D1256019

HOMINID EVOLUTION
PAST, PRESENT and FUTURE

The Taung child skull

HOMINID EVOLUTION
PAST, PRESENT and FUTURE

Proceedings of the Taung Diamond Jubilee
International Symposium, Johannesburg and Mmabatho,
Southern Africa, 27th January – 4th February 1985

Edited by

Phillip V. Tobias

Department of Anatomy
University of the Witwatersrand
Johannesburg, South Africa

With the assistance of
Valerie Strong
and
Heather White

Alan R. Liss, Inc., New York

Library of Congress Cataloging-in-Publication Data

Taung Diamond Jubilee International Symposium (1985:
  Johannesburg and Mmabatho, South Africa)
  Hominid evolution.

  Includes index.
  1. Human evolution—Congresses. 2. Australopithecines
—Congresses. I. Tobias, Phillip V. II. Strong,
Valerie. III. White, Heather. IV. Title.
GN281.T38   1985   573.2   85–15875
ISBN 0–8451–4202–X

To the Taung Child

At Taung, 3rd February 1985

When, in brief childhood here you used to play
In primal innocence, you could not know
The mission that awaited you or guess
That wayward Chance would cheat your early death
And bring you strangest immortality.

A million years, while Life around you grew
Into new forms to suit Earth's changing mood,
You slept, cocooned in stone, awaiting one
Who'd read your ciphered message and proclaim
Your signal place in Mankind's history.

Three-score years on, your infant face is known
And honoured round the world ... Heed not as we
Debate your lineage; though countless kin
May yet emerge to make their rival claims
As purer hominids, your place is sure:

You showed us first our cradle Africa.

True, first-born, we salute you, Child of Taung.

Jack P. Hutchings
Vancouver, British Columbia

Contents

PART I: ALL OUR YESTERDAYS

Taung: Retrospection and Perspective

Tongues in Trees. . . Sermons in Stones

PART II: THE VOICE OF TODAY

Ancestors of the Ancestors

Human Hominids That Say and Do

Humanization Con Brio: Accelerando, Largamente, Vivace

PART III: TOMORROW AND TOMORROW AND TOMORROW

Winding the Biological Clocks

New Shadow Pictures Beyond Röntgen's Wildest Visions

Sizing Up Brains and Human Nature

Present and Future Prospects for Dating the Past

Contributors

Emiliano Aguirre, Museo Nacional de Ciencias Naturales, Madrid 28006, Spain **[319]**

Harold J. Annegarn, Wits-CSIR Schonland Research Centre for Nuclear Sciences, University of the Witwatersrand, Johannesburg 2000, Republic of South Africa **[477]**

C.K. Brain, Transvaal Museum, Pretoria 0001, Republic of South Africa **[41]**

Timothy G. Bromage, Department of Anatomy and Embryology, University College London, London WC1E 6BT, England **[239]**

Brunetto Chiarelli, Institute of Anthropology, University of Florence, Firenze 50122, Italy **[397]**

J. Desmond Clark, Department of Anthropology, University of California, Berkeley, CA 94720 **[65]**

Ronald John Clarke, Department of Anatomy, University of the Witwatersrand Medical School, Johannesburg 2193, Republic of South Africa **[287]**

Glenn C. Conroy, Department of Anatomy and Neurobiology/Anthropology, Washington University Medical Center, St. Louis, MO 63110 **[419]**

Michael H. Day, Division of Anatomy, United Medical and Dental Schools of Guy's and St. Thomas's Hospitals, London SE1 7EH, England **[115]**

Yvette Deloison, Laboratoire d'Anthropologie, Muséum National d'Histoire Naturelle, Paris 75116, France **[135, 143]**

Robert B. Eckhardt, Department of Anthropology, The Pennsylvania State University, University Park, PA 16802 **[381]**

Jens Lorenz Franzen, Forschungsinstitut Senckenberg, Senckenberg-anlage 25, D 6000 Frankfurt am Main, Federal Republic of Germany **[255]**

Philip D. Gingerich, Museum of Paleontology, The University of Michigan, Ann Arbor, MI 48109 **[411]**

Leonard O. Greenfield, Department of Anthropology, Temple University, Philadelphia, PA 19122 **[465]**

Frederick E. Grine, Departments of Anthropology and Anatomical Sciences, State University of New York, Stony Brook, NY 11794 **[247]**

Otto-Joachim Grüsser, Physiologisches Institut, Der Freien Universität Berlin, 1000 Berlin 33, Federal Republic of Germany **[457]**

Phillip J. Habgood, Department of Anthropology, University of Sydney, New South Wales 2006, Australia **[367]**

Ralph L. Holloway, Department of Anthropology, Columbia University, New York, NY 10027 **[47]**

The number in brackets is the opening page number of the contributor's article.

William W. Howells, Peabody Museum, Harvard University, Cambridge, MA 02138 **[19]**

Jan Jelínek, Anthropos Institute, 659 37 Brno, Czechoslovakia **[341]**

Donald C. Johanson, The Institute of Human Origins, Berkeley, CA 94709 **[203]**

Reinhart Kraatz, Geologisch-Paläontologisches Institut der Universität Heidelberg, D-6900 Heidelberg 1, Federal Republic of Germany **[313]**

Jeffrey T. Laitman, Department of Anatomy, Mount Sinai School of Medicine, New York, NY 10029 **[281]**

Patrice P. Le Floch-Prigent, Laboratoire d'Anatomie, U.E.R. Biomédicale des Saints-Pères, Faculté de Médecine Paris-Ouest, 75270 Paris Cedex 06, France **[135]**

Jerold M. Lowenstein, Department of Medicine, University of California, San Francisco, CA 94143 **[401]**

Judy M. Maguire, Bernard Price Institute for Palaeontological Research, University of the Witwatersrand, Johannesburg 2001, Republic of South Africa **[151]**

Revil J. Mason, Archaeological Research Unit, University of the Witwatersrand, Johannesburg 2001, Republic of South Africa **[299]**

Henry M. McHenry, Department of Anthropology, University of California, Davis, CA 95616 **[221]**

Charles Oxnard, Department of Anatomy and Cell Biology, University of Southern California School of Medicine, Pasadena, CA 90033 **[271]**

Timothy C. Partridge, Department of Geography and Environmental Studies, University of the Witwatersrand, Johannesburg 2193, Republic of South Africa **[171, 303]**

Martin Pickford, Institut de Paléontologie, Muséum National d'Histoire Naturelle, Paris 75005, France **[107]**

Yoel Rak, Department of Anatomy and Anthropology, Sackler Faculty of Medicine, Tel Aviv University, Ramat Aviv 69978, Israel **[233]**

Charles A. Reed, Department of Anthropology, University of Illinois at Chicago, Chicago, IL 60680 **[89]**

Avraham Ronen, Department of Archaeology, The University of Haifa, Haifa 31 999, Israel **[329]**

Antonio Rosas, Departamento de Paleontologia, Facultad de Ciencias, Universidad Complutense, Madrid 28040, Spain **[319]**

Jeffrey H. Schwartz, Department of Anthropology, University of Pittsburgh, Pittsburgh, PA 15260 **[265]**

J.P.F. Sellschop, Wits-CSIR Schonland Research Centre for Nuclear Sciences, University of the Witwatersrand, Johannesburg 2000, Republic of South Africa **[477]**

Elwyn L. Simons, Department of Anthropology, Duke University Primate Center, Durham, NC 27705 **[101]**

Randall R. Skelton, Department of Anthropology, University of California, Davis, CA 95616 **[221]**

Geoffrey H. Sperber, Department of Oral Biology, University of Alberta, Edmonton, Alberta T6G 2N8, Canada **[443]**

T. Dale Stewart, United States National Museum of Natural History, Smithsonian Institution, Washington, D.C. 20560 **[335]**

Phillip V. Tobias, Department of Anatomy, University of the Witwatersrand Medical School, Johannesburg 2193, Republic of South Africa **[xix, 25]**

Russell H. Tuttle, Department of Anthropology, The University of Chicago, Chicago, IL 60637 **[129]**

Michael W. Vannier, Department of Radiology, Washington University Medical Center, St. Louis, MO 63110 **[419]**

John C. Vogel, National Physical Research Laboratory, Council for Scientific and Industrial Research, Pretoria 0001, Republic of South Africa **[189]**

Elisabeth S. Vrba, Department of Palaeontology, Transvaal Museum, Pretoria 0001, Republic of South Africa **[195]**

Sherwood L. Washburn, Department of Anthropology, University of California, Berkeley, CA 94720 **[3]**

Lutz-Rainer Weiss, Physiologisches Institut, Der Freien Universität Berlin, 1000 Berlin 33, Federal Republic of Germany **[457]**

M. Justin Wilkinson, Department of Geography and Environmental Studies, University of the Witwatersrand, Johannesburg 2001, Republic of South Africa **[165]**

Jan Wind, Institute of Human Genetics, Free University, Amsterdam 1007 MC, The Netherlands **[427, 437]**

Milford H. Wolpoff, Department of Anthropology, University of Michigan, Ann Arbor, MI 48109 **[355]**

Bernard A. Wood, Department of Anatomy and Biology, The Middlesex Hospital Medical School, London W19 6DB, England **[227]**

Ernst E. Wreschner, Department of Anthropology, The University of Haifa, Haifa 31 999, Israel **[387]**

Adrienne L. Zihlman, Department of Anthropology, University of California, Santa Cruz, CA 95064 **[213]**

Frans W. Zonneveld, Department of Diagnostic Radiology, Utrecht University Hospital, Utrecht 3511 GV, The Netherlands **[427, 437]**

Alessandro Zucchiatti, Wits-CSIR Schonland Research Centre for Nuclear Sciences, University of the Witwatersrand, Johannesburg 2000, Republic of South Africa **[477]**

Illustrations

Introduction

The impact of Raymond Dart's (1925) announcement of the discovery of the Taung child skull was immense and, ultimately, was to effect a revolution in human evolutionary studies. Quite justifiably, *Science 84*, in its Fifth Anniversary Issue in November 1984, included the revelation and aftermath of Taung among the journal's selection of "20 Discoveries that Changed our Lives".

Several colleagues and I deemed the 60th anniversary of that epochal event to be a most suitable moment at which to survey the state-of-the-art of palaeo-anthropology. It seemed fitting at this milestone to take stock of what had transpired during those 60 years in the quest for human origins, antecedents and descendants, to consider present-day trends and emerging directions in our discipline, and to contemplate and forecast possible future developments.

The idea of such a celebration was put to the University of the Witwatersrand, Johannesburg, which had been at the centre of the Taung activities of sixty years ago and which has since become an international centre for human evolutionary studies, as well as providing my own home-base for over 40 of those years. The plan was put also to the University of Bophuthatswana, within whose catchment area the old Buxton Limeworks at Taung is situated. Both universities, led by their respective Vice-Chancellors, Professor Karl Tober and Professor John Makhene, received the proposal with the greatest of enthusiasm and have throughout been most generous and supportive. When a similarly warm reception was accorded the idea by colleagues whom I consulted in various parts of the world and by scientific and financial bodies at home and abroad, we decided to proceed with the organising of the Taung Diamond Jubilee International Symposium. Happily we resisted the impulse to add another to the growing lexicon of ugly acronyms which have contaminated the language in a veritable onslaught ever since World War II; but in our planning office the project became known as "Taung 60". (But for the good judgment of our publisher, Alan Liss, this book might have been entitled *Taung 60!*)

Setting the Scene

At the Opening Ceremony of the Symposium, held on 28th January 1985, the large gathering was asked to project themselves back to that date 60 years earlier. "A mere 62 days have elapsed since the Taung child skull has come into the hands of Raymond Dart. Yet by 28th January 1925 the manuscript and illustrations of his paper have travelled by steam-train from Johannesburg to Cape

Raymond A. Dart and the Taung child skull, taken a few days after the announcement of the discovery

Town, a distance of some 1500 km (just over 900 miles), and by mail-boat from Cape Town to England, a two weeks' voyage,—and have reached the desk of Richard Gregory, the Editor of *Nature*, in London.

"On this day, the Editor is scratching his head over the astonishing and seemingly revolutionary article that lies before him. Sir Richard is plagued by a telegram from the Editor of the Johannesburg evening newspaper, *The Star*: Has Dart's article been received? What does *Nature* intend doing with it? But the Editor of *Nature* is a cautious man. He cables *The Star* that he has safely received all the submitted materials, but the discovery and Dart's claims are of so unprecedented a character that the paper has been referred to various experts in England, who have been asked to express their opinions on whether or not it should be published. One of the consultants is Sir Arthur Keith, who says— according to his *Autobiography*—'Go ahead and publish'.

"Meanwhile, Gregory's cable elicits a reply from *The Star*: we cannot with-hold publication later than the evening issue of 3rd February."

When the Taung Jubilee Symposium assembled at the University of Mmabatho on 1st February 1985, they were no longer 500 km from Taung: it was now a mere 250 km away! On that day 60 years earlier, there was a breathless hush of expectancy on several fronts.

"In Johannesburg, the young Professor of Anatomy sits waiting, we may assume rather nervously, for the outcome of the claims he is making and which are about to appear in print. In London, an editor sits waiting, we may assume rather uncertainly, for the response of the crowned heads of British science, to that same paper he is about to publish. In several offices of respected authorities in London and Cambridge, men of the ilk of Sir Arthur Keith, Sir Grafton Elliot Smith, Sir Arthur Smith Woodward and W.L.H. Duckworth are racking their brains over a set of proofs of Dart's paper that the Editor of *Nature* has sent to each of them and are musing over the choice of suitably wise and prudent remarks to offer the Editor in response to his puzzled enquiry. And in Johannes-burg, a newspaper editor is sitting impatiently waiting for 3rd February—to release the news to the world."

Then, on 3rd February 1985, a blistering hot Sunday, the Taung Jubilee delegates foregathered at the Buxton Limeworks, Taung, to see the famous site which had been visited by so few of the world's students of fossil man and to assist in the unveiling of a monument and plaque signalling for posterity the approximate position of the discovery. On that day, six decades ago, "*The Star* publishes the story and the name *Australopithecus* appears in print for the first time: on the same day, Gregory receives his proofs from the printer—and *Nature* carries the article on 7th February 1925. But by now four days have passed since the first public announcement and the strange names of Raymond Dart, Taung (or Taungs as it was known then) and that difficult one, Witwatersrand Univer-sity, and even more tricky *Australopithecus africanus*, have become known throughout the world, by means of the telegraph, the wireless and the press, which have not only sensed a good story, but have been excited by the idea that Africa, so long thoughtlessly dubbed the (metaphorically) Dark Continent, could have cradled mankind.

"There our tale begins and man's thoughts on his remote history have never been the same since."

Taung 60

The Taung Diamond Jubilee International Symposium took place between 27th January and 4th February 1985, with sessions mainly on the campus of the University of the Witwatersrand, Johannesburg, and some on the Mmabatho Campus of the University of Bophuthatswana. Close to 300 registered participants attended and they came to southern Africa for this purpose from some two dozen countries in all five continents. Fifty invited papers were given in the symposium proper, while 18 poster presentations were mounted. Two panel discussions and question-and-answer sessions, which had not been pre-arranged, were added to the programme by popular request of participants.

The aims of the Symposium were in keeping with its title, "The Past, Present and Future of Hominid Evolutionary Studies." Sessions were devoted to Taung 60 Years After; Hominid Evolutionary Studies in South Africa, East and North Africa, Asia, Australasia and Europe; The Evolution of the Upright Posture and Bipedal Gait; Molecules, Apes and Human Evolution; the Cultural Dimension and the Skeletal Dimension of Human Evolution; New Methods and Strategies in Hominid Evolution. These geographical and functional subdivisions cut across the historical, current and prospective categories: for such attempts to group subject matter under broader rubrics are essentially arbitrary. The same arbitrariness has beset the editor in devising the somewhat different categories followed in this, the book of the Taung 60 Symposium.

Thus, a number of papers in the Symposium took an unashamed backward glance at what has happened in palaeo-anthropology over the past 60 years, in the firm belief that an historical perspective is invaluable in the appraisal of where we stand today. The meeting endeavoured to glean some of the messages of the past which might help us to understand and to avoid the mistakes of yesterday in our planning for tomorrow. In Part I of this book are grouped 7 papers, the major thrust—though by no means the totality—of which is directed towards such historical reviews.

Inevitably, the major emphasis of the Symposium was on the present status of hominid evolutionary researches and concepts and the reporting of new results. No fewer than 33 papers are included in this section, Part II, under the title *The Voice of Today*, a phrase borrowed from those immortal words penned by Frederic W. Goudy for the anniversary of printing, "The Type Speaks: I am the Voice of Today, the Herald of Tomorrow."

A third aim of the Symposium was to explore some frontiers and spreading edges in our science. Where are we heading? What of tomorrow? Various speakers in this sector broached such diverse aspects as nuclear physics in the quest for fresh dating methods, new morphometric approaches, a diversity of cladistic analyses, the application of novel methods such as high-resolution computed tomography, the reconciliation of data provided by fossils with inferences from molecular studies, scanning electron microscopic studies on deposition and absorption of bone. Many of these papers explored and adumbrated innovating trends for the future of palaeo-anthropological studies. Ten of these offerings, in which the element of novelty seems to predominate, are grouped in Part III under the Shakespearean banner, *Tomorrow and Tomorrow and Tomorrow*: though it must at once be said that several other papers classified under Parts I and II also highlight pioneering approaches which will undoubtedly influence researches in our discipline over the remaining decade and a half of the 20th Century.

A Forum for Disputation

It was never the aim of the Taung Jubilee Symposium to achieve any kind of synthesis or even a consensus, on some of the burning issues that presently divide the fraternity of palaeo-anthropologists. On the contrary, it was a prime purpose to ventilate current arguments. Indeed, when I set forth the aims of the Symposium at the Opening Ceremony on Monday, 28th January 1985, I included this thought:

> "Here we shall encounter arguments, controversies, paradoxes. It should never be forgotten that that is a sign of health in a discipline. A branch of science which is not generating new hypotheses, contradictory viewpoints and dissentient voices—has ceased to grow, has passed its prime and is slipping into senescence prior to fossilization. So we should not be ashamed or embarrassed by our altercations. The role of dissent in the history of science is a tale that is waiting to be told. I return to this Orwellian theme. . . (in my contribution to this volume). . . for the gentleman sitting on my right (Professor Dart) has been throughout his life an arch dissentient, inclined to scientific heresies. And whether those outrageous Dartian views were upheld or not, they were generally of heuristic value, for they provoked counter-claims and catalysed new researches—even if the objective was only to shoot down those unconscionable ideas! So we shall and must argue our way through the Symposium. How dreadfully dull it would be if we all agreed on all points: we might as well pack up and go home right away!"

Whilst protesting that it is good for differences of opinion to be ventilated publicly, I found it necessary to sound a warning note:

> "Differing viewpoints may exist on such detailed matters as which species of *Australopithecus* gave rise to the genus *Homo*; to which group of hominids are those of Hadar most closely

related; which are ancestral and which belong to parallel, non-ancestral lineages; whether most evolutionary change in the hominids was of gradual nature with slower and faster times of change, or whether most change occurred at crisis points, sometimes called nodes or punctuations, with but little change occurring between successive nodes. The existence of differing views on such points of detail has been sometimes and rather thoughtlessly taken to indicate that one school of thought, by denying the interpretation of another school, is giving the lie to the very fundamental notion of the evolution of man. Nothing could be farther from the truth. This is perhaps a good moment to reaffirm that nothing in human biology makes sense except in the light of the evolutionary concept. To speak of the concept or hypothesis or theory of evolution is, in turn, often seized upon by anti-evolutionists as a sign of weakness in the evolutionary doctrine. Evolution, they are liable to declare, is only a theory. Thereby, of course, they are betraying their ignorance of the way in which science works—by the creating of hypotheses, the testing of them and the refuting or confirming or modifying of them. This approach, which is the essence of the scientific method, is no sign of weakness; it is surely the very strength of science.

"Having said that, let me hasten to add that there is probably no biological theory which is more thoroughly proven than the notion that living things have evolved. All the evidence goes to show that this is true of every part of the biosphere including man.

"So if we argue with one another over detailed interpretations, these are arguments about patterns of evolutionary change: they are not arguments about whether evolution has occurred."

(Address on the Aims of Taung 60:
28 January 1985: P.V. Tobias)

Indeed the differences of opinion *were* well and truly aired during the Symposium, but, withal, it was done without rancour or personal vendettas, as is the ideal way of science. On the whole, the symposiasts were able to avoid impassioned appeals to authority, personality cults and those interpersonal feuds with which our field has for too long been bedevilled. The moments of animosity which did surface were few and far between: and that, I believe, makes for better science.

This volume reflects some of the many cross-currents of thought in our discipline, particularly Part II which deals in the main with the current position and ongoing studies.

The Social Function of Palaeo-Anthropology

It would have been a rich bonus indeed if Taung 60, from its probings into the past history of mankind, had drawn lessons to guide us towards solving problems of man's complex and neurotic technocratic world of today and towards predicting in broad outline the world of tomorrow.

With a tinge of regret I note that these difficult and challenging questions did not receive the meeting's attention. Nonetheless, I take leave to set forth here a few remarks I made on this subject at the Opening Ceremony of Taung 60, because I believe that palaeo-anthropologists may not long escape addressing their minds to such issues:

> "Perhaps we may even permit ourselves to speculate a little on the future of human evolution. Have we any valid basis for predicting the nature, the pattern, of tomorrow's man from our profoundly expanding comprehension of today's, and our evergrowing, though still rudimentary, understanding of yesterday's hominids? We may even touch on *this* thorny question: What role does the student of hominid evolution have in planning for the future, as distinct from just sitting and waiting for the future to happen to us?
>
> "Do our studies of past and living human biology not give us grounds for pronouncing on such issues as race, races and racism (still contentious, ill-understood and obsessional problems here, as in some other parts of the world)? On the human population explosion, present and projected, and on the inversion of the demographic pyramid in modern industrialized and affluent communities? On the increasing urbanisation of *Homo sapiens*, a species which evolved under rural conditions but practically one in two of whose members are today living in urban areas (the figure is approximately 48 per cent)? On the accumulating evidence of problems in man's adaptation to city life, the rise of urban pathology?
>
> "I number myself among those who believe that the student of human evolution has not merely a basis, but a positive responsibility, not to eschew such knotty questions. It is possible that our evolutionary insights have provided us with a more secure groundwork on which to predict man's future, than those on which such initiatives as the *Club of Rome*, the *Committee on the next 30 Years* and the *Futuribles* project have been basing their somewhat gloomy prognostications."

These social functions and applications of palaeo-anthropology are perhaps a theme for another day and another symposium.

The Raymond Dart Lecture

Through the gracious cooperation of the Institute for the Study of Man in Africa (ISMA) and of Wits (our affectionate abbreviation for the University of the Witwatersrand), we were able to include the 1985 Raymond Dart Lecture in the programme of Taung 60. For more than two decades one of the academic highlights of the year at Wits has been the Annual Raymond Dart Lecture. This eponymous lectureship was established by ISMA, in association with the Wits University, in 1962. The first Raymond Dart Lecturers were the late Dr. Sidney H. Haughton F.R.S. (1964), whose address was entitled *The Australopithecine Fossils of Africa and their Geological Setting*, and the late Sir Wilfrid E. LeGros Clark F.R.S. (1965) whose lecture went under the title, *There is a Transcendence from Science to Science*. The 1985 Raymond Dart Lecture was the 23rd in the series and was delivered by Emeritus Professor Sherwood L. Washburn of the University of California, Berkeley, on Monday evening 28th January, in the Wits Great Hall, immediately after he had received the degree of Doctor of Science *honoris causa* from the Chancellor of the Wits University, Dr. Michael Rosholt. The lecture, called *Human Evolution after Raymond Dart*, is included in this volume by courtesy of ISMA, Wits and the Witwatersrand University Press.

Conferment of Honours

During the Symposium, the Wits University held a special Graduation Ceremony at which the degree of Doctor of Science *honoris causa* was conferred upon each of three of the most distinguished participants in Taung 60, Professor J. Desmond Clark, Emeritus Professor William W. Howells and Emeritus Professor Sherwood L. Washburn. On the same occasion an honorary degree was conferred upon the veteran excavator of ape-men and many-sided contributor to fossil hominid studies, Alun R. Hughes, on his retirement from the full-time service of the Wits Department of Anatomy to which he had dedicated 38 years of his life. A happy conjunction enabled one of my research students, Frederick E. Grine, to receive his Ph.D. degree in the same unusual ceremony: appropriately enough, his thesis dealt with the deciduous teeth of the Taung child and other hominids.

The Royal Society of South Africa, the country's oldest and senior scientific body, used the opportunity provided by Taung 60 to award its highest honour, Foreign Associateship of the Royal Society of South Africa, to Professor Yves Coppens, Dr. F. Clark Howell, Emeritus Professor W.W. Howells, Professor Ned Munger, Dr. T. Dale Stewart and Emeritus Professor S.L. Washburn, all of whom have carried out, promoted or encouraged work in South Africa.

The Book of the Symposium

This volume comprises 50 contributions. In harmony with the title, they have been arranged into three parts dealing, respectively, with the past (*All Our Yesterdays*), the present (*The Voice of Today*) and the future (*Tomorrow and*

Tomorrow and Tomorrow), of hominid evolutionary studies. Each part is further subdivided into sections.

Part I comprises two sections, *Taung: Retrospection and Perspective*, in which there are five papers of a mainly historical review character, focussed principally on Taung, including Professor Washburn's Raymond Dart Lecture; and *Tongues in Trees. . . Sermons in Stones*, containing two historical and conceptual papers on stone tools, "fossil behaviour" and adaptation.

In Part II, which is the largest portion of the book, there are six sections:

The first section, *Ancestors of the Ancestors*, consists of two papers on Oligocene and Miocene hominoids of Africa, some of which might have been ancestral to the Hominidae.

The second section, named *Rise of the Bipedalists*, contains four articles related to the anatomical basis of early hominid locomotion, as reflected both in comparative and palaeo-anatomy and in the famous Laetoli footprints.

In the third section, *Southern Sites*, five contributions present the results of recent studies on the southern African australopithecine fossil sites of Makapansgat, Sterkfontein, Swartkrans, Kromdraai and Taung, from the viewpoints of geology, geomorphology, geochronology, stratigraphy, palaeontology and palaeoecology.

Section 4 of Part II is devoted to *Australopithecus in the Forest of Hominid Trees*. The 10 articles explore various aspects of the morphology, cladistics, sexual dimorphism, ontogeny, zoogeography and phylogeny of the early hominids of South and East Africa—with glimpses at the hominids and apes of Asia.

The fifth section, *Human Hominids That Say and Do*, deals with hominid attributes that are apparently not features of *Australopithecus* but mark those taxa classified in the genus *Homo*: more particularly, the peripheral anatomical basis of speech and some early South African stone tool industries.

An accelerated tempo of change and dispersal of hominids are reflected in Section 6, under the arguably melodious style of *Humanization Con Brio: Accelerando, Largamente, Vivace*. In the eight offerings here later stages in human development are limned: new finds of fossil man are reported from Egypt and Spain; whilst other papers deal with phases of human biological and cultural evolution in Germany, Israel, eastern Asia and Australasia, symbolism and possible speed of evolutionary change.

Although a number of articles in Parts I and II look ahead in proposing new methodological approaches, Part III is dedicated more exclusively to newer and expected developments. Under the appellation, *Tomorrow and Tomorrow and Tomorrow*, we find four sections.

The first is called *Winding the Biological Clocks* and contemplates the respective roles in hominid evolutionary studies of cytogenetical data on living primates, of radio-immunoassay results on proteins from recent and fossil bones and, of the abandonment of the assumption of linearity in the devising of molecular evolutionary clocks, in favour of nonlinear clocks based on not one but several palaeontological divergence times.

Section 2 of Part III reveals some most striking recent radiological developments, just 90 years after Wilhelm Conrad Röntgen reported to the Würzburg Physical and Medical Society that he observed and photographed "shadow pictures" and possessed, for example, photographs of the shadow of the handbones. Under the banner *New Shadow Pictures Beyond Röntgen's Wildest Visions* are presented three articles on the exciting new technique of high-resolution computed tomography and first results obtained by its application; and one paper on radiographic palaeo-anatomy of teeth or "radiodontology".

In Section 3, entitled *Sizing Up Brains and Human Nature*, two papers consider the past and possible future growth of the hominid brain during phylogeny, and limitations of the degree to which human evolutionary data may throw light on human nature.

Finally, the fourth section of Part III presents a paper in which a group of nuclear physicists examine the feasibility of sensible modelling and chronological dating based on an array of isotopes, proton-induced X-ray emission and a new form of ultra-sensitive mass spectrometry using particle accelerators, which technology could be applied to both fossil bone and matrix, especially from the South African sites, where, until now, radio-isotopic dating determinations have not proved possible.

Some Usages and Spellings in the Text

m.y.BP—has been used for millions of years before the present

y.BP—years before the present

Neandertal, neandertalian, neandertaler—these spellings have been used consistently, in accordance with modern German orthography, though *neanderthalensis* is of course retained, when used as a trivial or subspecific name.

australopithecine, but *Australopithecinae*

hominid, but *Hominidae*

Acheulean, rather than Acheulian)

artefact, instead of artifact) for consistency

symphyseal, rather than symphysial)

American spellings (e.g. behavior, paleolithic) have generally been chosen, but where authors have favoured English spellings (e.g. palaeo-anthropology, behaviour) these have generally been left unchanged, in the camera-ready manuscripts.

Appreciation

As Organiser of the Symposium, as well as editor of these proceedings, I extend sincere thanks to all of the symposiasts and those whose articles appear in this work. Especially do I value their forebearance in the face of an editor inclined to be over-fastidious! For great help in the running of the Symposium special gratitude is due to the sessional chairmen and secretaries for their vital part.

Sincere appreciation is extended to Emeritus Professor Raymond A. Dart who not only served as Patron-in-Chief of the Symposium, but also honoured us with his presence on a number of occasions. Thanks are offered to all of the Patrons of the Symposium, including the Vice-Chancellors of the Universities of the Witwatersrand and of Bophuthatswana, and a dozen scientific bodies, societies and foundations from southern Africa and abroad.

The Symposium was made possible by the most generous financial support of the L.S.B. Leakey Foundation for Research into Man's Origin, Evolution and Behaviour, Dr. Hubert R. Hudson, The Chairman's Fund, Anglo-American Corporation, Barlow Rand Limited, the Wenner-Gren Foundation for Anthropological Research, Mr. Gordon P. Getty, the University of the Witwatersrand, Johannesburg, South African Airways, the University of Bophuthatswana, the Liberty Life Association of Africa Limited, The Holt Family Trust, Mrs. Diana B. Holt, The Institute of Human Origins, The Chairman's Fund, Anglo-Vaal Limited, Professor Raymond A. Dart, Sun International, Plasser Railway Machinery, National Co-operative Dairies Limited, South African Breweries Limited, The Coca-Cola Bottling Company and Stellenbosch Farmers' Winery.

I cannot begin to find words to express adequately my indebtedness to my Vice-Chairman, Professor Tim Partridge; Treasurer, Alun Hughes; Finance Officer, John Bunning; and the long-suffering but indefatigable Secretariat, consisting of Marie-Luise Betterton, Valerie Strong and Heather White, the latter two of whom have worked heroically in helping me edit the present work. Thanks be to all the members of the Main Organizing Committee, the Organizing Panel, the Ladies Committee, the Students' Committee and the Bookstall Committee; numerous staff members of the two Universities; and scores of persons and organisations who played their part in making the Symposium a social as well as an academic success.

My gratitude goes to Margo Crabtree for permission to use her magnificent photograph of the original Taung skull as the frontispiece, published here by courtesy of the American Association for the Advancement of Science; and to Jack Hutchings of British Columbia for his sensitive dedicatory poem apostrophizing the Taung child.

It is not often that an editor has the pleasure of dealing with so enthusiastic and devoted a publisher as Alan Liss. Not only did he agree to take on the task of publishing the proceedings of the Symposium without a moment's hesitation when first I approached him; but he insisted on attending the Symposium with his wife and I believe they did not miss a single session! It has been a great pleasure and a privilege to work with so dedicated, helpful and friendly a group of professionals as Alan Liss and his staff in the production of this book.

Phillip V. Tobias

Johannesburg
4th May 1985

PART I
ALL OUR YESTERDAYS

Taung: Retrospection and Perspective

Hominid Evolution: Past, Present and Future, pages 3–18
© *1985 Alan R. Liss, Inc.*

HUMAN EVOLUTION AFTER RAYMOND DART [*]

S. L. Washburn

Department of Anthropology
University of California
Berkeley, CA   94720

Sixty years ago, in 1925, Raymond Dart sent a paper to
Nature, probably the most widely read scientific periodical
of the time. The paper described a new kind of human being.
The fossil was that of a child with remarkably human teeth,
but with a very small braincase. Scientists had not been
expecting this combination. The situation was further
complicated by Dart's belief that this small-brained
creature had been bipedal. It has taken many years for
most scientists to accept the theory of a small-brained
African ancestor, rather than a large-brained Asiatic one.

The history of the discovery has been thoroughly
described by Dart himself (1959), and more recently by
Phillip Tobias (1984). Tobias has also organized this
international symposium honoring Dart and the discovery
of the little skull from Taung. I will not repeat the
facts which have been so clearly stated. What I hope to do
is to treat Australopithecus, as the species came to be
called, as a special case in the study of human evolution.
In dramatic form it illustrates the nature of science itself.

In the 1920s there were very few human fossils.
Scientists were eager to discover more, and one might
think that, granted this situation, the discovery of a
new kind of human being would have been greeted with joy

---

[*] *23rd Annual Raymond A. Dart Lecture which was delivered in
the Wits University Great Hall during the Jubilee Symposium,
on 28 January 1985, and is published here by kind permission of the Institute for the Study of Man in Africa and the
University of the Witwatersrand, Johannesburg. - Ed.*

and interest. But the actual situation was exactly the
reverse. Most of the leading scientists of the day criticized
Dart, both for the description of the specimen and for his
evolutionary conclusions. It was not that people did not want
more fossils to be found. They did, and Dart was advised to
gather more facts before he advanced far-reaching conclusions.

The peculiarity of the situation is well shown by comparing the reception of the discovery of fossil man near
Pekin with the discovery at Taung. The deductions based on
the Taung skull were criticized because it was only one
specimen and a juvenile at that. But the first find at Pekin
was one tooth and the first skull a juvenile. Dart's paper
was published years before Pekin was even found! The reaction
to the Taung skull was almost nothing but extreme criticism.

Clearly, the remarkable difference in the reception at
Taung and Pekin is not to be explained in any simple way.
The world of science was ready for the discoveries from
China; they fitted into what most scientists believed. Far
from the bones being objective facts to be judged as
evidence, there was an already established pattern of belief.
There was a climate of opinion which favored Davidson Black,
Asia, a general pattern of evolution, and not the silly
notion of small-brained bipeds from Africa. Raymond Dart
showed his imagination and courage in recognizing the meaning
of the fossil from Taung.

The situation will be clearer if we look at a few of the
very influential books that appeared shortly after the
discovery at Taung. Perhaps the most widely read was Hooton's
Up from the Ape (1931) in which Pekin is discussed and
illustrated, but there is nothing on Australopithecus,
Taung, or Raymond Dart. Leakey, in Adam's Ancestors (1934),
discussed Pekin but not Taung -- although the latter was found
sooner and published years before Pekin was found. Certainly
one of the most important evaluations of the Taung skull was
done by Keith (1931). Several chapters in New Discoveries
were devoted to Taung and it was concluded that Taung was an
"ape" (page 115), almost solely because of the size of the
brain.

At the time of the discovery at Taung there were two major theories of human evolution. The first was that the ancestors were primitive in a general way, that all parts of the body showed primitive features. Neandertals were pushed into this mold. The second theory was that the brain had evolved first, long before other parts of the body. A major factor in the proof of the brain-first theory was the Piltdown find. And Elliot Smith said that Piltdown is what should have been expected -- a modern-sized braincase and an ape-like face.

Taung fitted neither theory, and it is of interest to note that some of the strongest critics of Dart were advocates of the forgery known as Eoanthropus (Piltdown). If one believed that the large human braincase came first in evolution, then there was no place for Taung.

Dart's critics, scientists all, saw that the only way to settle the question of the nature of the Taung skull and its place in evolution was to find more specimens, particularly adults. Robert Broom set out with this goal in mind. In the 1930s and 40s he was successful, finding skulls of two kinds of Australopithecus and a nearly complete pelvis. The pelvis was so human that one scientist claimed that it must have "fallen in" from a much younger deposit. It clearly could not belong with the skull with which it was found.

By 1950 -- 25 years after the event we are celebrating -- many people found the accumulated evidence convincing. There had been two species of the genus Australopithecus in South Africa. Their brains had been no larger than those of the contemporary apes, and their limb bones showed that they had been upright. It had taken 25 years and numerous discoveries to provide the evidence which Dart's critics had demanded. Keith changed his mind and came to agree with Dart. Surely the accumulated "facts" showed that the small-brained bipeds had been a stage in human evolution.

Competing theories were eliminated. Piltdown was
shown to be a fake. Far from proving that the brain had come
first, it proved that the most highly regarded scientists had
been unable to tell a modern jaw from a fossil for over 30
years. The "facts" that had been used to "prove" the
traditional theories of human evolution were gone, but the
theories did not die.

The theory that the large brain evolved first and that
people like ourselves were very ancient was formulated
before 1900, supported by Piltdown, and this theory has
lasted to the present. Could the footsteps at Laetoli
have been made by an Australopithecus, or must a more advanced
form of human been present at that time? Must there be two
kinds of fossil humans in the collections from Afar?

The persistence of rival, contradictory theories (and
these are particularly common in the field of human evolution) depends on what has been called "emotional cerebration."
MacLean (1970) has outlined how the emotions, in science or in
politics, may stand on either side of any question. Let me
illustrate with a current, highly emotional controversy.
Raymond Dart suggested that at least some of our ancestors
had been highly aggressive. The playwright, Robert Ardrey,
dramatized this theory in his highly successful book,
African Genesis. The critics "knew" that our nearest
living ancestors were peaceful vegetarians, that Dart and
Ardrey were wrong. But recent studies show that chimpanzees
sometimes kill other chimpanzees. Gorillas may kill and
eat other gorillas. Male orangutans on occasion fight,
leaving great scars on the face. In the 1960s currently
held views by the scientific community were such that a
theory stressing aggression in human ancestry did not have
a chance. Yet, as was pointed out many years ago, the
male great apes had the anatomy for aggression. The
interpretation of the large canine teeth of the males was
recognized by Darwin.

Now field studies support the idea of some hunting and

intraspecies aggression in the great apes. It will be
exceedingly difficult to increase the role of fact and
decrease the role of emotional cerebration when evaluating
the role of aggression in human evolution.

Another widely held theory, contrary to the evidence, is
that continuous sexual attraction is what holds human society
together. Studies of the sexual behaviors of apes show that
orangutans are the most continually active sexually, gorillas
the least. Yet gorillas are the most social and orangs the
least. It has taken the combination of fieldwork and
carefully controlled studies in the laboratory to show the
interrelations of sexual behaviors with a wide range of
other social activities.

Field and laboratory studies of behavior and structure
have finally started to yield more probable conclusions.
But it is not likely that "facts" will be sufficient to
replace beliefs. In the matter of mating it is claimed
that the face-to-face position is unique to human beings
and has great psychological importance. But it is a
common position among gibbons, orangutans, pygmy chimpanzees
and gorillas. Schaller (1963, page 285) describes a
variety of mating positions used by gorillas, yet more than
20 years has not been enough to eliminate the myth that
humans are unique, and that face-to-face mating is a
major cause of human society.

In science there is a progress in which facts become
more clearly defined. For example, the discovery of the
pelvis of several specimens of Australopithecus strongly
supports Dart's original deductions based on the skull.
Surely one would think that the discovery at Laetoli
(Leakey 1981) would end this debate. But it is possible
to believe that the footprints were made by Homo, not
Australopithecus, and that the pelvis of Australopithecus
shows many ape-like features. My bias is that Dart's
reconstruction of the ilium from Makapansgat showing how
similar it was to a human ilium settled the matter. This
obviously did not convince one author who has managed to

write a whole book, a large part of which is devoted to Australopithecus, without mentioning Broom, Dart, Brain or Tobias in the long bibliography.

In glancing over the history of the study of human evolution, no matter how briefly, it becomes apparent that there have been two kinds of progress. In part, progress has come from the ideas of particular people, and, more often than not, such theories have been opposed by others. More importantly, progress has come from changes in the prevailing opinions from the cumulative effect of the actions of many individuals. Opinions are changed by the development of new techniques. Often the discoveries may seem to have little relation to one another, but in combination they build change.

The idea that humans had evolved rather than been created was a revolution of major proportions. Implementing this theory required the development of new techniques, and this process is still continuing. Evolution, as supported by comparative anatomy, embryology and a few fossils, was a very different science from the same theory supported by plate tectonics, physics and molecular biology. It should be remembered that the changes involve both the creative acts of individuals and changing climates of opinion.

Who could be so foolish as to believe that whole continents had drifted across the surface of the earth? Generations of college students were taught that Wegner exhibited a lack of common sense, and, as my geology professor remarked, "He wasn't even a geologist!"

In the next few minutes I want to speak of the scientific developments which have complicated the acceptance of the Taung fossil.

From the time of the initial discovery at Taung, the age of Australopithecus has been debated. The most common point of view was that it was too late to be in the ancestral line of human beings. The best it could have been was some sort of cousin, predominantly ape, with a few human features. But in the time of Piltdown and Taung there were no methods for objectively measuring geological time. The whole Age of the Mammals was thought to be only 3 million years. By 1931 the importance of radioactive time scales was recognized and the Age of the Mammals had been expanded twentyfold to 65 million years. In 1900 the age of the earth was supposed to be something on the order of only 65 million years! The short estimates of the age of the earth, plus natural human bias, had led to estimates that human beings must have been separated from the other primates for most of the Age of the Mammals. An Oligocene (30 million years) or early Miocene (20 million years) separation was common. The baboons and other mammals found with the Taung skull were surely late (Pleistocene), and so Taung was far too late to be a direct ancestor. The Galley Hill skeleton, the London skull, and Piltdown (a burial, a mistake and a fake) formed a prominent part of the evidence proving that Taung was too little and too late.

When looking back many years, it is easy to see why some scientists went wrong. Or is it? The idea that the separation between humans and other primates must be very early has continued to the present day. One might think that the radical change in the estimates of the age of the earth would have exerted a profound effect on the theories of human evolution. They did nothing of the sort.

The beginning of a new era in the dating of the stages of human evolution came with the determination of the age of a large Australopithecus in Olduvai Gorge. The potassium argon method showed that the deposit in which the skull was found was 1.8 million years. This was far longer than had been expected and alerted scientists to the importance of East Africa (long recognized by the Leakeys), and to the revolution in methods of determining dates. The dating of fossil humans ceased to be highly subjective, and over the last 20 years potassium argon,

fission track, and magnetic reversals have given promise
of making an objective framework for evolution (Curtis 1981).

It is the reduction of the human element that is so
important. It is not eliminated, as shown by recent controversies, but it is reduced, and the kind of research needed
to construct an accurate time scale is clearly defined.

Dart compared Taung with the African apes -- the
chimpanzee and the gorilla. Many scientists accepted this
point of view, but others thought that the closest living
relatives were prosimians, monkeys, or at least some
creatures more antiquated than any ape. Comparative anatomy
was regarded as a major science, a method that could resolve
the questions of relationships among the animals. But the
continuing controversies showed that well trained
anatomists could reach diametrically opposing conclusions.
Just as in the case of time, relations could only be
settled by some new techniques, and these depended on the
progress of science, NOT upon progress in anatomy as such.

Now molecular biology has settled the problems of
human relations. Just as Dart had thought, our closest
living relatives are the chimpanzee and the gorilla. The
study of a number of techniques -- immunology, sequences
of amino acids in proteins, electrophoresis, DNA
hybridization -- all show the close similarity of ape
and human, a similarity much closer than had been
expected. Ape and human differ in less than 1% of the
genetic material (Cronin 1983). The new techniques,
particularly DNA (Sibley and Ahlquist 1984), eliminated the
competing theories, something that comparative anatomy
had been unable to do.

Since solving the question of "Man's Place in Nature,"
a major problem for many years, one might think that the
discovery of techniques enabling the solution would have
been welcomed. Such was not the case. It has taken two
decades for the new methods to become moderately well

accepted, and as late as 1982 a book on human evolution
(Eldredge and Tattersall) appeared which contained no
discussion of the recent advances. Many scientists apparently
found the new to be threatening rather than enlightening. To
some, a well guarded intellectual territory seemed more
important than acceptance of an answer to the problems. The
speed and power of the new techniques must be illustrated to
be appreciated. Even after Piltdown had been proved to be a
fake, there remained the problem of the lower jaw. To what
kind of an animal did it belong? Lowenstein (1981), using
radioimmunoassay, a technique which permits immunology to be
done with microscopic amounts of material, has shown that it
belonged to an orangutan. The tests were performed in less
than two weeks, part time. Had it been available, this
routine test would have saved many years of controversy and
hundreds of scientific papers, and would have removed a
major block in the acceptance of Australopithecus! It
appears that the proof that Dart was right depends on recent
developments, and it is these developments that have fundamentally changed the climate of opinion.

How the current state of opinion contrasts with the old
is of interest. There is always a subjective element in the
study of evolution. This allows the existence of competing
theories. For a long time, uncertain comparative anatomy
and a very fragmentary fossil record placed almost no limits
on the theories. Brain size was enough to solve the problems
of human evolution.

These restrictions permit a reexamination of comparative
anatomy. Professor Dart would illustrate his lecture on human
evolution to his medical class with a graphic display. He
would walk to the back of the old medical amphitheater, grab
hold of a pipe, then swing back using the pipe as the arboreal
route to the stage. When he was over the podium he would call
out "And so he probably came to earth," let go of the pipe and
land right by his lecture notes. I am told no student ever
forgot this dramatic proof of human ape-like origins.

In his demonstration swing over the medical class,

Professor Dart used the anatomy which we share with the contemporary apes. The way humans swing is ape-like and very
different from the behavior of monkeys. It depends on the
anatomy of hand, wrist, elbow, shoulder and trunk. We stretch
to the side as the apes do, and this simple action summarizes
a great deal of anatomy. Monkeys stretch forward, in the
manner of a cat or a dog, summarizing a very different anatomy.

The late Adolph Schultz liked to point out that play-
grounds are arranged with rings, ladders and other devices,
so that the young may have the fun of climbing and swinging.
If you doubt our similarity to the apes, go to any play-
ground!

These similarities in behavior are based on complex anatomical patterns which form the basis for many hundreds
of scientific papers. In my opinion the methods of
traditional anatomy destroyed the similarities of ape and
human. The fundamental similarities were literally cut to
shreds by hundreds of eager anatomists. The pieces cannot
be reassembled into behavioral complexes. The reverse is
easy. It is easy to see the behaviors and then examine
their relation to the facilitating anatomy.

Next time you are standing on a crowded bus hanging by
one arm, consider what sort of a creature would design such
a system. Surely only an ape or a descendant of one! In the
zoo you can see apes take comparable positions, very like us
and very different from other animals. You have no problem
in using your arm as an ape does. The similarity in form and
function is apparent. Now take an anatomy book in which
bones, joints, ligaments, muscles, and proportions are all
described in different chapters and see how difficult it is
to compare ape and human.

The problems that arise when the attempt is made to
build theories of human evolution almost entirely on comparative anatomy have been reviewed by the late William
Straus (1949) in a paper entitled "The Riddle of Man's
Ancestry." The only reason I mention these problems here is

that Dart's "baby" and Broom's discoveries were evaluated in the context of the jumbled, troubled anatomy of their day.

Today we no longer have that problem. "Man's Place in Nature" is defined by molecular biology, especially by DNA hybridization and immunology. The new methods place limits on the old and enormously simplify the problems of human evolution. It is not that the old methods were necessarily wrong, but they did not convince people.

The emphasis on the African apes which stems naturally from their close similarity to humans leads to a reconsideration of some of the features of these apes. The early theories of human evolution stressed coming to the ground, the human descent from the trees. But the African apes, especially the gorilla, are on the ground much of the time. While they are skillful climbers, they are also successful ground livers. They have a most peculiar way of progressing on the ground -- they walk on the backs of their knuckles. The chimpanzee and the gorilla are the only animals that do this, and this raises the question of whether our ancestors were knuckle walkers. Is the problem of human origin descent from the trees or the evolution of a biped from a knuckle walker?

Sibley and Ahlquist's (1984) studies of DNA strongly suggest that the order of origin of African apes and humans is gorilla, chimpanzee, human. Granted short intervals of time (a point to which I will return), it is very difficult to see how gorillas and chimpanzees could have become knuckle walkers and how human beings could have avoided this adaptation. It should be noted that this is a rare and peculiar adaptation, not present in the non-African apes or in monkeys. Humans, obviously, cannot knuckle walk because their legs are too long, but they do occasionally adopt the position. One professor at Berkeley lectures standing behind a table and bearing his weight on his knuckles in a typical ape-like manner. For a few years this position was used in football. Unfortunately, this practice was given up and I can no longer lecture that knuckle walking is a position used by gorillas, chimpanzees, and professional football linemen.

The human wrist shows adaptations very like those of the knuckle walkers. For example, if I put my hand down flat in a monkey-like position, the end of the ulna is hardly articulated and bears no weight. If I want the ulna to function fully, I must turn the hand. The hand now faces my side. If I do this with both hands, the hands face each other and have the peculiar position they do in apes. This matter is not settled, but I think that the order of branching shown by the DNA would be taken as strong evidence for a knuckle walking ancestor if the ancestor in question were not our own. The least that can be said is that our hands are remarkably similar to those of knuckle walkers, especially considering that humans have been separated from the apes for at least 4 million years.

Fossils are necessary to determine just how long humans have been evolving after the separation from the apes. But estimates of the times of evolutionary separations may be made from the molecular biology. This development has been opposed by students of the fossil record, but the opposition is much less than it was a very few years ago. At the molecular level, evolution appears to proceed at remarkably uniform rates. The evolution of fibrinopeptides may be very fast, and that of cytochrome C very slow, but the rate is independent of that kind of animal whose fibrinopeptide is being studied. At an even more fundamental level, the rate of change in DNA seems to be the same in very distantly related animals. When Sarich first applied the idea of biological "clocks" to primate evolution his ideas were as welcome as the Taung skull was to Keith. Sarich (1983) has answered the objections to the "clock" approach, and I agree with his position. The essential point is that the molecular approach produces an internally consistent phylogenetic picture with approximate dates for the separation of the branches leading to the major kinds of primates, including human beings. It is far easier to approach human evolution with this framework than with the traditional one, which, incidentally, never produced agreement. Obviously, the arguments used in this paper will not hold at all if humans have been pursuing a separate evolutionary course for 25–35 million years (back to early Miocene or Oligocene).

At present we are still looking for some general scheme which will allow the conversion of molecular distances into years. Here is one that may be fun to consider. Alvarez and Alvarez (1980) suggest that a meteor hit the earth some 65 million years ago. This was the event which ended the domination of the dinosaurs and opened the way for the rapid evolution of the mammals. It had been known for a long time that the various orders of placental mammals all appeared at close to the same time. Immunology, particularly Goodman's (1976) surveys of the mammals, showed that the orders were all about equally distinct. If the orders are about 60–65 million years old then there is a date of origin in years, and all the different orders[1] have changed approximately the same amount immunologically. So the degree of anatomical change can be related to time.

If the order to which we belong, the primates, is some 60–70 million years old, then apes and humans as a combined group have been distinct for some 20 million years. Human beings have been separated from the apes for some 5 million years.

There are other ways of reaching approximately the same conclusions. My point is simply that just as anatomy needs other techniques to check its conclusions, so molecular biology needs controls for the measurement of time, controls which do not depend on the molecular evidence itself, no matter how good that may be.

----

1. Monotremes, marsupials and some insectivores are older, as traditionally believed.

Molecular biology does not eliminate the need for fossils or anatomy. Without fossils nothing would be known about the vast majority of the forms of life. What the molecular information does is to settle a very large number of problems and to provide objective information on issues which had been highly controversial.

In summary, Raymond Dart's recognition of the importance of the Taung skull opened a new era in the search for fossil humans. This search has proceeded at an accelerating rate until there are dozens of specimens and at least two species of the genus Australopithecus. The record shows that bipedalism is far older than large brains and stone tools. In fact, much of what we consider human evolved long after the separation from the apes.

Human intelligence, skills and language are products of uniquely human evolution. Isaac (1981) has shown how these different aspects of human behavior fit together, making a major new pattern. This human pattern evolved slowly for millions of years and then exceedingly rapidly for the last 40 or 50,000 years (Klein 1985). The latest acceleration is changing the world populated by a few hunters and gatherers into the crowded technical world of modern science.

Raymond Dart has contributed much to the modern formulation of the theory of human evolution. His contributions have been imaginative and courageous. It is a pleasure to honor him and his helper, the little child from Taung.

REFERENCES

Alvarez LW, Alvarez W, Asaro F, Michel HV (1980). Extraterrestrial cause for the Cretaceous-Tertiary extinction. Sci 208:1095.
Ardrey R (1961). "African Genesis." New York: Atheneum.
Ardrey R (1976). "The Hunting Hypothesis." New York: Atheneum.

Cronin JE (1983). Apes, humans, and molecular clocks: a reappraisal. In Ciochon RL, Corruccini AS (eds): "New Interpretations of Ape and Human Ancestry," New York: Plenum.

Curtis GH (1981). Establishing a relevant time scale in anthropological and archaeological research. In "The Emergence of Man," London: The Royal Society and the British Academy, p. 7.

Dart RA, Craig D (1959). "Adventures with the Missing Link." New York: Harper.

Eldredge N, Tattersall I (1982). "The Myths of Human Evolution." New York: Columbia U Press.

Goodman M, Tashian RE (1976) (eds). "Molecular Anthropology." New York: Plenum.

Hooton EA (1931). "Up from the Ape." New York: Macmillan.

Isaac GL (1981). Archaeological tests of alternative models of early hominid behaviour: excavation and experiments. In "The Emergence of Man," London: The Royal Society and the British Academy, p. 177.

Keith A (1931). "New Discoveries relating to the Antiquity of Man." London: Williams and Norgate.

Klein RG (1985). Breaking away. Nat Hist 94:4.

Leakey LSB (1934). "Adam's Ancestors." New York: Longmans, Green.

Leakey MD (1981). Tracks and tools. In "The Emergence of Man," London: The Royal Society and the British Academy, p. 95.

Lowenstein JM (1981). Immunological reactions from fossil material. In "The Emergence of Man," London: The Royal Society and the British Academy, p. 143.

MacLean PD (1970). The triune brain, emotion, and scientific bias. In Schmitt FO (ed): "The Neurosciences: Second Study Program," New York: Rockefeller U Press, p. 336.

Sarich VM (1983). Retrospective on hominoid macromolecular systematics. In Ciochon RL, Corruccini RS (eds): "New Interpretations of Ape and Human Ancestry," New York: Plenum, p. 115.

Schaller GB (1963). "The Mountain Gorilla: Ecology and
Behaviour." Chicago: U of Chicago Press.
Sibley CG, Ahlquist JE (1984). The phylogeny of the
hominoid primates, as indicated by DNA-DNA hybridization.
J Molec Evol 20:2
Straus WL (1962). The riddle of man's ancestry. In
Howells W (ed): "Ideas on Human Evolution," Cambridge:
Harvard U Press, p. 69.
Tobias PV (1984). "Dart, Taung and the 'Missing Link.'"
Johannesburg: Witwatersrand U Press.

Portrait of Raymond A. Dart and the Taung child, painted
by Alma Flynn of Johannesburg in 1983, on the occasion of
Professor Dart's 90th birthday

Hominid Evolution: Past, Present and Future, pages 19–24
© *1985 Alan R. Liss, Inc.*

TAUNG: A MIRROR FOR AMERICAN ANTHROPOLOGY

William W. Howells, Ph.D, D.Sc., F.S.A

Professor of Anthropology Emeritus
Peabody Museum, Harvard University
Cambridge, Mass., 02138 U.S.A.

The finding of the Taung child makes a telling reflection of the
nature of physical anthropology in the United States, then and for some
time after. The reception of Australopithecus in the Old World is well
known, from Professor Dart's own accounts and those of others. Reaction was
quite generally antagonistic. There was some slight sympathy in England but
only a German, Adloff, proclaimed it a hominid – a voice in the wilderness.
And what was the reception in America? Virtually none at all; a voice in a
vacuum. The New York Times put the find on the front page two days running,
with a couple of followups. The British journal Nature was available, of
course, and I have tear sheets from it of Dart's original 1925 article,
from Earnest Hooton's files at Harvard, showing that it was not simply
overlooked. Dart also was quoted in Science a couple of weeks later, and
had a second article in Natural History the next year.

But the anthropologists? Utter silence, except for Hrdlicka at the
National Museum. He was on a world skull-viewing trip in 1925, and detoured
to South Africa at Dart's invitation, where he studied the skull and the
site at first hand. Returning, he published his own account in the American
Journal of Physical Anthropology in the autumn of that year. The find was
important, he said, in extending the range of the anthropoid apes to South
Africa. He cited the British opinions, and gave as his own that the Taung
skull approached that of chimpanzees, with some reminiscences of a young
orang. As to its lack of supraorbitals, it was too young to have developed
any. In all probability, he wrote, "it is a new species, if not genus, of
the great apes."

In the next issue, early 1926, Hrdlicka approvingly reviewed two
foreign articles, one Portuguese and the other German, both of which
asserted that there was no justification for placing Taung anywhere in the
human phylum. A year later Hrdlicka commented on Dart's Natural History

article as one in which the author "very ingeniously, but, it seems obvious, more or less artificially, endeavors to humanize the 'Australopithecus'. It is not known that this effort thus far has found favor with any other student who gave truly earnest and critical attention to the otherwise very interesting and important Taungs relic."

After this, a void, with only one man taking some notice: William King Gregory, to whom I will return. As undergraduates of those days we heard nothing about it in courses. Naturally, there was no mention in Hrdlicka's 1930 volume, "The Skeletal Remains of Early Man". And no mention in Hooton's important book on human evolution, "Up From the Ape" in 1931. No scuttlebutt among enterprising or curious graduate students. By the same date Keith in England had relented somewhat, in his "New Discoveries Relating to the Antiquity of Man", accepting that Australopithecus, though a cousin of the chimp and gorilla, had developed in a human direction. And he allowed that Dart's assessment of Taung as a pre-human stage would call for serious consideration but for its probably too recent date — the entrance of this bugaboo which continued for some time to confuse the issue. At any rate, Keith devoted 80 pages of his book to the Taung fossil, while Americans devoted none at all. Australopithecus was not mentioned again in the pages of the American Journal of Physical Anthropology (and hardly elsewhere) until 1940. For that year, the whole volume was a commemoration of Hrdlicka's 70th birthday, and the committee of editors (not the birthday celebrant) invited Dart to contribute, which he did, writing on "The Status of Australopithecus." Thus was the other cheek turned.

Why this long neglect? Not because of any lack of interest in evolution and man's origins. Quite the reverse: in the United States there was intense interest in, and public controversy over, evolution during the 1920s. An Anti-evolution League was formed. In reaction scientists wrote letters to the papers. Clergymen chose up sides, with influential big-city preachers like John Roach Straton in New York and Aimee Semple McPherson in Los Angeles furiously denouncing evolution. Another trinomial divine, the liberal Harry Emerson Fosdick, lost his New York pulpit by preaching an anti-fundamentalist sermon, only to have a still grander church bestowed on him. Ex-president Wilson expressed his belief in evolution as would, he wrote, any man of education and intelligence. The champions of the two camps were Henry Fairfield Osborn and William Jennings Bryan. Osborn, the august, aristocratic and impressive president of the American Museum of Natural History would, I feel sure, have abandoned hope of Heaven could he have been present at last spring's celebration of ANCESTORS, the great reunion of hominid fossils at the Museum. Osborn was a paleontologist by trade, but also a superb publicist, who kept the Museum in the newspapers constantly while contributing his opinions to Science magazine several times a year. He was deeply interested in human origins and believed that they lay in Asia. Hoping to find the evidence he sent Roy Chapman Andrews on richly caparisoned expeditions to the Gobi Desert where Andrews, as the world knows, found dinosaur eggs instead.

Bryan was the three-time unsuccessful candidate for presidency of the nation, known as the Great Commoner, a man of the people, and always referred to as silver-tongued, for his oratorical powers. Now, in the 1920s, he was threatening a campaign to bar evolution from schools and, by one quote, to put the Bible into the Constitution. The whole controversy boiled over in 1925, the year of Taung, in the Great Monkey Trial in Dayton, Tennessee. A school teacher, John Thomas Scopes, had broken the new state law against the teaching of man's descent from a lower form. Anyone old enough must still have vivid memories of the newspaper stories and cartoons. The president of Harvard, the president of Yale, leading clergymen and a legion of scientists offered to help the defense. Evolution, in the favored parts of the country, was not only persuasive, it was chic. In opposition, Bryan put his fame and his faith at the service of the prosecution; he arrived in Dayton proclaiming that the battle between evolution and Christianity was to be a duel to the death. The defense had selected a panel of experts to testify for evolution, but to their dismay the judge barred them from testifying. Bryan himself became the chief witness, as an expert on the Bible. Scopes was convicted. But Bryan brought ruin on himself and his cause by getting trapped in problems of interpreting Genesis. And his ghastly proclamation bore fruit; he himself died of a heart attack, still in Dayton, just after the trial ended. The impression made by the trial on the public was immense, and smoothed the progress of popular understanding of human evolution from then on. Many years were still to pass before radio broadcasting would dare to deal with evolution on the airways, but six decades have changed the climate.

In 1922 Osborn had announced, in a lead article in Science , the discovery in Nebraska of the fossil molar tooth of an anthropoid primate, which he named Hesperopithecus haroldcookii (in honor of the finder.) On examination by Gregory and Milo Hellman it was pronounced to be in fact more similar in character to teeth of Pithecanthropus than to those of anthropoid apes. A North American ape-man was sensational, of course. However, when more such teeth turned up later in the same deposits, Gregory had to conclude in 1927 that all belonged to a Pliocene peccary, not to a primate. But in 1922 any stick would do to beat a fundamentalist.

In this atmosphere, when Osborn could use a single badly worn tooth to bedevil the anti-evolutionists with the improbable notion of a Pliocene apeman in America, it is difficult to see why Dart's explicit, detailed and highly suggestive paper was not avidly seized on as a better club to bash the bumpkins. But it was not. Again, why the long neglect? It is hard to believe that other writers were simply intimidated by Hrdlicka. Rather, there existed virtually no hands-on paleoanthropology in the United States. Germany and France, with plenty of fossil specimens, had their own schools of study and of phylogenetic interpretation well established, and in England there was the strong tradition among anatomists of interest in human fossils, together with forums for discussion like the Geological Society. However, this tradition was deflected from interest in Australopithecus because of Taung's provincial origins, as well as because of the confusion of dates and the obsession with Piltdown. But in America only Hrdlicka, with his Old World experience, showed real concern with phylogeny, as he argued his "Neanderthal stage", or lineal evolution from Java Man direct to Homo sapiens via the Neandertals It is true that Osborn in 1915 and MacCurdy in 1924 published books on prehistory in which the known human fossils were described at length but more compendiously than critically, and with phylogeny shown almost as an afterthought, following Keith in making all early fossil forms side branches without issue. The same tree appears in my course lecture notes of a few years later, and in fact it continued to flourish until the arrival of Weidenreich. In this field physical anthropology was then rather unfocussed, aside from having no Old World fossil materials in hand.

Assemble a physicist, a mammalogist, a classicist, an Africanist, a demographer and some others equally diverse, and what have you got? The American Association of Physical Anthropologists, at its first meeting in 1930 in Charlottesville, Virginia. The above are the fields of doctoral study respectively of Franz Boas, W. K. Gregory, E. A. Hooton, Melville Herskovits and Frederick Osborn. These, along with anatomists, prehistorians, general anthropologists and sundry others met there, under the scepter of Ales Hrdlicka, a graduate in homeopathic medicine. But Hrdlicka had studied at Broca's school in Paris as well. This dean of American physical anthopology had already founded the journal of that ilk in 1918, and had been trying for over thirty years to create some kind of institute devoted specifically and solely to the field.

Physical anthropology already had a long tradition, established especially by Samuel George Morton a century before. Another tradition of anthropology in the United States was the subsuming under that name of all fields — physical anthropology, archaeology, cultural anthropology and linguistics — instead of the separation among them more typical of European institutions. The current or recent presence of the American Indians may have fostered a tendency to look at them from all angles together. At any rate, the tradition led to the concept of a general anthropologist, one who had some acquaintance with most of these subjects, and to an expectation that all of them would be offered in a university, well before a specialist in physical anthopology was likely to be present.

Among the 83 charter members there were in fact few professed physical anthropologists. These were, apart from Hrdlicka, Stewart and Krogman, and Hooton, with a couple of his students including Shapiro. There were also anatomists whose central interest was physical anthropology and who remained prominent in the field, like Todd, Schultz, Straus, Terry, Mildred Trotter, Davidson Black and Dudley Morton. And, always important, Gregory the paleontologist and comparative anatomist. For some years the annual meetings of the Association took place in a comfortable small room, generally at the Wistar Institute in Philadelphia, where they heard something fewer than 30 papers.

In this framework, physical anthropology was not yet much of a discipline, and the only real teacher was Earnest Hooton at Harvard, who was himself largely self-taught, following his own interests. Hrdlicka and Todd gave some instruction but had no body of students. And Franz Boas, the most sophisticated of them all in the understanding of genetic variation and its mathematical measurement, was too occupied with other kinds of anthropology to be effective. Nevertheless the very openness of this self-developing field was its strength. Hooton's students, always encouraged to follow their individual bents, brought variety into the field as they gradually got positions in university departments. By the time of World War II they were bringing in the necessary special training and tools — genetics, serology, experimental anatomy, mathematical control and analysis — to make the discipline a science. After the War the great expansion of departments of anthropology — unique to the United States and Canada — also gave employment to really large numbers of physical anthropologists of this next generation. At the same time the largesse of the National Science Foundation and the Wenner-Gren Foundation supported work on a scale that allowed the adoption and use of ever more refined techniques. By last year, the most recent annual meetings of the American Association of Physical Anthropologists, its membership more than 1200, were holding, not one session but 38 or 39, with a total menu of over 300 papers. A highly visible outcome has been the fact that several leading American universities now have team investigations of <u>Australopithecus</u> and his relatives and forbears as a major research investment of their anthropology departments. And many of the techniques have become those of a hard science. This is the state to which 60 years of physical anthropology have come in North America. We should accordingly venerate two men. Hrdlicka, in spite of his waywardness, was indefatigable and effective in promoting the subject, founding the Journal and insisting that physical anthropology should have its own institution or society, as in France. Hooton, with his compelling teaching and openness, drew students to him over a long generation and sent them on their ways to follow their own chosen lines.

And how was Taung, now grown up into <u>Australopithecus,</u> faring in this hemisphere? It had one clearheaded friend throughout, Gregory at the

American Museum in New York. In the 1920s Gregory was fighting on two
flanks: against Wood Jones and his tarsian hypothesis on one side, and his
own Museum's president Osborn on the other. The latter, like Wood Jones,
favored a long separate, non-anthropoid ancestry for man. Gregory took up
the cudgels in 1930 against Osborn, refuting his hypothesis point by point
and, as useful evidence, specifically citing Australopithecus as a
progressive ape with a prevailingly human dentition.

In 1937 there was held in Philadelphia a major symposium on Early Man.
Of the 36 papers in the published volume many were highly parochial and of
little present interest. However, Broom made an early report on the new
finds of Plesianthropus and Paranthropus. And Gregory took full account
of Australopithecus, placing it as a very late member of the
Dryopithecus phylum and, on the basis of the first lower molar, ranking it
as one of two fossils particularly close to man, the other being one of the
Siwalik fragments later assigned to Ramapithecus. Gregory and Milo
Hellman then traveled to Johannesburg at Dart's invitation and in 1938 and
1939 published several papers on the dentition stressing the human features
and also creating a subfamily, Australopithecinae, for the now expanding
South African forms. They did not, however, place it in Hominidae. Of
course, in the late 1940s, with Broom and Robinson's discovery of pelvic
and spinal parts, and Dart's finding of revealing occipital evidence of
posture, it was all over. In 1947 Le Gros Clark in England argued that the
australopithecines were hominids and Keith, now in his eighties, handsomely
admitted that Dart had been right and he wrong. At the Cold Spring Harbor
Symposium in 1950 the hominid status was a matter of common consent, and
Mayr went so far as (temporarily) to incorporate Australopithecus in the
genus Homo.

Another decade was to pass before the entrance on the scene of the
varied East African forms. It had already taken a quarter of a century
after Taung before its hominid status was recognized — slow work. Of
course, there were handicaps: the young age, the solitary specimen, and the
constantly expressed idea that its date was too late for an actual human
ancestor. Nevertheless, at the present time, with so much attention
focussed on all these African hominids, it is curious to look again at
Dart's 1925 paper and see how well and simply he had assessed the meaning
of his find. Dart has been called many names in his long joint career with
the Taung child: just last November Phillip Tobias has called him
ebullient, emotional, imaginative, charismatic and also, although inclined
to scientific heresies, perceptive and prescient. How very true. Let us add
tenacious and courageous. What price heresy now? Galileo had his
Inquisition; Dart had his Establishment. But on the American side there
was, in 1925, no establishment critically equipped to assess
Australopithecus one way or the other, and the sin seems to have been mere
inattention.

Hominid Evolution: Past, Present and Future, pages 25–40
© *1985 Alan R. Liss, Inc.*

THE FORMER TAUNG CAVE SYSTEM IN THE LIGHT OF CONTEMPORARY
REPORTS AND ITS BEARING ON THE SKULL'S PROVENANCE: EARLY
DETERRENTS TO THE ACCEPTANCE OF AUSTRALOPITHECUS*

Phillip V. Tobias, MBBCh., PhD., DSc., DSc.h.c.

Head, Department of Anatomy, University of the Witwatersrand
Medical School, York Road, Parktown, Johannesburg 2193.

Honorary Professor of Palaeo-anthropology, Bernard Price
Institute for Palaeontological Research

Buxton near Taung is a point on the cliff face of the
Kaap (or Ghaap) Plateau where immense aprons of limestone have
formed by the action of springs and streams issuing from the
underlying rocks of the Campbell Rand. It is not the only
such lime fan along the face: others are at Witkrans, Thoming,
Boetsap, Klein Boetsap Fossil Hill, Ulco Grootkloof and Gar-
rokop (Young 1925; Peabody 1954; Partridge 1973, 1982; Butzer
1974, 1980; Marker 1975).

Two features of the lime tufas are noteworthy. First,
they dissolve readily, so forming solution cavities, some of
which, in turn, have become filled with sand, stones, vegetation and bones. Secondly, they are excellent preservatives of
the remains of living things that were carried or dragged into,
or trapped in the cavities. During fossilization, the chemicals in the limestone penetrated the bones and gradually replaced the organic matter, mainly collagenous fibres. Hence
a number of deposits of fossil bones have been preserved along
the escarpment. There are examples at Boetsap and Ulco - and
at the Buxton Limeworks at Taung. One skull that came out of
the Buxton Limeworks in November 1924 outshone all the rest.

THE BUXTON LIMEWORKS

Some time before 1916, the rich limestone deposits at
Taung were first revealed by Maurice Nolan - but then it was

* Shortened version of an address delivered at the University of Bophuthatswana, Mmabatho Campus, on 1st February 1985.

to the world of commerce and not to the world of science. He laid claim to 1000 acres of Buxton's lime deposits, limestone being a substance essential to the process of gold extraction. His brother, Herbert Gladstone Nolan, in 1896-7, had played a part in uncovering the Sterkfontein caves near Krugersdorp. H. G. Nolan set up the Nolan Lime Works and started to quarry limestone at Buxton about the beginning of 1916.

Quarry 1 was on the east face of the most southerly and oldest of the four major lime tufas at Buxton, namely the Thabaseek (from Thaba Sige, Black Mountain, in the language of the descendants of the 18th century Chief Tao, the Lion, and his followers). According to Gordon (1925a, b), the geologist to Nolan Lime Works, bodies of sandy limestone within caverns and fissures were exposed in quarry 1 from the outset. Some of these cave fillings contained bones.

THE BABOONS OF TAUNG

A geology student, S. Taylor, from the University of Cape Town, had the honour of recovering the first fossil baboons from Taung in 1919, about 3 years after quarrying had begun. Had the turn of the wheel been otherwise, it might have been that great university which would have had the ecstasy - and the agony - of revealing Africa's first ape-man. As things turned out, another baboon fossil, in the hands of J. Salmons, led to the Taung skull being unveiled to the world by the Witwatersrand University. Yet Taylor of Cape Town's engineering school deserves a place in the annals of Taung.

Not long after, more fossil baboons were found at Buxton by C. Urbach, S.J. Botha and W.J. Wybergh (Tobias 1984). This Wybergh published the first geological report on Taung, a note on the Buxton deposits in "The limestone resources of the Union" (Wybergh 1920). On 19 May 1920, S.H. Haughton read a paper on these baboons before the Royal Society of South Africa in Cape Town. His paper was not published at the time, though he later made his notes available to Broom, Dart and James Gear. After Dart's (1925a) announcement, Haughton (1925) published a note on the baboons. He recognised that they were different from, and smaller than, the Chacma baboon living in southern Africa today.

The presence of these extinct baboons led to the first statements about the age of the Buxton deposit. They moved

Wybergh (1920) to infer that the Buxton lime formation was
"relatively of great age and hardly to be classed as recent",
while Haughton (1925) expressed the view that his proposed
new species of small fossil baboon "may extend back in point
of time to a level contemporaneous with the early and possibly
pre-Pleistocene of Europe". These were the first of many attempts to find, guess, estimate or guesstimate the age of the
Thabaseek tufa and its contained fossils (Vogel 1984).

Towards the middle of 1924, E.G. Izod, a director of the
Northern Lime Company, visited Taung. From A.E. Spiers, the
Works Manager of the Buxton Limeworks, he obtained a fossilized baboon skull. Thinking it would make an appropriate, if
unusual, paperweight, Izod took it back to Johannesburg. His
son 'Pat' Izod, a student at Wits, showed the specimen to his
friends including Josephine Salmons. She was a science student in the Anatomy Department of the University's embryonic
Medical School, under its recently-appointed Australian head,
Raymond A. Dart. She showed the skull to Dart and he confirmed that it was a primate fossil akin to the extant African
baboons (Dart and Craig 1959; Tobias 1984).

Dart discussed the Buxton baboon with Robert Burns Young,
the Professor of Geology at the Witwatersrand University*. In
response, Young asked A.F. Campbell, General Manager of the
Northern Lime Company, to arrange for any further fossils to
be kept. By coincidence, Young had just been commissioned to
inspect the mineral deposits at Thoming, south of Buxton. He
promised to call on Spiers and look out for further specimens.

THE BUXTON CAVE SYSTEM

About that time, the Buxton quarrymen were blasting out
the calcified sandy filling of a cave in the limestone. The
directors did not like those pinkish, sandy deposits because
they yielded an impure lime, whereas the rest of the Buxton
limestone was nearly 100 per cent pure. In fact, from 1919,
occasional queries were raised at the Lime Company's board
meetings, as to how long it would be before the quarrymen got
through the pink earth. The cave filling had yielded the

---

* R.B. Young had, in 1908 published "The Life and Work of George William
  Stow", the geologist and ethnologist who had conducted the first geolog-
  ical survey in southern Africa and written a monograph on the geology of
  Griqualand West, just south and south-west of Taung.

baboons of 1919-1920. After 5-6 more years, the quarry face had been displaced about 100 metres (Hrdlička 1925) from the original rockface at, or close to, the eastern toe of the Thabaseek tufa. For all of this distance, the quarrymen had been observing - and reporting to the directors - the pinkish, sandy cave-deposits, which were commercially less valuable. Gordon (1925a, 1925b), who had been consultant geologist to the Lime Company from the beginning of operations at Buxton early in 1916, states that the cave system originally reached "right to the east front of the deposit" (1925b, p.xxxviii). According to Hrdlička (1925), in a report soon after he had excavated at Buxton quarry no. 1 in 1925, the quarrymen reported that petrified bones had been found in the pink sandy lime repeatedly. He adds, from information given on the spot, that "The majority of these specimens were lost in the operations, others were taken away by employés" (Hrdlička 1925, p.383)."

It is a remarkable point, which has hitherto escaped comment, that the small baboons found in 1919-1920 were of essentially the same forms as those found by Peabody in 1947-1948. Peabody excavated baboons near the positions of the hominid discovery of 1924 and of Hrdlička's disclosure of fossil baboons in August 1925. This suggests that the quarried part of the Thabaseek tufa had been riddled with a number of approximately contemporaneous cavities, or with a single cave system with tunnel-like galleries. This immense cave system had by August 1925 been shown to extend in the tufa for at least 100 metres in an almost south-north direction, since this quarry was being worked northwards, parallel to and not far from the original eastern margin of the Thabaseek tufa.

In horizontal extent the Buxton cave system seems to have borne comparison with the extensive cave systems of Sterkfontein and Makapansgat. The latter caves were formed in dolomite ribbed with chert and oolite, whereas those of Buxton were dissolved in a limestone tufa. The end-point was roughly the same: great underground cavities which at intervals opened to the surface, such openings allowing access of pinkish or reddish sand and animal bones, in a situation ideal for their preservation. At Taung, Broom (1946) "for convenience" spoke of Dart's Cave, Hrdlička's Cave and also of Spiers' Cave: at least the first two were probably no more than ramifications of a single labyrinthine cave system (Peabody 1954). The virtual identity of the baboon fossils in breccias at opposite ends of this Buxton cave system, separated horizon-

tally by some 100 metres, is in keeping with this interpretation.

THE APE-MAN OF TAUNG

Late in 1924, a Buxton quarryman, M. de Bruyn, possibly
instructed by A.F. Fleming, collected fossils, especially cer-
copithecoid skulls, that were being blasted out from the pink
sandy deposits in the face of quarry no. 1. He recognised the
baboon skulls and endocranial casts. One day, in November
1924, he found an endocast that was much bigger in size than
those of baboons. He delivered the specimen and other pieces
of rock containing bones to Spiers' office. There the pile of
rocks, including the oddly large brain-cast, lay in the Works
Manager's office for a week or two. Spiers was not to know
that he was temporarily sharing his office with the most critical ancestral skull ever discovered! On a November day in
1924, Young arrived. Despite a protracted search, I have not
found any record of the dates of de Bruyn's discovery or of
Young's visit, save that Young arrived in November and that
the skull had lain in the office for at least one week and,
at most, several weeks.

Young brought the large endocast and several chunks of
pinkish rock to Johannesburg and gave them to Dart. From a
letter sent by Dart to Sir Arthur Keith on 26 February 1925,
we learn that the specimen came to Dart on 28 November 1924 :
that is one date of which we can be certain. The other date
mentioned in the same letter was 6 January 1925, on which day
Dart's paper on the skull went away to London. Thus, only 40
days elapsed between the time the skull was placed in Dart's
hands and the despatch of his manuscript. (The reference in
Dart and Craig (1959) to 73 days prior to 23 December 1924 is
clearly an error (Tobias 1984).) What a 40 days they were!
In that time Dart, using improvised tools, extracted the skull
from the rock, studied, measured and described its morphology,
compared it with other hominoid skulls, had it drawn and
photographed, wrote his paper, had it typed, and despatched
his manuscript to Nature. It was an incredible high-speed ach-
ievement, even by the standards of 1985.

During those memorable 40 days, Dart, with imagination
and insight, discerned the skull's distinctive traits (despite
the fact that it was the skull of a young child). He appreciated those of its features now considered to be hallmarks of

early, small-brained hominids. With incisive neuro-anatomical acumen, and not a little courage, he pointed out features of the brain cast, which he believed indicated advancement of the cerebrum in an "ultra-simian" or "pre-human" direction, despite the small brain-size. He taught that quality rather than quantity was important in early stages of hominid brain evolution. The small canine tooth he likened to that of man rather than to the fanglike canine of modern apes. On the base of the cranium and endocast, he detected subtle pointers to the carriage of the head: it was poised, he asserted, on a nearly upright spinal column, in contrast with the oblique backbone of the apes. He went further: if the creature went on two legs instead of on all fours, the hands must have been freed for human-like sensory probings and motor skills.

To Dart, the Taung skull comprised an amalgam of apelike and manlike characters. Thus, he inferred, it represented an intermediate stage between apehood and manhood. It was, he felt, a higher primate that had belonged to an extinct ape species possessing features which were 'ultra-simian' and 'pre-human'. He called the new genus and species to which he assigned the specimen, Australopithecus africanus (Dart, 1925a).

Dart had the insight to realise that the Taung skull was relevant to the discussion of hominid origins and evolution. He apprehended also that the skull revealed something about the sequence of hominid evolution, that was quite at variance with what was expected on the teaching of his mentor, Elliot Smith. For if Taung represented a grade of anthropoid that was approaching hominid status, it showed that during hominidization, upright posture, bipedal gait, man-like canine reduction and qualitative making-over of the brain, were attained before brain enlargement. Moreover, Dart (1925a) had the historical sense to remind the world of that half-forgotten prediction, made by Darwin in 1871, that "... it is somewhat more probable that our early progenitors lived on the African continent than elsewhere".

In one respect, Dart's initial affirmation was modest, as wisdom after the event permits us to acknowledge. He was emphatic that A. africanus fell outside the range of the pongids. Yet he could not bring himself boldly to enrol it as a member of the Hominidae. Even the name, Australopithecus, points to this hesitation, for it means 'southern ape' (not southern man). It is interesting that W.J. Sollas of Oxford, after studying and comparing a median section through the Taung cranium that Broom had made for him, found it to be so human-like that he declared, in a letter to Broom dated 25 May 1925, that he would have named the Taung creature Homunculus (little man) (Broom 1946:17)! Years later, with hindsight, Heberer (1965) proposed to re-name it Australanthropus (southern man). Of course, on the rules of the game, the name once given must generally remain.

Dart even declined to look upon A. africanus as an ape-man. In 1925, Sir Harry Johnston, the celebrated British explorer, administrator and writer, cabled Dart that he thought the specimen was like an ape-man. Dart (1925b) referred to the cable at a meeting of the Medical Association of South Africa in Johannesburg in February 1925. He begged to differ from Sir Harry: it was, he said, 'a humanised type of ape'.

Instead of aligning Australopithecus with the hominids, Dart (1925a) proposed a new family to accommodate the genus. Homo-simiadae was the name suggested, literally man-apes: but the proposal failed to gain acceptance.

Yet, modest as Dart's initial claims might now appear to have been, they flew in the face of the fixed ideas of the time and aroused a storm of controversy and contumely. Dart was damned or dismissed by almost all his fellow anthropologists and by numerous members of the lay public. He was a heretic. He was impetuous and ebullient. He did not respect the grey-headed authorities who knew better! Even the name he gave Australopithecus was deemed etymologically unsound and barbarous (Woodward 1925; Bather 1925)!

Dart's willingness to place his head on a block over the status of his fossil child illustrates a feature of his make-up: he was a born rebel.

DART AND THE RÔLE OF DISSENT IN SCIENCE

Re-reading Nineteen Eighty-Four in the year which, fortuitously, Orwell chose as the setting of his macabre tale, I construed it as a lament over what Orwell saw as the impending decline and death of the freedom to dissent. Forcefully, I was struck by the rôle and value of dissent in scientific inquiry. How crucial it is, if science is to seek out and cross new frontiers, that some few thinkers should rebel against the conventional wisdom. Such a dissident was Dart. Perhaps he was even more of a rebel than the late Louis Leakey. For back in 1925 Dart had to topple more preconceived notions of the day than Leakey or any other student of hominid evolution was subsequently to confront.

To dissent needs originality, inventiveness, imagination, luck - and, above all courage. In withstanding the opposition and calumny which dissent elicits, Dart showed rocklike inner strength. Yet it is not hostility alone that the scientific rebel has to face: he has to contend with the unutterable loneliness of his position. Such a terrible solitariness did Darwin face 125 years ago when he published "The Origin of Species". Such a desolation did Dart confront when, for decades after the discovery of the Taung skull, he was virtually alone in his belief that Australopithecus had stood upon the threshold of humanity.

Even before the Taung skull was found, Dart's heterodoxies had been evident. In 1919, Dart and Shellshear tried to overthrow a long entrenched theory of neuro-anatomy, namely Wilhelm His's 19th century principle that all nerve-cells in the body arose from the embryo's neural tube and neural crest. Their challenge to His's doctrine (Dart, Shellshear 1922) was followed by Dart's (1923) paper on the brain of the ancestral Zeuglodont whales, in which he reached heterodox conclusions about the trigeminal nerve and the cerebellum.

So, even before he was out of his twenties, Dart was acquiring a reputation for spurning authority and jumping to conclusions. Of himself he wrote much later, "Such a person, I can see now in retrospect, was not only controversial, but upsetting and potentially dangerous." (Dart 1973). When Keith, that powerful figure in British anatomy and anthropology, was asked whether Dart was a suitable person for the Chair of Anatomy at the University of the Witwatersrand, he wrote: "I was one of those who recommended him for the post, but I did so, I am now free to confess, with a certain degree of trepidation. Of his knowledge, his power of intellect, and of imagination there could be no question; what rather frightened me was his flightiness, his scorn for accepted opinion, the unorthodoxy of his outlook." (Keith 1950:480).

Dart was but 29 years of age when he was appointed to the Witwatersrand Chair. He turned 30 only days after he set foot in the 37-year-old mining town of Johannesburg. How much long-

er would the appreciation and appraisal of Australopithecus
have taken, if it had fallen into the hands of a more conventional and timid scientist than Dart?

If I speak feelingly, it is because to a modest degree,
something like this has befallen my own career on several oc-
cations. For example, I experienced such aloneness when,
from 1955, following early studies by the Wits Kalahari Research Committee on the San (Bushmen), I tried to convince
colleagues that the San were not a "dying race" or "a degener-
ate product of desert conditions", as respected text-books
were proclaiming, but a 50 000-strong, highly adapted and adept people (Tobias 1956). Again, for 15 years I strove to establish that the taxon, which Leakey, Napier and I (1964) had
named Homo habilis, represented a good species of hominid.
Although both notions are now widely accepted, there were
years of intellectual isolation before the points were gained.

SOME DETERRENTS TO THE ACCEPTANCE OF AUSTRALOPITHECUS

On the 60th anniversary of the discovery of the Taung
child, we recall that it was the first of all Africa's testi-
monies to the dawn of humankind. That skull of an ancient
child of humanoid aspect and all that flowed from its revelation were to change dramatically our conception of human origins. Inevitably, it is counted among the major scientific
discoveries of the 20th century. Yet, the skull could not
have been unveiled under less auspicious circumstances. The
dice were heavily loaded against its acceptance.

Geographical obstacle. For one thing it seemed to be in
the wrong part of the world! 'Java Man' (Pithecanthropus
erectus) had been brought to light by E. Dubois in 1890-92.
Asia came to be seen as the probable cradle of humanity by
leading investigators. These included H.F. Osborn and W.D.
Matthew of the American Museum of Natural History, Davidson
Black of Canada and later China, and A.E. Grabau, described
as "the father of Chinese palaeontology" (Koenigswald 1956).
North of the Himalayas was the favored area and in the early
1920's Roy Chapman Andrews led expeditions to the Gobi Desert
on the Central Asian High Plateau. This Asia-centered view
was reinforced by the finding of apparently ancient, human-
like teeth in China early in the 20th century. These teeth
led Grabau to use, for the first time, the term "Peking Man"
in the 1920s. Hence, on the ideas of the time, the Taung

child was geographically inconvenient.

Dating deterrent. While men like D.M.S. Watson and A.W.
Rogers held that "there was no probability of the age of the
(Buxton) deposit being determined", others opined that the
Taung skull was too recent to have been ancestral to man. For
example: "The Taungs ape is much too late in the scale of time
to have any place in man's ancestry" (Keith 1925b) and " ...
one can be certain that early and true forms of men were already in existence before the ape's skull described by Prof.
Dart was entombed in a cave at Taung" (Keith 1925c); "The
antiquity of the find is not necessarily very great ..."
(Hrdlička 1925); "The geological evidence ... seems to indicate that Australopithecus africanus was not the predecessor,
but contemporary of man." (Gordon 1925a); " ... The Taung
skull is not geologically of great antiquity - probably not
older than Pleistocene ..." (Broom 1925), which view, Broom
(1930) was later to retract, after identifying a number of
mammals found in the breccia with the hominid skull (" ...
the evidence is strongly in favour of the deposit being Pliocene, and I think quite likely it will prove to be Lower Pliocene" op. cit.:814). Although some of these judgments went
beyond the available facts, collectively they created a climate of belief that Taung was of too recent a geological age to
permit Dart's claims to be supported.

Morphological obstacle. The structure of the Taung skull
was not in accord with one of the prevailing concepts of a
human ancestor. A strong body of opinion, supported especially
by Elliot Smith, held that the expansion of the absolute size
of the brain (which is so staggering a feature of modern man)
must have started early in the advent of hominids, perhaps
when jaws and teeth were still quite ape-like. The Piltdown
forgery seems to have been an attempt to translate this concept into hard 'fact' - for that infamous fraud was based on
a large-brained human skull and an ape's carefully doctored
lower jaw and teeth! Taung turned this notion on its head:
for it represented a creature whose teeth were human-like,
whilst its natural endocast betrayed a brain-size no bigger in
absolute terms than that of some modern apes. On the fixed
idea of the day, the child of Taung was anatomically the wrong
kind of creature.

Juvenile obstacle. A fourth drawback was that the skull
represented a child, as is inferred from the presence of all
20 deciduous teeth, whilst the first permanent molars had

erupted, the lowers completely, the uppers incompletely. At such a stage of dental eruption, the modern human child would be nearly 6 years and the modern ape child rising 4 years (Keith 1925a). That the first discovery of an 'ape-man' was of a child was a decided disadvantage, as Keith (1925a), Elliot Smith (1925), Woodward (1925) and Duckworth (1925) pointed out. Its juvenile status left two questions unanswered: How much of the humanoid aspect of the child was simply the consequence of immaturity? Secondly, into what kind of adult would the Taung child have grown, if it had lived longer?

Dartian deterrent. Fifthly, it seemed to some that the fossil fell into the wrong hands. Dart was young (31 3/4 years), rather inexperienced, inclined to scientific heresies, flighty, fanciful, a product of those troubled wartime years of the previous decade. Some would have said that these were a most improbable combination of qualities for the man who was destined to describe and evaluate the world's first discovered, small-brained, bipedal hominoid. Yet, in the event, Dart proved to be the right man in the right place at the right time (Tobias 1985).

TAUNG A PREMATURE DISCOVERY

Although the discovery and Dart's claims constituted a breakthrough in the history of man's probings for ancestral roots, the world treated Dart's claims with scepticism, even scorn. This opposition, and the intervention of World War II, delayed the acceptance of Dart's claims for over 25 years.

It is easy to describe Dart's discovery as ahead of its time or premature. Yet, when we look more closely at that statement, what does it mean? Are we simply saying, it was before its time, because the time was not ripe for it? Such a statement explains nothing; it simply says the same thing over twice in different words; it is a pure tautology and it rings hollow. Yet is there no sense in which a discovery may be described as premature - and the description ring true?

Stent (1972) raised this very question in speaking of the prematurity of the Avery et al. (1944) proof that DNA was the basic hereditary substance. He sought a criterion of a premature discovery, other than its failure to make an impact at the time. He proposed that "A discovery is premature if its implications cannot be connected by a series of simple logical steps to canonical, or generally accepted, knowledge." The 1944 study of Avery and his co-workers failed to be connected with the accepted knowledge of the day because the view still prevailed that DNA comprised a regular sequence of repeating items. On this view, the specificity of the genes could not reside in the DNA and had been assigned to the chromosomal protein. Avery's claim could not be connected logically to this paradigm. In this sense, the 1944 claim was premature.

Similarly, Mendel's concept of the laws of inheritance could not be appreciated in 1865, as no component of cells was known to behave in the manner shown by Mendel's "essences". Moreover a statistical approach was foreign to biological research. So Mendel's 1865 concept of heredity was premature (Stent 1972). In 1883 and 1884 (the year of Mendel's death) the chromosomal phenomena of meiosis were demonstrated; but by then Mendel's paper had been forgotten. Only after its rediscovery nearly 20 years later was it realised by Sutton and Boveri that the behaviour of Mendel's essences paralleled that of chromosomes in meiosis. Also, by 1900, the use of statistics in biology was common.

Perhaps another example of a premature discovery is provided by acupuncture. Although practised in the Orient for centuries, it failed to be appreciated in scientific medicine, for its reults could not be logically related to canonical knowledge. Only recently is scientific medicine taking a serious interest in acupuncture, since the discovery of the endogenous opioids has provided a rational basis for it.

Dart's claims for the Taung skull clearly fall into the category of a premature discovery, as defined by Stent, since they could not be logically connected to the canonical knowledge of the day. The geographical zone was wrong; the geological age seemed to be too young; its morphology was at variance with the concept of the primacy of brain enlargement; whilst the acceptance by a number of authorities of the veracity of Piltdown militated against the acceptance of Taung.

On this view, the measure of Mendel's prematurity was 35 years; that of Avery and his colleagues 6-8 years, and that of Dart 22-28 years - according to which event and date one regards as marking the ultimate acceptance of Dart's claims.

THE FATE OF TAUNG

Although Dart did not place <u>Australopithecus</u> in the hominids, others eventually did classify it as such. But this
could not happen until many more African hominid fossils, including adult specimens, had been unearthed and investigated
by Dart, Broom, Schepers, Robinson, Le Gros Clark and others.
It could happen only after hundreds of skeletons of great apes
had been studied by Zuckerman and others, so that the limits
of variability of the apes could be statistically defined. One
had to know what <u>was</u> an ape in order to decide what was not!

There had to be time for the old Asia-centered concept of
human origins to lapse into disuse, after evidence had accumulated that the oldest African man-like fossils were more ape-
like and more ancient than the oldest Asian hominid fossils.
Another concept that had to be shed was the idea that the increase in absolute brain-size had been an early feature in
hominid emergence. Evidence to support this theory had been
furnished by the big-brained Piltdown skull.

TAUNG AND THE PILTDOWN FORGERY

The Piltdown remains, found in a gravel pit on the Sus-
sex Downs of England between 1908 and 1913, showed the astonishing combination of a large-brained cranium, of rather
modern aspect, with an ape-like jawbone* and lower canine
tooth. As long as Piltdown, was accepted as genuine and considered an ancient human precursor, it was impossible to accept that <u>Australopithecus</u> was ancestral to man. One of
Dart's former students, J.S. Weiner, suggested in the early
1950s that a hoax had been perpetrated. K.P. Oakley, through
more precise fluorine tests than those previously applied,
showed that the jaw and canine tooth were modern and of a different age from the cranium. They had clearly not belonged to
the same individual. Weiner, Oakley and Le Gros Clark (1953)
then detected and laid bare the signs of the intentional forgery.

It was none too soon; for as their report (1953) stated,
"... the elimination of the Piltdown jaw and teeth from any
further consideration clarifies very considerably the problem
of human evolution. For it has to be realised that 'Piltdown

---

* Now known to have belonged to an orang-utan (Lowenstein, this volume).

man' (Eoanthropus) was actually a most awkward and perplexing element in the fossil record of the Hominidae, being entirely out of conformity both in its strange mixture of morphological characters and its time sequence with all the palaeontological evidence of human evolution available from other parts of the world." (op.cit: 145-6). When the hoax had been perpetrated more than 40 years earlier, its features had been in conformity with the then fixed ideas about human evolution.

The exposure of the Piltdown remains as fraudulent dealt a final, fatal blow to the notion that the increase of absolute brain-size had been first in the field. Piltdown had helped to delay the acceptance of Dart's claims for Australopithecus. In some peoples' minds, it produced a hold-up of 28 years - from 1925 to 1953!

AFTERWORD

By the early 'fifties, almost all obstacles to the acceptance of Australopithecus had disappeared. The first 30 years of the Taung debate, 1925 to 1954, were over and in the second 30 years, from 1955 to 1984, Australopithecus gained pretty well universal acceptance as a member of the hominids and as a genus, one or more of whose species were on the direct lineage of modern man. The 30 years of loneliness, the lean years of rejection, had given way to 30 years of acceptance, the fat years of recognition. Happily Dart, now 92 years old, has lived to see his claims vindicated.

The message of Taung and Australopithecus points powerfully to the non-human or animal origins of humankind. Australopithecus may be looked upon as the animal phase of the hominids and Homo as the human hominids. Thus, Australopithecus presents itself as the veriest crystallization of the evolutionary concept as·applied to man.

Finally, the available evidence signals Africa as the cradle of mankind. Nearly 25 per cent of the Earth's habitable surface, Africa straddles the Equator; about two-thirds of its landmass lies between the Tropics. This area has been the crucible which on present evidence generated the unique biological and cultural elements that became Man. Those who disparage Africa and its achievements may need reminding of its supreme benefaction : Africa gave the world the first hominids and the first human culture.

My warm thanks are extended to Valerie Strong.

Avery OT, McLeod CM, McCarty M (1944). Studies on the chemical nature of the substance inducing transformation of pneumococcal types. J Exp Med 79:137.

Bather FA (1925). The word Australopithecus and others. Nature 115:947.

Broom R (1925). Some notes on the Taungs skull. Nature 115: 569.

Broom R (1930). The age of Australopithecus. Nature 125:814.

Broom R (1946). The Occurrence and General Structure of the South African Ape-men. Part 1, The South African Fossil Ape-men, the Australopithecinae. Transvaal Mus Mem 2:11.

Butzer KW (1974). Palaeoecology of South African australopithecines: Taung revisited. Curr Anthrop 15:367.

Butzer KW (1980). The Taung australopithecine: contextual evidence. Palaeont afr 25:59.

Dart RA (1923). The brain of the Zeuglodontidae (Cetacea). Proc Zool Soc (London), p 615.

Dart RA (1925a). Australopithecus africanus, the man-ape of South Africa. Nature 115:195.

Dart RA (1925b). Australopithecus africanus: demonstration before SAf Medical Association, Johannesburg, February 1925. Med J S Afr 20:183.

Dart RA (1973). Recollections of a reluctant anthropologist, J Hum Evol 2:417.

Dart RA, Craig D (1959). "Adventures with the Missing Link". New York: Harper.

Dart RA, Shellshear JL (1922). The origin of the motor neuroblasts of the anterior cornu of the neural tube. J Anat 56:77.

Duckworth WLH (1925). The fossil anthropoid ape from Taungs. Nature 115:236.

Gordon GB (1925a). The Taungs discovery: geological point of view. The Cape Times, 27 February 1925:11.

Gordon GB (1925b). Comments on R.B. Young's paper on the calcareous tufa deposits of the Campbell Rand. Trans Geol Soc S Afr 28:xxxvii.

Haughton SH (1925). Note on the occurrence of a species of baboon in limestone deposits near Taungs. Trans Roy Soc S Afr 12:68.

Heberer G (1965). "Menschliche Abstammungslehre". Stuttgart: Gustav Fischer: 313.

Hrdlička A (1925). The Taungs ape. Amer J Phys Anthropol 8:379.

Keith A (1925a). The fossil anthropoid ape from Taungs.

Nature 115:234.

Keith A (1925b). The Taungs skull. Nature 116:11.

Keith A (1925c). The Taungs skull. Nature 116:462.

Keith A (1950). "An Autobiography". London: Watts.

Koenigswald GHR von (1956). "Meeting Prehistoric Man". London: Thames & Hudson.

Leakey LSB, Tobias PV, Napier JR (1965). A new species of the genus Homo from Olduvai Gorge. Nature 202:7.

Marker ME (1975). On Taung revisited. Curr Anthrop 16(2):295.

Partridge TC (1973). Geomorphological dating of cave opening at Makapansgat, Sterkfontein, Swartkrans and Taung. Nature 246:75.

Partridge TC (1982). The chronological positions of the fossil hominids of southern Africa. Cong Int de Pal Humaine, Nice. Prétirage 2:617.

Peabody FE (1954). Travertines and cave deposits of the Kaap escarpment of South Africa, and the type locality of Australopithecus africanus Dart. Bull Geol Soc Amer 65:671.

Smith GE (1925). The fossil anthropoid ape from Taungs. Nature 115:235.

Stent GS (1972). Prematurity and uniqueness in scientific discovery. Sci Am 227:84.

Tobias PV (1956). On the survival of the Bushmen. Africa 26:174.

Tobias PV (1984). "Dart, Taung and the 'Missing Link'". Johannesburg: Witwatersrand University Press for The Institute for the Study of Man in Africa.

Tobias PV (1985). The life and times of Emeritus Professor Raymond A. Dart. S Afr Med J 67:134.

Vogel JC (1984). Did Australopithecus and Homo co-exist? Nuclear Active 31:19.

Weiner JS, Oakley KP, Le Gros Clark WE (1953). The solution of the Piltdown problem. Bull Br Mus (Geol) II: No. 3

Woodward AS (1925). The fossil anthropoid ape from Taungs. Nature 115:235.

Wybergh WJ (1920). The limestone resources of the Union. In "Limestones of Natal, Cape and Orange Free State Provinces". Mem S Afr Geol Survey No. 11.

Young RB (1925). The calcareous tufa deposits of the Campbell Rand, from Boetsap to Taungs native reserve. Trans Geol Soc S Afr 28:55.

Hominid Evolution: Past, Present and Future, pages 41–46
© *1985 Alan R. Liss, Inc.*

INTERPRETING EARLY HOMINID DEATH ASSEMBLAGES: THE RISE OF
TAPHONOMY SINCE 1925

C.K. Brain

Transvaal Museum
P.O. Box 413, Pretoria
South Africa

Within four years of his description of the Taung child
skull, Professor R.A. Dart published speculations on the
origin of the bone accumulation found in association with
the skull. He wrote (Dart 1929:648):

"Examination of the bone deposit at Taungs shows that it
contains the remains of thousands of bone fragments. It is
a cavern lair or kitchen-midden heap of a carnivorous beast.
It is not a water-borne deposit and the Taungs remains could
not have been washed into the cavern from the surface. The
bones are chiefly those of small animals like baboons, bok,
tortoises, rodents, bats and birds. Egg shells and crab
shells have also been found. This fauna is one which is not
characteristic of the lair of a leopard, hyena or other
large carnivore, but is comparable with the cave deposits
formed by primitive man. The deposit was, therefore, formed
by primitive man or by Australopithecus, an advanced ape
with human carnivorous habits. As no human remains have
been found there, and as no Australopithecus remains have
been found elsewhere in known Pleistocene deposits, I am
of the opinion that the deposit was formed by the Taungs sub-
man himself".

From this statement it is obvious that even at this early
stage, Dart was starting to come to grips with basic taphonomic concepts, with the overall composition of a bone
assemblage and with identification of the bone-collecting
agencies involved. Shortly afterwards he started to speculate on the origin of depressed fractures that he observed
in the cranial vaults of fossil baboon and hominid skulls

from Taung and Sterkfontein, surmising that these were the
results of deliberate hominid violence. Turning his attention to the bone accumulation in the Makapansgat Limeworks
cave, he produced a remarkable series of 39 papers between
1949 and 1965 (for a full bibliographic listing, see Brain
1981) which set forth his hypothesis of predatory, canniba-
listic australopithecines who made use of an *osteodontokeratic* (bone, tooth, horn) culture. His concepts and statements were highly provocative and stimulated a great deal
of further related research, including all the taphonomic
contributions which I have subsequently made. In fact, in
my case, it was Dart's Makapansgat work that motivated my
taphonomic investigations of the other South African australopithecine assemblages, leading to the publication of a
book *The Hunters or the Hunted?* (Brain 1981). The results
reached were not the same as Dart's but there is no question
as to the source of inspiration for the work.

During the 1970s, interest in the interpretation of bone
assemblages had developed to the point that a Wenner-Gren
symposium on Vertebrate Taphonomy and Paleoecology was held
at Burg Wartenstein, Austria in 1976, organised by A.K.
Behrensmeyer, A.P. Hill, A. Walker and myself, with generous
support from Lita Osmundsen. It was here that the science of
Taphonomy may be said to have crystallised and the symposium
led to a book *Fossils in the Making* (Behrensmeyer and Hill
1980). Since then taphonomic research is being vigorously
conducted by a number of people, leading to the publication
of such books as *Life History of a Fossil* (Shipman, 1981)
and *Bones: Ancient Men and Modern Myths* (Binford, 1981). In
1984 the First International Conference on Bone Modification
was held in the United States, organised by Robson Bonnichsen
and Marcella Sorg. It attracted a large number of delegates
and its proceedings are now in the press.

Although it would be incorrect to claim that taphonomic
speculations started with Raymond Dart and the Taung skull,
it was Dart's imagination and impetus that promoted the
discipline into a flourishing field of research.

TAPHONOMY AS APPLIED TO THE TAUNG SITE

The geological context of the Taung fossils was described
by Young (1925) soon after the finding of the hominid skull.
He assumed that the Buxton travertine, in which the cave deposit lay, formed a single entity that had resulted from

continuous deposition. In his later study, Peabody (1954) concluded that the travertine mass consisted of four separate units, of different ages, and that the Taung skull came from an infilling in the oldest of these, the Thabaseek travertine. Each travertine mass was found to have gently sloping basal layers upon which were built up hemispherical carapaces of pure calcium carbonate whose layers were frequently very steeply inclined. It is between such concentric carapace layers that secondary caves are prone to form.

In view of the extensive mining which has been done in the Buxton travertines, it is difficult to reconstruct the situation as it may have been when the remains of the now-fossil animals were accumulating in the cave. Recently I have had the opportunity of examining a series of very similar but undamaged travertine deposits set in a wild and natural environment of the Namib-Naukluft Park in S.W. Africa/Namibia. Here it is possible to observe travertines forming and eroding, while natural taphonomic processes may be studied in the numerous secondary cavities within the travertine masses themselves.

Fig.1. A massive deposit of cliff travertine, about 70 m high, built out from a dolomite valley in the Naukluft Mountains, S.W. Africa. Before mining, the Taung deposit was of this kind.

Figure 1 shows a characteristic, massive block of travertine about 70 m high, built out by spring action in a steep valley of Precambrian dolomite. Several such travertines are forming at present in the Naukluft mountains and, in such places, it is possible to see how closely the formation of carapace travertine is linked to the presence of moss and algae. The steeply-inclined carapace layers are formed where water flows or seeps slowly over living moss, often in the form of hanging curtains. Quite apart from natural evaporation of this lime-rich water, it is the photosynthesis of the moss and algae that removes carbon dioxide from the water, leading to the precipitation of calcium carbonate. Moss-banks in various stages of calcification may be observed in such places and it is not unusual to find natural cavities behind hanging curtains of moss. Such cavities, enlarged by subsequent erosion (see Fig. 2), are probably the sort of secondary hollow in which the Taung fossils were deposited.

Fig. 2. A characteristic cavity in the steeply-inclined "carapace" lime of a Naukluft travertine. The cave where the Taung child skull came to rest may have had a similar appearance.

Examination of the bones currently accumulating in the

Naukluft travertine caves show that the expected taphonomic accumulating processes are at work there. Some bones have been collected by porcupines; others show unmistakable signs of leopard feeding activity.

In the light of current taphonomic criteria, what might one say about the circumstances of death of the Taung child? The skull came to light through mining and, although many baboon skulls were blasted out of the travertine face at about the same time, stratigraphic and contextual relationships are not known. In the late 1940s the University of California Africa Expedition excavated fossils at Taung from a filling known as Hrdlicka's Cave, thought to be a continuation of the infilling that yielded the original skull. At least one large block of limestone, numbered 56831, was preserved intact (Fig. 3) and contained numerous relatively complete baboon skulls, together with skeletal fragments showing extensive carnivore damage.

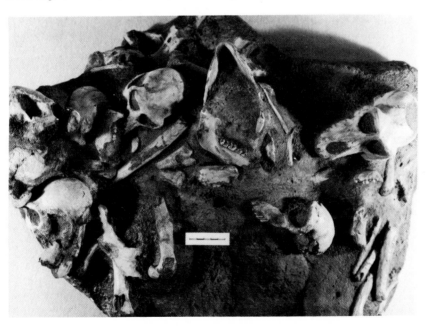

Fig. 3. A block of sandy limestone from a cave filling adjacent to that which yielded the Taung child skull. Extensive carnivore damage to the bones suggests that we are dealing here with the food remains of a large cat which fed on baboons in a cave lair.

I would have no hesitation in attributing these baboon remains to the feeding activity of a large carnivore, probably a leopard, which regularly consumed its prey in a cave lair. Both Peabody (1954) and Butzer (1974) have pointed out that there were two stratigraphic horizons in the Taung cave, - a lower sandy level indicating dry conditions, and an upper calcareous one, source of the hominid skull, reflecting a moister environment. It is not known if the taphonomic conclusions drawn from the baboon-skull block from Hrdlicka's Cave, coming probably from the lower dry levels of the cave, would apply equally well to the context of the hominid skull itself. My personal suspicions are that they would and that the Taung child fell prey to a large cat in the same way as the baboons did.

ACKNOWLEDGEMENTS

Particular thanks are due to Professor P.V. Tobias for his initiative in organising the Taung Diamond Jubilee International Symposium and for his invitation to contribute to the proceedings. My thanks go also to the S.W. African Dept. of Nature Conservation, in particular Dr E. Joubert, for permission to do research in the Namib-Naukluft Park.

REFERENCES

Behrensmeyer AK, Hill AP (1980). "Fossils in the Making. Vertebrate Taphonomy and Paleoecology".Chicago: University of Chicago Press.
Binford LR (1981). "Bones: Ancient Men and Modern Myths". New York: Academic Press.
Brain CK (1981). "The Hunters or the Hunted? An Introduction to African Cave Taphonomy". Chicago: University of Chicago Press.
Butzer KW (1974). Paleoecology of the South African australopithecines: Taung revisited. Current Anthropology 15: 367.
Dart RA (1929). A note on the Taungs skull. S Afr J Sci 26: 648.
Peabody FE (1954). Travertines and cave deposits of the Kaap Escarpment of South Africa, and the type locality of *Australopithecus africanus* Dart. Bull Geol Soc Amer 65: 671.
Shipman P (1981)."Life History of a Fossil". Cambridge: Harvard University Press.
Young RB (1925).The calcareous tufa deposits of the Campbell Rand, from Boetsap to Taungs native reserve. Trans Geol Soc S Afr 28:55.

Hominid Evolution: Past, Present and Future, pages 47–62
© *1985 Alan R. Liss, Inc.*

THE PAST, PRESENT, AND FUTURE SIGNIFICANCE OF THE LUNATE
SULCUS IN EARLY HOMINID EVOLUTION

Ralph L. Holloway

Department of Anthropology
Columbia University
New York, New York 10027

INTRODUCTION

This Jubilee of the Taung discovery, and Prof. Raymond
Dart's extraordinary prescience in realizing its
significance, provide an opportunity to look once again at a
controversial issue of great importance to our nascent
understanding of human brain evolution. There is a certain
irony to it all: 60 years ago, the position and importance
of the lunate sulcus on the Taung endocast were integral
parts of the debates of those times; 60 years later, the
issue would appear to be unresolved, surrounded by a
veritable penumbra of doubt, confusion, and alas, ignorance.

Several "anniversaries" are worth mentioning at the
outset of this paper, which is unabashedly an homage and
tribute to one man so instrumental in setting my own career
in the perfidious endeavours of paleoneurology. 1986 will
mark the 60th anniversary of Dart's (1926) re-assertion
that the Taung lunate sulcus was in a non-pongid posterior
position. 1986 will be the 50th anniversary of the Clark et
al (1936) paper which attempted to rebut Dart's earlier
claims, and as will be shown below, holds an important key
to the puzzle of both Taung and Hadar interpretations
(Holloway 1981a, 1983,a,b,c, 1984). 1987 will be the 40th
"anniversary" of Clark's (1947) classic paper which, while
not definitively identifying the position of the cerebral
landmarks on the Taung specimen, nevertheless supported most
of Dart's original perceptions, including the lunate sulcus.
1986 will mark the 40th anniversary of Schepers' (1946,
1950) attempts to interpret those convolutional details

which brought him to an unusual statement of positive accord. 1986 will mark the 30th anniversary of Dart's (1956) paper which recognized the reorganization of brain structures in early hominid evolution as something equally important as mass, or size (if not more so).

THE ISSUE OF THE LUNATE SULCUS

"Whatever the total dimensions of the adult brain may have been, they are not lacking evidences that the brain in this group of fossil forms was distinctive in type and was an instrument of greater intelligence than that of living anthropoids... The pithecoid type of parallel sulcus is preserved, but the sulcus lunatus has been thrust backwards toward the occipital pole by a pronounced general bulging of the parieto-temporo-occipital association areas...

"The expansion in this area of the brain is the more significant in that it explains the posterior humanoid situation of the sulcus lunatus. It indicates... the fact that this group of beings... had profited beyond living anthropoids by...a...much larger area of cerebral cortex to serve as a storehouse of information concerning their objective environment... They possessed to a degree unappreciated by living anthropoids the use of their hands and ears and the consequent faculty of associating with the colour, form, and general appearance of objects, their weight, texture, resilience, and flexibility, as well as the sounds emitted by them... They had laid down the foundations of that discriminative knowledge of the appearance, feeling, and sound of things that was a necessary milestone in the acquisition of articulate speech." (Dart 1925:197-198; see also Dart 1926:324-327 for comments regarding the size of the Taung brain and expansion of posterior parietal association cortex, and McGregor 1925.)

Perhaps the language is a bit more optimistic or piquant than modern descriptions of cerebral regions and functions, but Dart's (1925) quotation above compares favorably with conclusions drawn from modern neurobiology (c.f. Geschwind 1965; Damasio, Geschwind 1984; Hyvarinen 1982; Angevine, Smith 1982; Jones 1984, Eidelberg, Galaburda 1984), regarding both the sensory and integrative functions of anterior occipital and posterior parietal cerebral cortex, i.e., areas 18 and 19 per Brodmann, or peristriate cortex.

The lunate sulcus is the anterior boundary of primary visual striate cortex, which has a particular cytoarchitectonic appearance, as it contains the stria of Gennari, densely composed of stellate neurons. This is koniocortex, and is the first way-station in the cerebral neocortex for the termination of the radiating visual fibers leaving the lateral geniculate body of the thalamus. It is a crescentic furrow or sulcus, concave posteriorly, and is invariably in an anterior position in Anthropoidea. The function of striate cortex is almost totally sensory, although the anterior margins have some integrative functions. The striate cortex's nearest anterior neighbor is parietal "association" cortex, better known for its integrative functions in cognition, although its posterior margin also subserves visual sensory tasks. Damage within the striate cortex produces scotopias, or areas of visual blindness. Ablations in parietal "association" cortex do not normally produce scotopias.

In Pan, Pongo, Gorilla and Hylobates, the lunate sulcus is readily visible, always anterior to the lateral calcarine sulcus, and always posterior to the interparietal sulcus whose posterior portion abuts against the lunate sulcus (see Holloway 1983b for diagrams). In modern Homo sapiens, the lunate is often present, and when visible is in a decidedly posterior position (see Levin 1936, and Connolly 1950 for frequencies). Interestingly, Smith (1904a) found an unusual exception to this general condition in one Egyptian woman's brain. Smith (1904a, 1904b, 1907a,b, 1927, 1929) spent years studying the distribution, development and comparative aspects of the lunate sulcus. This does not mean that invoking his name is, as Falk (1983a: 487) described it, "argumentum ad verecundiam", but rather an indication that as Dart did study under Smith, he (Dart) had probably heard of the lunate sulcus, and did not cast his eyes upon the Taung endocast as an unknowledgeable virgin who would confuse the lambdoid suture with the lunate sulcus! But in the interest of Latin scholarship, Falk missed this essential point. Falk (1982:86) claims that one looks in vain for any reorganization in the human brain, despite her earlier (1980a:104) examples from Armstrong's (1979, 1980) work. Passingham and Ettlinger (1973:241-242), using Stephan's (1969) data showed that the human value for primary visual striate cortex fell well beneath the log-log equation of striate cortex vs. brain volume. In 1979, I showed the same results with 4 independent samples, and had discussed such findings in 1976 as a response to Jerison's (1973) critique of the reorganization concept.

On Stephan et al's (1981) newer data base, the Homo sapiens value for visual striate cortex is 66% less than expected for a primate of its brain size. The lateral geniculate body is 106% less than expected! (These are for the Anthropoidea only. If prosimians are included, the deviations are 121% and 147% respectively.) Sometime in the course of human evolution, primary visual striate cortex became relatively reduced, and posterior parietal cortex enlarged. This is not some mere "packaging problem" (Radinsky 1979) whom Falk cites (1983a: 488) "argumentum ad verecundiam", without a single empirical observation to prove it, but rather a true reorganizational feature important to cognitive functioning. Results with stereoplotting techniques (Holloway 1981b, 1984) point to the same conclusion. The questions are, when did this happen, and how can it be demonstrated unambiguously?

BACK TO THE ENDOCASTS

The problem with the Taung endocast, as both Dart and Clark (and others) knew , was that the lunate sulcus could not be unambiguously demonstrated, given the position and openness of the lambdoid suture. Schepers (1946), too, realized this, but opted for placing the lunate sulcus in the approximate position followed by the lambdoid sutural ridge.

"The identification of the lunate sulcus (18) in Australopithecus africanus, originally made by Dart, provoked a great deal of adverse criticism on the part of many scientists, notably Keith (1929). Being ordinarily situated relatively far forward on the dorsolateral surface in living anthropoids... and just behind the parallel sulcus (24), the suggestion that it may be represented by the curved sulcus near the tip of the occipital pole, which lay almost opposite the lambdoidal suture-line, was widely taken to signify over-enthusiasm on the part of the sponsors of the hominid status of the Taungs fossil. As far as is known, none of these critics have had access to the original cast. Even Dart studied a partially cleaned cast. Since the adherent cortical lamellae have been removed there can be little doubt as to the precise identification of the various occipital sulci, and Dart's original homology for the lunate sulcus must be sustained." (Schepers 1946:192).

Falk's (1980b: 531-532) response was as follows:
"Sutures are readily distinguished from sulci on endocasts (Falk 1978c), since sulci are represented by grooves and sutures are represented by protruding lines that often look like sutures. The landmark that Dart (1925) and Schepers (Broom and Schepers 1946) identified as the lunate sulcus is definitely the lambdoid suture, as suggested by Clark et al (1936:268)" (Falk 1980b: 531-532.)

It comes as a surprise to me that Clark (1947) should reverse his earlier position (Clark et al 1936) after working on the originals rather than a cast and six chimpanzee brains and associated endocasts. In fact, as Clark et al (1936) indicated, the lambdoidal suture and the lunate sulcus did overlap (approximately) in one chimpanzee specimen, and Hirschler (1942:17,31,54) in his review and studies found this conjunction to occur occasionally. Incidentally, there is nothing in Schepers' (1946) discussion to suggest he confused the lambdoid suture with the lunate sulcus, as alleged by Falk (1980b).

Falk's (1980b) solution to this problem was to define a small sulcal depression well anterior to where it would occur on any pongid brain, as the lunate, even though its superior portion was interrupted by a longitudinal gyrus just lateral to the midsagittal plane, and without any inferior crescentic morphology. The normal Pan position, based on 6 Pan brain casts, violated all cerebral morphology when placed on the Taung endocast (Holloway 1981a). To rebut this demonstration, Falk (1983) introduced some undefined "shape factor" (p. 487) to account for the discrepancy, based on ratios taken from arc measurements on unscaled photographs of Pan brain casts. The actual measurements (Holloway 1984) indicate Falk's lunate position on Taung to be 2.52 S.D's anterior to that in a typical pongid pattern (Fig. 1).

In 1983c, in a preliminary report on the AL 162-28 posterior endocast portion of a reputed A. afarensis (Holloway 1983c), groove or furrow "B" was defined not as the lunate sulcus, but as a depression caused by the inferior lip of the parietal bone. Falk (1985) now regards this as a true lunate and, by incorrectly orienting her

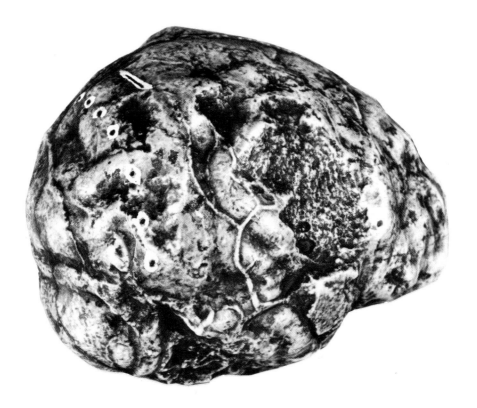

Fig. 1

     Lateral view of the Taung endocast, slightly rotated
from the midsagittal plane to show the convolutional
patterns more thoroughly. The small line marked at top left
is Falk's (1980b) placement of the lunate sulcus. To the
left is an arc of white dots containing black dots, which
represents the average location of the lunate sulcus in <u>Pan</u>,
based on six <u>brain</u> casts (c.f. Holloway 1981a). Falk's
"line" falls 2.5 standard deviations outside of the <u>Pan</u>
average (Holloway 1984). Note how the typical <u>Pan</u> placement
violates the transversely disposed convolutional detail
between the lambdoid suture and Falk's line.

drawing by an anterior rotation of the endocast fragment, claims it (the furrow) to be a lunate sulcus in an anterior position. As shown below, her conclusions are erroneous, and her placement actually strengthens the argument for a posterior position of this sulcus.

CLARK ET AL (1936) AND CLARK (1947)

In this paper, Clark and his colleagues studied six chimpanzee brains, and drew their convolutional details (their Figs. 1-6). They also made endocasts from the same crania whence came the brains. Soot was deposited on the endocast surfaces and then wiped off with a cloth stretched over a flat board. Thus soot was erased from prominences (gyri) and left intact in the furrows (sulci).

They compared the two configurations and discovered that gyri and sulci were seldom replicated on the endocasts by the impressions left by the once pulsating brains on the endocrania. This was particularly true of the lunate sulcus! They used this demonstration to argue forcibly that Dart was incorrect in perceiving the Taung lunate as non-pongid:

"The lunate sulcus, surprisingly enough, produces only ill-defined depressions on the cast. With the possible exception of the left hemisphere on specimen No. 1, and the left hemisphere of No. 2, it is very doubtful whether the course of the sulcus could be inferred correctly from the casts. On the other hand, the lambdoid suture always produces a conspicuous furrow which might readily be mistaken for a lunate sulcus, the more so because the occipital pole bulges rather conspicuously behind this furrow..."

"The furrow taken by Dart to represent the sulcus lunatus resembles precisely in its position and appearance the furrow on all the chimpanzee endocranial casts caused by the lambdoid suture. There can be little doubt, therefore, that this is the correct interpretation..." (Clark et al, 1936:268; (see also Cunningham 1892).

An important clue appears to have been ignored in their paper. In each case, a groove "X" was found in a posterior position, and it was caused by the posterior inferior lip of the parietal bone in the lambdoidal sutural margin! Groove

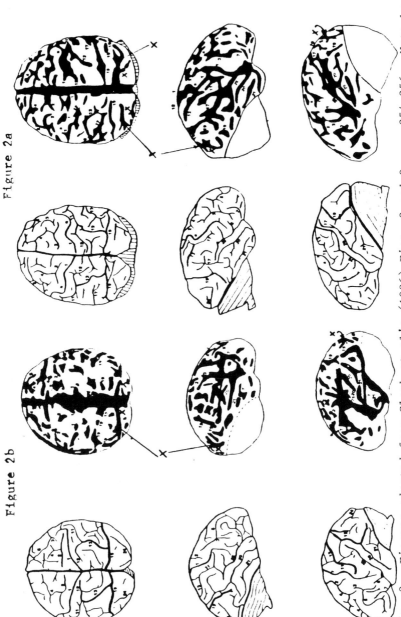

Figure 2a

Figure 2b

Fig. 2. Figures adapted from Clark et al's (1936) Figs. 2 and 3, pp. 254-256. Note how furrow "x" is clearly posterior to sulcus 19, the lunate, on both dorsal and lateral views.

"X" always transects the lateral calcarine sulcus and always appears well posterior to the lunate sulcus (Fig. 2). It is a striking morphological pattern and can be readily seen on almost every Pan brain that this author has examined, as well as on the endocasts.

Unpublished observations on 36 Pan troglodytes endocasts in my laboratory show this groove to present in pronounced form in 30 cases, and less pronounced in the remainder. It is seldom if ever coincident with the lunate sulcus on these endocasts, which can seldom be seen, an observation fully in accord with Clark et al's (1936) study, and Hirscher's (1942) review.

The AL 162-28 specimen shows this groove ("X" of Clark et. al.) clearly and the interparietal sulcus abuts this groove posteriorly. Falk (1985) agrees with my identification of the interparietal (IP). When the endocast fragment is oriented properly, the groove is quite posterior, as is the caudal end of the interparietal sulcus. Unless the AL 162-28 A. afarensis specimen represents a freak of hominoid brain morphology as yet undiscovered in any living or fossil specimens, one must conclude that if the lunate sulcus did exist, it did so in a posterior non-pongid position. But let us return to Clark:

"A close study seems to make it fairly clear that, if the sulcus lunatus is farther forward than the depression numbered 18 in Schepers' diagram, it must be at least as far anterior as the depression marked 26 (for between 18 and 26 the convolutions are disposed in an approximately antero-posterior direction, and they are clearly enough marked on the cast to exclude the possibility of a transversely disposed sulcus lunatus in this position. But a sulcus lunatus in the position of the depression marked 26 would be much farther forward (relative to the distance between frontal and occipital poles of the cerebral hemisphere) than in any modern ape, and would indicate an area striata of an extent and complexity beyond reasonable expectation. Thus it must be inferred that a sulcus lunatus, if present in typical form, must have been at least as far back as sulcus 18. On the other hand, the impression on the cast which has been interpreted as sulcus 18 is very ill-defined, and may simply be related to an elevation of bone along the line of the lambdoid suture and not to a cortical sulcus at all. Whichever interpretation may be correct, the evidence provided by the cast can certainly be taken to indicate that the brain belonging to the Australopithecus skull did not possess a lunate sulcus of the typical simian type. (Clark 1947: 312, emphasis mine).

Falk (1980b), in error, misread Clark (1947), and proclaimed Schepers' (1946) groove (21, parieto-occipital fissure) to be the lunate sulcus. In the stereoplotting (Holloway 1981b) and arc-chord measurements (Holloway 1984) papers, both, techniques concurred in showing Schepers' (1946) grooves (21) and (26) to be far more anterior than any in the chimpanzee. (The latter paper found Falk's 1983a placement by traditional arc-chord techniques to be 2.5 S.D.'s anterior to a true pongid position of the lunate sulcus). I believe Clark's studies of the original specimen, along with Schepers' diagrams, led Clark to regard the lunate as either not present, or posteriorly positioned, as a lunate in the intervening regions would completely violate the remaining cerebral morphology (Fig. 1).

Clark could be wrong in his acceptance of some of Schepers' (1946) convolutional details, but Clark was guarded over accepting some of them. In particular he was hardly confusing the lambdoid suture with the lunate sulcus.

SOME NEWER EVIDENCE FROM HADAR

Falk's (1985) criticism of my study of the Hadar AL 162-28 endocast (Holloway 1983) provides a basis for further hypothesis testing. As she agrees with my depiction of the posterior end of the IP sulcus, it is simple to measure the distance from OP (occipital pole) to IP (interparietal) and compare that distance with values from chimpanzee brains, most of which are smaller in volume than the Hadar 162-28 specimen. The results are quite dramatic (Table I), as the distance from OP to IP in AL 162-28 is roughly 1/2 that found on chimpanzee brains. Even the infant case, with a brain weight of 136 grams, has a larger value! Similarly, the distance from OP to what Falk regards as lunate sulcus is on AL 162-28, roughly 1/2 of the distance found on chimpanzee brains. Clearly, if there was a lunate on AL 162-28, it was in a very posterior position compared to the chimpanzee, and, one assumes, any other pongid brain. These measurements thus disprove her contention of an anteriorly-placed, pongid-like lunate sulcus for AL 162-28. Falk's orientation of the endocast is almost 40° in error from proper anatomical placement of the cranial fragment (Kimbel et al 1982; Holloway and Kimbel, in press).

Table 1

SPECIMENS	VOLUME (ML)	DISTANCE FROM OP TO IP(mm)
Hominid		
AL 162-28	375-400	16
Chimpanzees		
infant	140	25.5
1	407	34
2	320	30
3	290	30
4	295	29
5	270	28
6	250	29.5
7	300	33.5
8	400	32.5
9	275	27
Chimpanzee Mean $\bar{x}$		30.39
S. D.		2.43

Number of S.D.'s of AL 162-28 from chimpanzee
    average = 5.92.

Table 1

   Figures showing OP (occipital pole) to IP (interparietal
sulcus, caudal end) distance in mm (adapted from Holloway
and Kimbel, in press).   Note that most chimpanzee brain
casts are smaller than the AL 162-28 endocast, yet the OP-IP
chord distance is approximately double in chimpanzee.

DISCUSSION

There is a considerable difference between saying that a brain shows some human-like features and saying it is human, fully human, or does not retain any pongid features. In my papers on human brain evolution, I have not said that australopithecine brains were human. A more human-like disposition of the primary visual striate cortex is a far cry from a fully human brain. After all, these creatures lived about 2-3 m.y ago, and it would be amazing if <u>no</u> primitive features of cerebral morphology were retained from ancestral pongid-like forms.

More recently, Falk claims to have discovered a pongid-like morphology of the frontal lobe in australopithecines (Falk 1982, 1983a, b, 1984). She claims a fronto-orbital sulcus is visible in Taung, STS60, and KNM-ER 1805 from Kenya. With all due respect, the following must be said: (1) The inbedded frontal lobe region of the Taung endocast renders such an observation impossible. (2) The crushed and distorted type 2 endocast from Sterkfontein does not show that region intact. (3) STS 60 similarly lacks any clear indication of that sulcus, given the broken surface morphology in that region. (4) The original KNM-ER 1805 endocast was made by me. Both frontal and temporal lobe regions are damaged and eroded, and on the original endocast, the fronto-orbital sulcus <u>cannot</u> be seen on either side. The line drawings in Falk's (1983b) article are at odds with photographs of the original. Indeed, her fronto-orbital sulcus could be interpreted as an inferior frontal or horizontal limb of the Sylvian sulcus.

In conclusion, a plea is made for newer and more objective methods to be developed, to locate convolutional details. Careful metric and stereoplotting tests of hypothesized hominid sulci are an apparent improvement, but newer approaches would be welcome.

Acknowledgements

I am grateful for the support of the L.S.B. Leakey Foundation, and to NSF (BNS-79-11235 and BNS-84-18921) which permitted me to cast the chimpanzee brains. I am very grateful to Mrs. Beverley March for her patience and skill in typing this paper. I owe countless thanks to Drs. P.V. Tobias, R.A. Dart and Mr. Alun Hughes for all their help, interest and hospitality since 1969.

References

Angevine JB Jr., Smith MS (1982). Recent advances in
    forebrain anatomy and their clinical correlates. In:
    Thompson RA and Green JR (eds), "New Perspectives in
    Cerebral Localization." New York: Raven, p.1.
Armstrong E (1979). A quantitative comparison of the
    hominoid thalamus. I. Specific sensory relay nuclei.
    Am J Phys Anthropol 51: 365.
Armstrong E (1980). A quantitative comparison of the
    hominoid thalamus. II Limbic nuclei anterior principalis
    and lateralis dorsalis. Am J Phys Anthropol 52: 43.
Broom R, Schepers GWH (1946). "The South African Fossil
    Ape-Men: The Australopithecinae," Transvaal Mus Mem 2:1
Clark WE LeGros (1947). Observations on the anatomy of
    the fossil Australopithecinae. J Anat 81: 300.
Clark WE LeGros, Cooper D Zuckerman S (1936). The
    endocranial cast of the chimpanzee. J Roy Anthrop Inst
    66: 249.
Connolly CJ (1950). "External Morphology of the Primate
    Brain," Springfield Ill: CC Thomas.
Cunningham DJ (1892). Contribution to the surface anatomy
    of the cerebral hemispheres. Roy Irish Acad Sc Cunningham
    Memoirs, No. VII, Dublin.
Damasio AR, Geschwind N (1984). The neural basis for
    language. Ann Rev Neurosci 7:
Dart RA (1925). Australopithecus africanus: the man-ape
    of South Africa. Nature 115: 195.
Dart RA (1926). Taungs and its significance. Nat Hist. 26:
    315.
Dart RA (1956). The relationship of brainsize and brain
    pattern to human status. S Afr J Med Sci 21: 23.
Eidelberg D, Galaburda AM (1984). Inferior parietal lobule.
    divergent architectonic asymmetries in the human brain.
    Arch Neurol 41: 843.
Falk D (1978). External neuroanatomy of Old World monkeys
    (Cercopithecoidea). Contrib Primatol 15: 1.
Falk, D (1980a). Hominid brain evolution: the approach from
    paleoneurology. Yearbook of Phys Anthropol 73: 93.
Falk, D (1980b). A reanalysis of the South African
    australopithecine natural endocasts. Amer J Phys
    Anthropol 53: 525.
Falk, D (1982). Primate neuroanatomy: an evolutionary
    perspective. In Spencer F (ed): "A History of American
    Physical Anthropology, 1930-1980," New York: Academic

Press, p. 75.

Falk, D. (1983a). The Taung endocast: A reply to Holloway. Am J Phys Anthrop 60: 479.

Falk, D (1983b). Cerebral cortices of East African early hominids. Science 222: 1072.

Falk, D (1985). Hadar AL 162-28 endocast as evidence that brain enlargement preceded cortical reorganization in hominid evolution. Nature 313, 45.

Geschwind N (1965). Disconnexion syndromes in animals and man. Brain 88: 237, 585.

Hirschler P (1942). "Anthropoid and human endocranial casts." Amersterdam: N.V. Noord-Hollandsche Uitgevers Maatschappijen.

Holloway RL (1976). Paleoneurological evidence for language origins. New York Acad Sci 280: 330.

Holloway RL (1979). Brainsize, allometry, and reorganization: toward a synthesis. In Hahn ME, Jensen C, Dudek BC (eds): "Development and Evolution of Brain Size: Behavioral Implications." New York: Academic Press, p 59.

Holloway RL (1981a). Revisiting the South African Taung australopithecine endocast: the position of the lunate sulcus as determined by the stereoplotting technique. Am J Phys Anthropol 56: 43.

Holloway RL (1981b). Exploring the dorsal surface of hominoid brain endocasts by stereoplotter and discriminant analysis. Philos Trans R Soc Lond B 292: 155.

Holloway RL (1983a). Human brain evolution: a search for units, models and synthesis. Canad J Anthro 3: 215.

Holloway RL (1983b). Human paleontological evidence relevant to language behavior. Human Neurobiol 3: 105.

Holloway, RL (1983c). Cerebral brain endocast patterns of the AL 162-28 Hadar A. afarensis hominid. Nature 303: 420.

Holloway, RL (1984). The Taung endocast and the lunate sulcus: a rejection of the hypothesis of its anterior position. Amer J Phys Anthropol 64: 285.

Holloway, RL, Kimbel, WH (in press). Brain reorganization in the Hadar AL 162-28 endocast: A reply to Falk. Nature, N.D.

Hyvarinen J (1982). "The Parietal Cortex of Monkey and Man," (Studies of Brain Function, Vol. 8) Berlin: Springer-Verlag.

Jerison HJ (1973). "Evolution of Brain and Intelligence." New York: Academic Press.

Jones EG (1984). History of cortical cytology. In: Peters

A, Jones EG (eds) "Cerebral Cortex," Vol. 1. New York Plenum p. 1.

Keith A (1929). "The Antiquity of Man." London: Williams and Norgate.

Kimbel WH, Johanson DC, Coppens Y (1982). Pliocene hominid cranial remains from the Hadar Formation, Ethiopia. Am J Phys Anthropol 57: 453.

Levin G (1936). Racial and "inferiority" characters in the human brain. Am J Phys Anthropol 22: 345.

McGregor JH (1925). Recent studies on the skull and brain of Pithecanthropus erectus. Nat Hist 25: 544.

Passingham RE, Ettlinger G (1973). A comparison of cortical functions in man and other primates. Int Rev Neurbiol 16: 233.

Peter A, Jones EG (eds) (198)4. "Cerebral Cortex. Vol. 1 Cellular Components of the Cerebral Cortex." New York; Plenum.

Radinsky LB (1979). The fossil record of primate brain evolution. James Arthur Lecture, New York: Amer Mus Nat Hist.

Schepers GWH (1946). The endocranial casts of the South African Ape Men. In Broom R, Schepers GWH "The Australopithecinae." Transvaal Mus Mem 2.

Schepers GWH (1950). The brain casts of the recently discovered Plesianthropus skulls. Part II of Broom R, Robinson JT, Schepers GWH. "Sterkfontein Ape-Man Plesianthropus." Transvaal Mus Mem 4.

Smith GE (1904a). Studies in the Morphology of the human brain with special reference to that of the Egyptians. No. 1. The occipital region. Records. Egyptian Govr School Med 2. 725.

Smith GE (1904b). The morphology of the occipital region of the cerebral hemispheres in man and apes. Anat Anz 24: 436.

Smith GE (1907a). A new typographical survey of the human cerebral cortex. J Anat 41: 237.

Smith GE (1907b). New studies on the folding of the visual cortex and the significance of the occipital sulci in the human brain. J Anat & Phys 41: 198.

Smith GE (1927). "Essays on the Evolution of Man." Oxford: University Press.

Smith GE (1929). The variations in the folding of the visual cortex in man. Mott Memorial volume. "Contributions to Psychiatry, Neurology, and Sociology." London.

Stephan H. (1969) Quantitative investigations on visual

structures in primate brains. Proc. 2nd Internat. Congr.
Primates 3: 34.
Stephan H, Frahm H, Baron G (1981). New and revised data
on volumes of brain structure in Insectivores and
Primates. Folia Primatol 35: 1.

Tongues in Trees . . .Sermons in Stones

Hominid Evolution: Past, Present and Future, pa
© *1985 Alan R. Liss, Inc.*

LEAVING NO STONE UNTURNED: ARCHAEOLOGICAL ADVANCES AND
BEHAVIORAL ADAPTATION

J. Desmond Clark

University of California
Berkeley, CA 94720

INTRODUCTION

In 1971 a young American archaeologist, James Gallagher,
had been mapping a number of surface assemblages of obsidian
artifacts near an important source of the raw material in the
highlands of Ethiopia. The assemblages were clearly of
several different ages and, apparently, all of some antiquity
since the local villagers denied working any of the stone.
Much was his surprise, therefore, on revisiting the site to
find a *new* assemblage of blades that had not been there
before. The villagers vehemently denied any responsibility
but their examination of the new assemblage revealed that a
man from a more distant village had been passing the outcrop,
decided his beard was a bit rough, and sat down, struck a few
blades, selected one of them and, having completed a successful shave, went on his way. If further proof were needed it
lay in the *hairs* still adhering to the blade used as a razor
(Gallagher 1972)! Quite a neat and, as it turned out, accurate, piece of detection which, of course, opened up all sorts
of lines of enquiry for the archaeologist. I cite it here to
illustrate the emphasis that archaeologists place today on
trying to reconstruct, through the evidence of the present,
what it was that the early hominids whose buried assemblages
they excavate were *doing* and to understand what a collection
of stone artifacts and bones - which too often is all that
survives - means in terms of the behavioral activities of
which they once formed a part. Gallagher's experience also
shows how ethnological analogues can, if correctly used, be
made to provide invaluable understanding of past human behavior and technology; this example serves also to emphasize the

new direction that palaeo-anthropology is taking today in
seeking to identify recurring patterns of behavior through
the use of comparative evidence from the present.

During the past thirty years and, in particular, since
the 1960s, method and theory in archaeology have advanced
out of all recognition now that they are freed from the
restricting parameters of the earlier years of the century.
And today the prehistorian is increasingly able to contribute
significantly to understanding the evolving complexities of
hominid technology and behavior. The fossils themselves show
us the general pattern of behavior - bipedality, forelimbs
able to manipulate simple tools and so on - but it is the
archaeological residues in their context that are the clues
to the many phenomena affecting the hominids' way of life.

EARLIER DEVELOPMENTS

The groundwork laid by earlier archaeologists has largely
made possible what we are able to do today. It was they who
built the framework within which we still work and they who
evolved the terminological and taxonomic system that still
permits communication between prehistorians all over Africa
and, indeed, in the Old World generally (Bishop, Clark
1967). Since the late 19th century the impetus in African
prehistory came from two directions: in northern Africa it
came from the northwest from the work of French palaeontologists and archaeologists, because, in Egypt, perhaps under-
standably, archaeologists were too engrossed in searching for
unrifled Dynastic tombs to spend much time on the intrinsical-
ly far less rewarding study of prehistoric sites. In both
northwest and northern Africa, the French terminology provided
the framework for a cultural chronology based on geological
contexts with associated faunal and archaeological assemblages
(Balout 1955:10-20).

At the other end of the continent there was no early
historical archaeology and the prehistoric stone artifacts,
first recognized in the 1850s, were so numerous that European
prehistorians declared that there were too many for them to
be very old! (Goodwin 1935:311, 318). The terminology developed in South Africa is now in general use throughout the
continent south of the Sahara. Its foundations were laid by
two men - John Goodwin who was the first professionally
trained archaeologist to work in South Africa, mostly in the

Cape, and 'Peter' Van Riet Lowe, an engineer by profession
working chiefly in the Free State and Transvaal. Their
findings were first published in 1929 in the classic *Stone
Age Cultures of South Africa* (Goodwin,   Van Riet Lowe 1929).
They divided the Stone Age sequence into Earlier, Middle and
Later stages within which major cultural entities were given
local names as had been done earlier in Europe, on the basis
of the so-called "type fossils" regarded as characteristic
artifacts of a particular ethnic population. The only means
of dating was *relatively* based on stratigraphic superposition
and, occasionally, by associated faunal remains, so that
chronologies were totally without precision and archaeologists
were mostly searching for and excavating stratified assemblages, very often in secondary contexts. Each succeeding or
different set of artefacts encountered was regarded as the
product of a discrete ethnic population and so as the result
of migration - an hypothesis that is still sometimes with us
in spite of the findings of the last two decades.

Goodwin and Lowe - correctly - considered the makers of
the Later Stone Age cultures to be modern humans like ourselves - the ancestors of the Khoi San peoples - and they also
appreciated the extent to which raw material can affect the
form of artefacts made from it.

Goodwin's excavation of the Oakhurst Shelter on the south
coast was the first systematic study of such a sequence and
shows his identification of cultural entities later known to
be widely distributed in the sub-continent (Goodwin 1938).
At the same time, Van Riet Lowe and the geologists Sohnge and
Visser (Sohnge, *et al.* 1937) did the first systematic study
of a river valley sequence - the Vaal basin - and for many
years their work was a yardstick, against which similar studies
were measured and interpreted in terms of the Pluvial/Inter-
pluvial Climatic Hypothesis developed in the late '20s and
'30s in East Africa by Wayland (1934) and Leakey (1936).

This, then, was the foundation upon which a new era of
archaeological research began and the late 1940s saw the beginnings of the interdisciplinary, team-oriented research that
is the very basis of early man studies today. One of the
first examples was the work of Raymond Dart and Robert Broom
who gave us *Australopithecus* and showed that Africa was the
original home of man; and the cooperation between Van Riet
Lowe, the founder of the South African Archaeological Survey,
and the Abbé Breuil, then the acknowledged world authority on

the cultures of the Palaeolithic. The potential for team-
work was obviously there but it was not until 1956, when
C.K. Brain (1958:72-3) found the first stone artifacts at
Sterkfontein, that archaeologists also joined in the search
for the australopithecines and other hominids being undertaken in the Transvaal cave breccias with such dramatic success by Dart, Broom and Robinson.

Previous to the 1950-60 decade, prehistorians in Africa
were very thinly scattered over vast distances and were
working in virtual isolation. Then, in 1947, Louis Leakey
organized in Nairobi the first of the international Pan-
African Congresses on Prehistory and Quaternary Studies to
which came most of the Quaternary geologists, palaeontologists and prehistorians working in the continent to exchange
information and to view many of the sites already made famous
by the work of Louis and Mary Leakey in Kenya and Tanzania.
The Congress has now met nine times and done much to break
down barriers of isolation and to foster international collaboration that is the mainstay of our research today (Clark
1980).

Then, in 1950, Willard Libby gave us radiocarbon chronology and we finally saw the true time-depth for the last
40,000 years and more. In the 1960s still other methods of
dating carried the chronology even further back and there
emerged a framework giving in radiometric years before the
present the time taken for the respective hominid species,
sub-species and associated cultural manifestations to evolve
(Bishop,   Miller 1972). By the mid-1950s, Dart (1957) had
carried out his pioneer analysis of the faunal assemblage
from the australopithecine breccias at Limeworks in the
Makapan valley and Revil Mason (1957) had begun to apply
quantitative analysis to complete collections of stone artifacts which is, of course, an essential basis for all archaeological work today. This progress in radiometric time
scales accompanied by the abandonment - due to the work of
Basil Cooke (1958) and Dick Flint (1959) - of the Pluvial/
Interpluvial hypothesis, at last freed the prehistorian to
look at what his artefacts might mean - no longer as *"types"*
- but in terms of the *behavior* of the hominid populations
that made them. Also in the early 1960s, ecological and
functional studies were begun.

NEW TECHNIQUES

The main new advance in prehistoric studies came first
in the Great Rift Valley in East Africa where deep sedimentary sequences with rich fossil and artifact contents lay in
relatively undisturbed contexts. Here Mary Leakey (1971)
adopted techniques never before applied to the excavation of
earlier Pleistocene sites and, in the late 1940s at Olorgesailie (Leakey 1952; Isaac 1977), a new dimension was added
to the vertical stratigraphic sequence namely *"horizontal
archaeology"* and early sites were now excavated with the care
in uncovering, plotting and recording in common use today.
However, the significance of the new finds was really only
fully appreciated after the discovery in 1959 in Bed I at the
Olduvai Gorge of *Zinjanthropus boisei* in relatively undisturbed context with associated behavioral evidence (Leakey
1959). Now hominid fossils began to be recovered by systematic survey instead of chance surface or mining finds.

During the '60s and '70s activity proliferated especially
in East Africa and Ethiopia, recently referred to by Glynn
Isaac as "the Klondike" of Palaeoanthropology with the Rift
Valley as the "Motherlode" (Isaac 1984). The approach is now
multi-disciplinary and the pooled researches are resulting in
increasing knowledge of the palaeo-climatic and ecological
conditions under which the hominid lineage evolved and of
some of the more significant events that triggered these major
biological and cultural developments; all based upon sequences
securely correlated by stratigraphic "marker beds" and faunal
assemblages within a dependable chronological framework.

To a great extent, the archaeologist relies on the work
of other disciplines for evidence of the palaeo-climate and
environment but several recent studies are more specifically
archaeological in their objectives. Work by Behrensmeyer
(1982), Schick (1984) and Kaufulu (1983) has shown the circumstances and history of an archaeological accumulation and also
the conditions and sites preferred as hominid occupation
places - particularly the drier savanna close to stream banks,
riparian forest and point bars.

Today the approach is more problem-oriented and theoretical, complicated by the greater caution now needed in forming
hypotheses because of the increasing corpus of our knowledge.
The earlier, too literal approach sometimes gave rise to un-
justifiable conclusions which we now try to avoid as we real-

ize that knowledge of the processes and agencies involved in
the formation of a site is essential to its interpretation.
The relatively new sub-discipline of taphonomy traces the
history of a carcass from the time of death and seeks to
identify the agencies responsible for its dispersal
(Behrensmeyer, Hill 1980). Experimental studies in bone
fracture and modification and in stone tool manufacture are
providing important evidence of the thought processes behind
butchery techniques, lithic technology and the reduction sequence in stone tool manufacture. Nicholas Toth (1982) and
Peter Jones (1981) with their replicative experiments can now
identify normal and abnormal flaking assemblages and so show
what has been removed - as in the case from Koobi Fora where
the study of conjoining flakes showed that the end-product,
the core/chopper, had been removed, probably by the makers,
thereby suggesting that the hominids transported some of their
stone equipment from place to place (Bunn *et al.* 1980). The
advantages and limitations of various raw materials were
clearly perceived by the hominids as demonstrated for the
Acheulean at Olduvai by Peter Jones's experiments (Jones 1978).
At Koobi Fora, from the order of removal of cortex flakes
from an Oldowan core, Toth was able to suggest that the makers
of Oldowan artefacts may have been predominantly right-handed
(Toth: in press).

Experiments in possible functional uses of stone tools
(Jones 1980; Toth, Schick 1983) show their appropriateness
for different purposes and Lawrence Keeley's (1980) more sophisticated laboratory techniques can identify the polishes
that result from using them on different kinds of material.
The method has growing potential and it has recently been
possible to identify wood, soft plant and meat polish on chert
and ignimbrite artefacts from 1.5 m.y. old activity areas at
East Turkana (Keeley, Toth 1980).

Bone analysis now goes far beyond mere taxonomic identification. Aging, sexing and quantitative patterning of
species and parts provide information on hominid foraging and
hunting preferences and abilities (Klein 1979; Klein *et al.*
1983). Brain's (1982) experiments have shown that polish and
striations resulting from digging bulbs with bone fragments
closely parallel those on the points of bone tools from
Swartkrans and Sterkfontein. Researchers can now distinguish
carnivore bone assemblages from those made by hominids
(Shipman 1981) and even scavenged from hunted bones (Vrba
1980; Binford 1981, 1984; Brain 1981). Surface modification

of bone has been the subject of several specialist studies
and criteria have been determined for distinguishing cut-
marks resulting from using stone tools (Potts, Shipman
1981; Bunn 1982, 1983). At least one Plio-Pleistocene assemblage from Koobi Fora with no stone artifacts has been identified as hominid on the basis of stone-tool cut-marks on the
bone (Bunn: pers com.). One somewhat startling discovery
made by Tim White (White: in press) is the series of seven-
teen cut-marks, made with a stone tool, on the face and forehead of the Bodo hominid fossil first described by Glenn Conroy
(Conroy *et al*. 1978) from Middle Pleistocene sediments in the
Middle Awash section of the Ethiopian Afar Rift. Now we can
say that Bodo man was defleshed but the main question remains
- why?

These are some of the new approaches, techniques and
methodology stemming from our expanded horizons in the
present decade. It is the empirical evidence from the archaeological sites themselves and the steadily expanding information from primate behavior studies and ethnography that enable
us to generate hypotheses and form models that we then test
to arrive at a "best fit" explanation. Perhaps we are as
close to answers as we will ever get, but there is reason to
hope, for, as Thoreau said, "Some circumstantial evidence is
very strong, as when you find a trout in the milk."

EARLY HOMINID BEHAVIORAL MODELS AND PROBLEMS

Pre-Toolmaking Hominoids/Hominids

Archaeologists are primarily concerned with hominid
activity after the appearance of stone tools some 2.5-2.0
m.y. ago. We have evidence of hominids for 3.0 m.y. before
that but no identifiable tools, though almost no buried surface
with hominid remains has yet been excavated to record everything in immediate association that might make it possible
to identify relatively undisturbed "living places" (Dechant-
Boaz 1982:236-250). As yet, therefore - and not forgetting
the excellent circumstantial evidence in this time range from
the Transvaal caves - behavioral hypotheses and interpretations remain essentially theoretical and probably untestable
in this time range as, for example, Lovejoy's (1981) hypothesis based on pair-bonding, sexual division of labor and base-
camp living. The pronounced sexual dimorphism in early

australopithecines might also suggest alternative models as that in Fig. 1, for example, based on aggressive group behavior not all that far removed from Dart's original suggestion. This would at least seem worth investigating, now that so much more is known about great ape aggressive behavior (Harding, Teleki 1981).

PAIR-BONDING HYPOTHESIS

(After SHIPMAN 1983 based on LOVEJOY 1981)

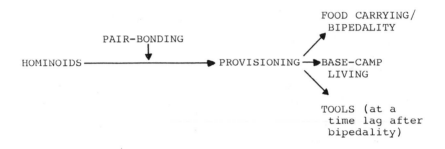

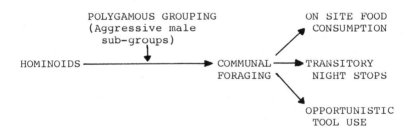

Fig. 1.

The Earliest Tool-Makers: 2.5-1.8 m.y. ago

Intentional stone tool manufacture almost certainly began, relatively suddenly, between 2.5 and 2.0 m.y. ago

with most of the earliest sites between 2.0 and 1.8 m.y.
(Harris 1983). At the same time, the first larger brained
hominid *(Homo habilis)* appeared at Olduvai (Leakey *et al*
1964) and Koobi Fora (Leakey, Walker 1973). Also, broken
bones, some with cut-marks, are now consistently found with
artefacts on relatively small, single context, activity areas.
It would thus appear likely that meat was now a new and
important item of diet (Isaac 1971) - there is no such evidence as yet from earlier periods. The enlarged cranium and,
by inference, intelligence, have been assumed - no doubt
correctly - to be concomitant with these behavioral innovations and some significant modification of social organization is also likely to have occurred - for example food-
sharing based on the central place foraging theory first
developed by Isaac (1978). The Oldowan stone tool-kit is
found together with several different kinds of bone accumulations, but the original interpretation that early hominids were
efficient hunters has now been considerably modified by comparative studies of bone-waste resulting from hunting and
scavenging and, in particular, by quantitative surveys carried
out last year in the Serengeti by Robert Blumenschine
(Blumenschine: in press and pers. com.). It now appears probable that early hominids obtained their meat mostly by
scavenging and this applies also to the remains of single
large animals associated with Oldowan artefacts - for example
the *Elephas recki* and *Deinotherium* from Beds I and II at
Olduvai Gorge (Leakey 1971:64-66, 85-6). Using a sharp flake
hominids were able to cut through the thick, tough hide of a
large carcass and obtain quantities of meat unavailable to
even the large carnivores until the carcass had sufficiently
decomposed. It was this advantage, I believe, that occasioned
the intentional manufacture of stone tools.

By this time also there were two hominid forms in the
fossil record - the gracile *Homo habilis* and the robust
australopithecine (Leakey, Walker 1976). Biologically,
either *could* have made tools and there are contexts where
both are found with artefacts. However, the rapid biological
evolution of the gracile and the unchanged nature of the
robust, probably vegetarian, form until it disappeared about
1.0 m.y. ago, suggest that the gracile one was probably the
only tool-maker. If so, then it was probably the addition of
significant amounts of meat and bone marrow to the diet that
made this already omnivorous form into an efficient, larger-
brained tool-maker and scavenger.

Two events are likely to have brought about important changes in hominid behavior. The first is the very significant lowering of world temperatures between 2.6 and 2.3 m.y. ago (Shackleton and Kennett 1975) when the east African vegetation belts were lowered by more than 1000m (Bonnefille 1983 and in press; Gasse 1980) and cooler, drier climatic conditions with forest/savanna mosaic ecology predominated in the Rift (Vincens 1979, Cerling *et al.* 1977). The second event is the appearance of the robust australopithecine, *A. boisei*, around 2.2 m.y. ago in the Rift (Howell 1982:95-100). It seems not impossible that this may have been a higher altitude, forest-related creature that moved with other animals down from the high plains with the vegetation communities on which they depended (Assefa *et al.* 1982). Such a situation as in Fig. 2 might have arisen which ultimately led to the extinction of *A. boisei* due to unsuccessful competition - with *Homo erectus*. Such an hypothesis could probably best be tested through the fossil record which, at Koobi Fora, has already dramatically demonstrated the contemporaneity of these two forms.

Unspecialized Hunters: 1.6-0.2 m.y. ago

From about 1.5 m.y. BP to around 200,000 years ago, a second kind of tool assemblage was present in the archaeological record. This Acheulean tool-kit, containing the first standardized tools - handaxes and cleavers, mostly made from large flakes - appears to supplement the Developed Oldowan and, at present, the most convincing of several suggested explanations for these two parallel tool-kits is that they are task-specific with some forms being more general purpose tools. However, in spite of an infinite quantitative variety, these assemblages all fit within a uniform pattern met with in the late Lower to Middle Pleistocene time range in Africa, Europe, the Middle East and India. Seven or eight different kinds of artefacts are identifiable on the basis of functional morphology and, although a single part of the equipment (Developed Oldowan or Acheulean) sometimes occurs in isolation, in Africa, more often than not, a broad range of artefacts combining both traditions is present on the activity areas (Clark 1975a).

One enigma has long been the large numbers of bifaces at many Acheulean sites - apparently many more than would have been needed unless the hominid groups were now always

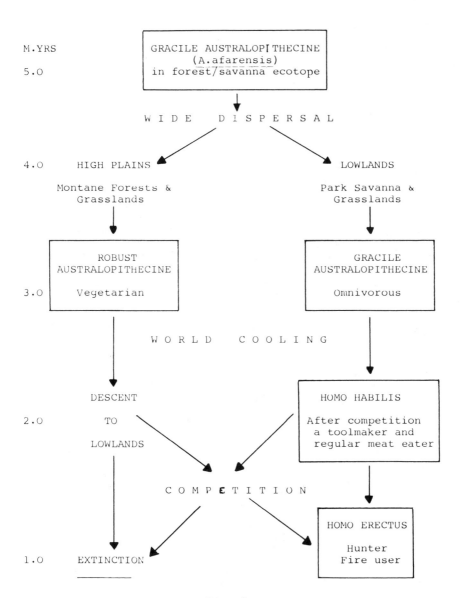

Fig. 2.

much larger. As yet, we lack any reliable means of estimating for how long these Acheulean sites were occupied but, more often than not, they occur in situations where prevailing climatic conditions could have favored the persistence of the same topography for scores or even hundreds of years without much modification. They may also have been preferred locations often revisited before burial or changed conditions made them uninhabitable.

A very different situation exists in the Middle Awash in the Ethiopian Afar Rift where concentrations on surfaces in Middle Pleistocene, fine grained, fluviatile deltaic deposits can have provided only very temporary resting places for Acheulean hominids, since they were rapidly covered by the next layer of sediment, possibly in the same season (Clark *et al.* 1984). The same kinds of bone concentrations are found as occurred with the Oldowan. These are all quite small assemblages such as can be expected to have been left by small transitory groups of hominids carrying out a limited set of activities for a brief time before moving on. The light duty flake component is usually found with the more abundant bone concentrations, while the relatively discrete biface assemblages did not have very much bone associated. Large concentrations of bifaces do occasionally occur but in channel sands and not in primary context. Middle Pleistocene hominid groupings must, indeed, sometimes have been larger than those of the makers of the earlier Oldowan complex, but the Middle Awash sites suggest that highly mobile, small but cohesive groups may still have been the basic unit of Acheulean social patterning. A very wide-ranging mobility is also suggested in the high plains of Gadeb, for example, by the rare obsidian handaxes some 100km distant from the nearest source (in the Rift) of that rock which can only have been carried onto the site by the makers of the Acheulean (Clark, Kurashina 1979).

By later Acheulean times, some 1.0 m.y. ago, *Homo erectus* had spread widely throughout Africa except into the rainforest, and the first move out of the continent appears to have been into the Middle East and the Asian tropics between 1.5 and 1.0 m.y. ago; later archaic forms of *Homo sapiens* moved into temperate Europe and the cold north. *Sapiens* characteristics are widely recognized in African fossils from about 300,000 or more years ago and it is thought that this biological change may have taken place first in Africa (Day,   Stringer 1982; Rightmire 1984; Brauer 1984) during the later Acheulean, though it remains to be determined quantitatively whether it was accompanied by any significant cultural development. At present it would seem that the success of the spread of *H. erectus* and archaic *H. sapiens* may have been due to their greater efficiency as hunters. In cold climates large mammal carcasses would have had particular importance in view of the restricted range of plant foods. The first evidence of wooden spears are a stabbing spear at Clacton around 300,000 BP (Oakley *et al* 1977) and a throwing spear at Lehringen (Movius 1950) about 100,000 BP and at Kalambo Falls we find wooden digging equipment and a club (Howell, Clark 1963). Also with the Acheulean at this time claims have been made for dwelling structures (Clark 1967, de Lumley 1969) and the first faint signs of symbolism and abstract thought (Oakley 1981; Edwards 1978).

It has also been suggested that it was the ability to make effective use of fire that enabled *H. erectus* and his successors to live in the cold and temperate lands of Eurasia. Evidence of fire in the form of charcoals and burned bone occurs in Europe and the Far East by 300,000 BP and not much later in Africa (Oakley 1956). However, at more ancient open sites in the African tropics, where charcoal is not preserved, only indirect evidence survives - burned earth or fire-fractured rocks. Some of the most compelling evidence is the fire-reddened earth from burned tree boles and termite nests, comparable to that observable today. In the Middle Awash such burned clay occurs in sediments of Later Pliocene to Middle Pleistocene age, with stone artifacts and bone in juxtaposition to some of the latter (Clark *et al* 1984). Natural fires are quite common and, if the advantages of fire had been perceived by early hominids, there is no reason why they should not have used it. Fig. 3 (after Clark and Harris: in press) shows some of the ways fire could have been used by them without any knowledge of how to *make* it; so that fire could have been in regular use long before *Homo erectus* erupted into Eurasia, but future research is needed to verify this.

The Emergence of Modern Man 200,000-35,000 BP

Until radiometric dating showed the contemporaneity of the Middle Palaeolithic of Eurasia and the Middle Stone Age of Africa, it was thought that Africa was a biological and cultural backwater after the Middle Pleistocene (Clark 1975b).

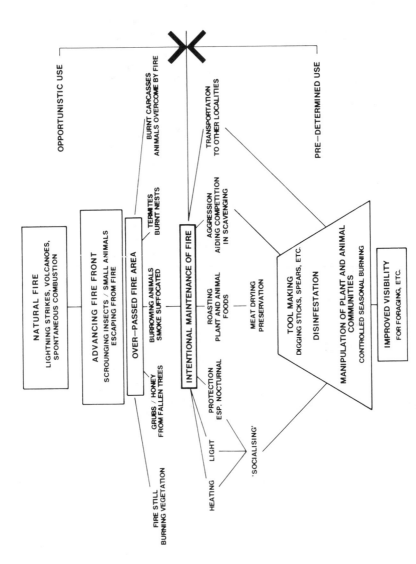

SOME POSSIBLE USES OF NATURAL FIRE BY EARLY HOMINIDS

Fig. 3.

Now we know that human skeletal remains with modern characteristics, associated with Middle Palaeolithic/Middle Stone
Age industries, were widely distributed in north, east and
south Africa in situations perhaps earlier than in the Middle
East: for example, those from Dar es-Soltan II (Debenath
1975; Ferembach 1976a), Temara (Roche, Texier 1976;
Ferembach 1976b), Ngaloba (Laetoli) (Day *et al* 1980), Omo
(Kibish) (Day 1972), Klasies River Mouth (Singer, Wymer
1982) and Border Cave (Beaumont *et al* 1978) over 100,000
years old. Certainly by 30,000 years ago, Modern Man with a
regionally variable Later Stone Age technology was present
throughout the continent. The most significant research on
the emergence of these later cultures comes from recent work
on the South African south and southwest coastal and some
inland sites (Deacon HJ 1976, 1979; Deacon J 1978; Klein
1983; Parkington 1984). Now, it can be seen that the classic
Upper Palaeolithic technology of Western Europe is but one of
several distinctive technologies all used by modern forms of
man with equal success and it is clear that the superiority
of modern man is not shown by the morphology of his tool-kits,
but by the ways these were adapted to make the most efficient
use of the resources being exploited (Clark 1981). This successful early diversity, in semi-isolation over many millennia, finally gave rise to the rich and varied cultures we
know of from history and ethnography. The chart (Fig. 4)
summarizes some of the main cultural advances prehistory shows
to have taken place, firstly in Africa alone and then more or
less contemporaneously throughout the later Pleistocene world.

THE FUTURE

   In the last few years a good beginning has been made in
recognizing some of the problems and pitfalls in understanding
cultural evolution and we can now plan research strategies to
reduce the number of alternative models suggested by the circumstantial evidence. We still need to identify the best
potential localities and to carry out in-depth, team-oriented,
empirical studies of activity areas in their contexts so that
our corpus of reliable data is expanded. Some workers consider that the theoretical base should take precedence over
the fieldwork data but, especially in contexts that predate
the appearance of Modern Man, empirical data are the only
sure evidence on which to construct hypotheses. We need to
avoid what Leonardo da Vinci called "the supreme misfortune
when theory outstrips the performance".

AGE AND MAIN DIVISION	STONE TECHNOLOGY		SOME OTHER TOOLS *	BEHAVIOURAL TRAITS
	GENERAL	SPECIFIC		
Present LSA 0.035my	SPECIALISED COMPOSITE TOOLS AND MANY ARTIFACT SETS ABSTRACTIONS	MANY REGIONAL TRADITIONS AND FUNCTIONALLY SPECIFIC TOOL SETS		POPULATION EXPANSION SMALL BANDS INTENSIVE SEASONAL RESOURCE USE IN SMALLER AREAS.COHESION FOR CEREMONIES AND EXCHANGE
0.04myr MSA 0.15myr	SIMPLE COMPOUND TOOLS AND FEW ARTIFACT SETS (Hafting) (Begins)	SEVERAL REGIONAL TRADITIONS AND CHRONOLOGICAL FACIES		SEASONAL TRANSHUMANCE RESOURCE AND TASK SPECIFIC CAMPS REGULAR REOCCUPATION BY LARGE BANDS IN LARGE TERRITORIES
0.2 myr ESA II ACHEULIAN and D.OLDOWAN 1.8 myr	FIRST STANDARD TOOLS FORMS	LARGE CUTTING BIFACE TOOLS AND CHOPPER/LIGHT DUTY COMPLEX		ORGANISED HUNTING LARGE ANIMALS SCAVENGING AND PLANNED COLLECTING. WIDELY ADAPTED. SITES REOCCUPIED
2.0 myr ESA I OLDOWAN 2.5 myr	SELECTED OPPORTUNISTIC TOOLS	CHOPPERS SPHEROIDS LIGHT DUTY HEAVY DUTY SMALL CUTTING TOOLS		SCAVENGING AND HUNTING SMALL ANIMALS COLLECTING INSECTS, EGGS; PLANT FOODS FROM BRIEF STOP OVERS
NO KNOWN FLAKED STONE TOOLS 5.0 myr	SIMPLE OPPORTUNISTIC TOOLS: STICKS ETC.	MANUPORTS ?		COLLECTING PLANT FOODS INSECTS AND SMALL MAMMALS

Vertical labels in the "SOME OTHER TOOLS" column: SHAPED — DIGGING STICKS — CARRYING TRAYS ETC. — SPEARS — THROWING STICK — OPPORTUNISTIC — FIRE ? — BONE TOOLS

* - - -. - - - INFERRED

_____ CONFIRMED

Fig. 4.

   Fieldwork is becoming more costly and sometimes more difficult so that funds and expertise need to be concentrated where maximum results can be expected. Also most of the fossils and cultural remains in which we are interested occur in Third World countries where it is essential that work be done in close collaboration with local institutions and scientists. Perhaps too, the training of professionals and technicians and the provision of laboratory and storage facilities should be an integral part of all long-term research programs by expatriate teams. Although palaeoanthropologists now have at their disposal a greatly increased aggregate of data with which to help solve their basic problems, nevertheless much more specific research is needed especially - on prehistoric diet - like that of Sept (1984), Vincent (1983) and Stahl (1984) into the plant foods used by early hominids before and after the use of fire; studies in bone chemistry, such as Andrew Sillen's on strontium/calcium analysis which promises to reveal the extent of hominid dependence on plant and animal foods (Sillen, Kavanaugh 1982). Stable carbon and nitrogen isotope ratios are showing the nature of prehistoric diet (Van der Merwe, Vogel 1978; Van der Merwe 1982; Schoeninger *et al* 1983) and how much meat was eaten and, if Ambrose's research (pers. com.) extends the usefulness of these techniques to *fossil* bone, it will be of great significance. At the same time, initial studies on organic trace elements on the edges of stone tools hold out promise that haemoglobin traces identified on recent prehistoric artefacts may be identified also on ancient tools (Loy 1983). These are but some of the new techniques now being developed.

   One of our most pressing problems is the need to use archaeological methods to locate and recover associated material on single-context, pre-tool-making hominid find-sites in the open. Another, of great significance, is systematic investigation of localities with *groups* of sites to show how hominids may have made use of space; as Isaac (1975) has put it, to study "the scatters between the patches," in addition to the concentrations themselves, and how these combine to provide models for the "catchment base". We also need to know how far back in time fire, "that great civilizer", has been significantly associated with hominid activity areas. And, since research over the past two decades has largely been concentrated on understanding earliest human origins and society, systematic study of the circumstances and cultural manifestations of the appearance of Modern Man is long overdue.

Prehistorians working on cultural evolution in the Pleistocene are now much more aware of the complexity of the problems of interpretation. Many of these seem insurmountable but the input from many directions is now much greater than ever before. We are now more cautious about our ability to provide answers but there is also real enthusiasm for the search for the evidence, the comparative approach, the identification of recurring patterns and the elimination of red herrings. I am sure that the findings of the next sixty years will far transcend those of the past sixty we have just been considering. In the future, public interest in the quest for human ancestors, regularly stimulated by provocative writings from the entrepreneurs of anthropology, will certainly not diminish in extent, as each new fossil find is greeted with the recognition it deserves and takes its place in the human family tree. But we need to remember that each new evolutionary advance to be seen in the fossil record was made possible only by the cultural heritage that is unique to ourselves. If tools are a product of the human brain it is also tools that "maketh man".

REFERENCES

Assefa G, Clark JD, Williams MAJ (1982). Late Cenozoic history and archaeology of the upper Webi Shebele basin, east-central Ethiopia. Sinet: Ethiop J Sci 5(1):27.

Balout L (1955). "Préhistoire de l'Afrique du Nord: Essai de Chronologie." Paris. Arts et Métiers Graphiques.

Beaumont PB, de Villiers H, Vogel JC (1978). Modern man in sub-Saharan Africa prior to 49,000 years BP: a review and evaluation with particular reference to Border Cave. S Afr J Sci 74:409.

Behrensmeyer AK (1982). The geological context of human evolution. Annual Review Earth Planetary Sci 10:39.

Behrensmeyer AK, Hill A (1980). "Fossils in the Making." Chicago. Chicago University Press.

Binford LR (1981). "Bones: Ancient Men and Modern Myths." New York. Academic Press.

Binford LR (1984). "Faunal Remains from Klasies River Mouth." New York. Academic Press.

Bishop WW, Clark JD (eds) (1967). "Background to Evolution in Africa." Chicago; London: Chicago Univ Press.

Bishop WW, Miller JA (eds) (1972). "Calibration of Hominoid Evolution." Edinburgh, Toronto: Scottish Academic Press.

Blumenschine RJ (in press). Characteristics of prehistoric hominid scavenging opportunity. Nature.

Bonnefille R (1983). Evidence for a cooler and drier climate in the Ethiopian uplands towards 2.5 m.yr. ago. Nature 303:487.

Bonnefille R (in press). Environment of early hominids in Africa. Paper presented at the 8th International Conference of Ethiopian Studies, Addis Ababa, 1984.

Brain CK (1958). The Transvaal Ape-man-bearing cave deposits. Transvl Mus Mem 11.

Brain CK (1981). "The Hunters or the Hunted? An Introduction to African Cave Taphonomy." Chicago: Chicago Univ Press.

Brain CK (1982). The Swartkrans site: Stratigraphy of the fossil hominids and a reconstruction of the environment of early *Homo*. Cong Int de Paléontologie humaine, Premier Congrès, Prétirage T.2. Nice:676.

Brauer G (1984). A craniological approach to the origin of anatomically modern *Homo sapiens* in Africa and implications for the appearance of modern Europeans. In Smith FH, Spencer F (eds): "The Origins of Modern Humans: A World Survey of the Fossil Evidence," New York: Alan Liss p 327.

Bunn HT (1982). Meat-Eating and Human Evolution: Studies on the Diet and Subsistence Patterns of Plio-Pleistocene Hominids in East Africa. Ph.D. dissertation, Department of Anthropology, University of California, Berkeley.

Bunn HT (1983). Evidence on the diet and subsistence patterns of Plio-Pleistocene hominids at Koobi Fora, Kenya and Olduvai Gorge, Tanzania. In Clutton-Brock J, Grigson C (eds): "Animals and Archaeology: Hunters and Their Prey," Oxford: B A R p.21.

Bunn H, Harris JWK, Isaac G, Kaufulu Z, Kroll E, Schick K, Toth N, Behrensmeyer AK (1980). FxJ j50: an early Pleistocene site in northern Kenya. World Archaeology 12(2): 109.

Cerling TE, Hay RL, O'Neil J (1977). Isotopic evidence for dramatic climatic changes in East Africa during the Pleistocene. Nature 267:137.

Clark JD (1967). The Middle Acheulian occupation site at Latamne, northern Syria (second paper). Further excavations (1965): General results, definition and interpretation. Quaternaria 10:1.

Clark JD (1975a). A comparison of the Late Acheulian industries of Africa and the Middle East. In Butzer KW, Isaac G (eds): "After the Australopithecines," The Hague: Mouton p 605.

Clark JD (1975b). Africa in prehistory: peripheral or paramount?   Man 10(2):175.

Clark JD (1980).   Thirty years of the Pan-African Congress, 1947-1977.   Proc 8th Pan-Afr Cong Prehist and Quat Studies, Nairobi, 1977.   Nairobi: p 3.

Clark JD (1981).   New men, strange faces, other minds: An archaeologist's perspective on recent discoveries relating to the origin and spread of Modern Man.   Proc Brit Acad 67:163.

Clark JD, Asfaw B, Assefa G, Harris JWK, Kurashina H, Walter RC, White TD, Williams MAJ (1984).   Palaeoanthropological discoveries in the Middle Awash Valley, Ethiopia.   Nature 307:423.

Clark JD, Harris JWK (in press).   Traces of fire: A critique of studies of Lower Pleistocene sites in East Africa.   The African Archaeological Review.

Clark JD, Kurashina H (1979).   Hominid occupation of the East-Central Highlands of Ethiopia in the Plio-Pleistocene. Nature 282:33.

Conroy G, Jolly C, Cramer D, Kalb J (1978).   Newly discovered fossil hominid skull from the Afar Depression, Ethiopia. Nature 275:67.

Cooke HBS (1958).   Observations relating to Quaternary environments in East and Southern Africa.   Trans Geol Soc S Afr 61(Annex):1.

Dart RA (1957).   The Makapansgat australopithecine osteodonto-keratic culture.   Proc 3rd Pan-Afr Cong on Prehist.   Livingstone, 1955, London:161.

Day MH (1972).   The Omo human skeletal remains.   In Bordes F (ed):   "The Origin of *Homo sapiens*," Paris: UNESCO p 31.

Day MH, Leakey MD, Magore C (1980).   A new hominid fossil skull (LH 18) from the Ngaloba Beds, Laetoli, northern Tanzania.   Nature 284:55.

Day MH, Stringer CS (1982).   A reconsideration of the Omo Kibish remains and the *Erectus-Sapiens* transition.   In "Congrès int de Paléont Humaine, 1ier Cong Nice, 1982." Prétirage, Vol 2, Nice: p 814.

Deacon HJ (1976).   "Where Hunters Gathered."   S Afr Archaeol Soc Monograph Ser 1:1-231.

Deacon HJ (1979).   Excavations at Boomplaas Cave - a sequence through the Upper Pleistocene and Holocene in South Africa. World Archaeology 10:241.

Deacon J (1978).   Changing patterns in the late Pleistocene/early Holocene prehistory of southern Africa as seen from the Nelson Bay Cave stone artifact sequence.   Quat. Research 10:84.

Débenath A (1975).   Découverte de restes humains probablement atériens à Dar-es-Soltan (Maroc). C R Hebd Séanc Acad Sci (Paris) 281-D:875.

Dechant-Boaz D (1982). Modern Riverine Taphonomy: Its Relevance to the Interpretation of Plio-Pleistocene Hominid Palaeoecology in the Omo Basin, Ethiopia. Ph.D. dissertation, Department of Anthropology, University of California, Berkeley.

Edwards SW (1978). Non-utilitarian activities in the Lower Palaeolithic: a look at the two kinds of evidence. Current Anthropology 19:135.

Ferembach D (1976a). Les restes humains de la Grotte de Dar-es-Soltan 2 (Maroc) Campagne 1975. Bul Mem Soc Anthrop Paris Ser 13, 3:183.

Ferembach D (1976b). Les restes humains atériens de Temara (Campagne 1975). Bul Mem Soc Anthrop Paris Sér 13, 3:175.

Flint RF (1959). Pleistocene climates in eastern and southern Africa. Bull Geol Soc Amer 70:343.

Gallagher JP (1972). A preliminary report on archaeological research near Lake Zuai. Annales d'Ethiopie 9:13.

Gasse F (1980). Les diatomées lacustres Plio-Pléistocènes du Gadeb (Ethiopie). Rev Algologique, Mem hors Sér 3.Paris.

Goodwin AJH (1935). A commentary on the history and present position of South African prehistory with full bibliography. Bantu Studies 9(4):293.

Goodwin AJH (1938). Archaeology of the Oakhurst Shelter, George. Trans Roy Soc S Afr 25 pt 3:229.

Goodwin AJH, Van Riet Lowe C (1929). Stone Age cultures of South Africa. Ann S Afr Mus 27:1.

Harding RSO, Teleki G (eds) (1981). "Omnivorous Primates: Gathering and Hunting in Human Evolution." New York, Columbia Univ Press.

Harris JWK (1983). Cultural beginnings: Plio-Pleistocene archaeological occurrences from the Afar Rift, Ethiopia. African Archaeol Review 1:3.

Howell FC (1982). Origins and evolution of African Hominidae. In Clark JD (ed: "The Cambridge History of Africa, Vol I From the Earliest Times to c.500 BC," Cambridge Univ Press p 70.

Howell FC, Clark JD (1963). Acheulian hunter-gatherers of sub-Saharan Africa. In Howell FC, Bourlière F (eds): "African Ecology and Human Evolution," Chicago:Aldine,p 458.

Isaac GLI (1971). The diet of early man: aspects of archaeological evidence from Lower and Middle Pleistocene sites in Africa. World Archaeol 2:278.

Isaac GLI (1975). The scatter between the patches. Kroeber Anthrop Soc Papers, Berkeley, California.

Isaac GL1 (1977). "Olorgesailie: Archaeological Studies of a Middle Pleistocene Lake Basin." Chicago: Univ Press.

Isaac GL1 (1978). The food-sharing behavior of proto-human hominids. Sci Amer 238:90.

Isaac GL1 (1984). The archaeology of human origins: Studies of the Lower Pleistocene in East Africa, 1971-1981. In Wendorf F, Close A (eds): Advances in World Archaeology, 3:1.

Jones PR (1978). Effects of raw materials on biface manufacture. Science 204:835.

Jones PR (1980). Experimental butchery with modern stone tools and its relevance for Palaeolithic archaeology. World Archaeol 12:153.

Jones PR (1981). Experimental implement manufacture and use: a case study from Olduvai Gorge, Tanzania. Philos Trans R Soc Lond, B 292:189.

Kaufulu ZM (1983). The Geological Context of Some Early Archaeological Sites in Kenya, Malawi and Tanzania: Microstratigraphy, Site Formation and Interpretation. Ph.D. dissertation, Department of Anthropology, University of California, Berkeley.

Keeley LH (1980). "Experimental Determination of Stone Tool Uses: A Microwear Analysis." Chicago:Univ Press.

Keeley LH, Toth N (1980). Microwear polishes on early stone tools from Koobi Fora, Kenya. Nature 293:464.

Klein RG (1979). Stone Age exploitation of animals in southern Africa. Am Sci 67:151.

Klein RG (1983). The Stone Age prehistory of southern Africa. Ann Rev Anthropol 12:25.

Klein RG, Allwarden K, Wolf CA (1983). The calculation and interpretation of ungulate age profiles from dental crown heights. In Bailey GA (ed): "Hunter-Gatherer Economy in Prehistory," Cambridge Univ Press p 47.

Leakey LSB (1936). "Stone Age Africa." Lond: Oxford Univ Press.

Leakey LSB (1952). The Olorgesailie prehistoric site. In Leakey LSB, Cole S (eds): "Proc Pan Afr Cong on Prehist 1947," Oxford: Blackwell, 209.

Leakey LSB (1959). A new fossil skull from Olduvai. Nature 184:491.

Leakey LSB, Tobias PV, Napier JR (1964). A new species of the genus *Homo* from Olduvai Gorge. Nature 202:7-9.

Leakey MD (1971). "Olduvai Gorge, Vol 3." Cambridge Univ Press

Leakey REF, Walker AC (1973). New australopithecines from Lake Rudolf, Kenya III. Am J Phys Anthropol 39:205.

Leakey REF, Walker AC (1976). *Australopithecus, Homo erectus* and the single species hypothesis. Nature 261:572.

Lovejoy CO (1981). The origin of man. Science 211:341.

Loy TH (1983). Prehistoric blood residues: detection on tool surfaces and identification of species of origin. Science 220:1269.

Lumley H de (1969). A Palaeolithic camp at Nice. Sci Amer 220:42.

Mason RJ (1957). The Transvaal Middle Stone Age and statistical analysis. S Afr Archaeol Bul 12(48):119.

Movius HL (1950). A wooden spear of Third Interglacial age from Lower Saxony. South-West J Anthrop 6(2):139.

Oakley KP (1956). Fire as a Palaeolithic tool and weapon. Proc Prehist Soc 21:36.

Oakley KP (1981). Emergence of higher thought 3.0-0.2 Ma. BP. Philos Trans R Soc Lond B 292:205.

Oakley KP, Andrews P, Keeley LH, Clark JD (1977). A reappraisal of the Clacton spearpoint. Proc Prehist Soc 43: 13.

Parkington JE (1984). Changing views of the Later Stone Age of South Africa. Advances in World Archaeology 3: 90.

Potts R, Shipman P (1981). Cutmarks made by stone tools on bones from Olduvai Gorge, Tanzania. Nature 291:577.

Rightmire GP (1984). *Homo sapiens* in sub-Saharan Africa. In "The Origins of Modern Humans: A World Survey of the Fossil Evidence", New York: Alan Liss, p 295.

Roche J, Texier JP (1976). Découverte des restes humains dans un niveau atérien supérieur de la Grotte des Contrebandiers à Temara (Maroc). C R Hebd Séanc Acad Sci, Paris 282-D: 45.

Schick KD (1984). Processes of Palaeolithic Site Formation: An Experimental Study. Ph.D. dissertation, Department of Anthropology, University of California, Berkeley.

Schoeninger MJ, DeNiro MJ, Tauber H (1983). Stable nitrogen isotope ratios of bone collagen reflect marine and terrestrial components of prehistoric human diet. Science 220: 1381.

Sept JM (1984). Plants and Early Hominids in East Africa: A Study of Vegetation in Situations Comparable to Early Archaeological Site Locations. Ph.D. dissertation, Department of Anthropology, University of California, Berkeley.

Shackleton NJ, Kennett JP (1975). Late Cenozoic oxygen and carbon isotope changes at DSDP Site 284: implications for general history of the northern hemisphere and Antarctica. In Kennett JP, Hout RE *et al.*: "Initial Reports of the Deep Sea Drilling Project," Vol 29. Washington: US Govt Print Office, p 801.

Shipman P (1981). Application of scanning electron microscopy to taphonomic problems. Ann N Y Acad Sci 376:357.

Shipman P (1983). Early hominid lifestyle: hunting and gathering or foraging and scavenging? In Clutton-Brock J, Grigson C (eds): "Animals and Archaeology, 1. Hunters and Their Prey," Oxford: B A R p 31

Sillen A, Kavanaugh M (1982). Strontium and palaeodietary research: a review. Yearbk Phys Anthropol 25:67.

Singer R, Wymer J (1982). "The Middle Stone Age at Klasies River Mouth in South Africa." Chicago:Univ Press.

Sohnge PG, Visser DJL, Van Riet Lowe C (1937). "The Geology and Archaeology of the Vaal River Basin." Pret: Govt Printer.

Stahl AB (1984). Hominid dietary selection before fire. Curr Anthropol 25:151.

Toth NP (1982). The Stone Technologies of Early Hominids at Koobi Fora, Kenya: An Experimental Approach. Ph.D. dissertation, Department of Anthropology, University of California, Berkeley.

Toth NP (in press). Evidence for preferential right-handedness in the Early Stone Age. J Hum Evol.

Toth NP, Schick KD (1983). The cutting edge: an elephant butchery with stone tools: interim evidence. Foundation for Research into the Origin of Man 5:8.

Van der Merwe NJ (1982). Carbon isotopes, photosynthesis and archaeology. Amer Sci 70:596.

Van der Merwe NJ, Vogel JC (1978). ^{13}C content of human collagen as a measure of prehistoric diet in woodland North America. Nature 276:815.

Vincens A (1979). Analyse palynologique du site archéologique FxJ j50: formation de Koobi Fora, East Turkana (Kenya). Bul Soc Géol France 21(3): 343.

Vincent A (1983). Tuberous plants in the evolution of human diet. Anthroquest, LSB Leakey Found 26:15.

Vrba ES (1980). The significance of bovid remains as indicators of environment and predation patterns. In Behrensmeyer AK, Hill AP (eds): "Fossils in the Making: Vertebrate Taphonomy and Palaeoecology," Chicago:Univ Press p 247.

Wayland EJ (1934). Rifts, rivers, rains and early man in Uganda. J Roy Anth Inst 64:333.

White TD (in press). Cutmarks on the Bodo cranium: a case of prehistoric defleshing. Nature.

Hominid Evolution: Past, Present and Future, pages 89–97
© *1985 Alan R. Liss, Inc.*

ENERGY-TRAPS AND TOOLS

Charles A. Reed

Department of Anthropology
University of Illinois at Chicago
Chicago, Illinois, U.S.A. 60680

As a farmer's son growing up during the 1920's in the Cascade Mts. of Oregon, U.S.A., I was familiar with tools; we used them all the time. Some forty kinds quickly come to memory. A tool was an object, usually of metal but of wood sometimes, which was hand-held and allowed the user to accomplish a task which could not have been done without the use of the tool, or perhaps allowed the task to be done more easily. Machines powered by horses, electricity, or gasoline engines were not tools. A hammer was a tool but a nail was not. Bottles, jugs, buckets, boxes, etc., were containers but not tools. Ladders or lawnmowers, which could be moved by hand but rested on the earth when used, were not tools. A whip was not a tool but a curry-comb was. Yes, we farmers knew precisely which objects were tools and we knew too that tools were man-made, man-used. We would have laughed at anyone who suggested otherwise.

As I grew older I learned that occasionally some objects were used by some animals (I would now say some other animals) as we would use tools, but this knowledge in no way changed my concept that only man made tools. Indeed, college courses of that time and later emphasized the 'fact' that only humans made tools.

Man the tool-maker was a cultural tradition of thousands of years originating in our human regard for ourselves as expressed in Genesis 1:27 on the religious side of our culture and in the Aristotelian concept of the Scala Naturae philosophically.

Anthropologists thought much as did the farmers among whom I was reared; tools were made by man. Definitions of the Hominidae often if not usually included the unique tool-making ability of 'man'as a criterion of the family ("Man the Tool-maker," Oakley 1949 and subsequent editions, for instance). Those anthropologists had little reason to think otherwise, and that little was ignored or easily explained away.

In 1954 I became by chance an osteo-archaeologist, identifying the broken bones, mostly non-human, excavated from prehistoric archaeological sites. I was soon asked "Are domestic animals human tools?" I didn't know then the answer to that question, but based on my experience as a farmer I thought not. I had of course learned that stone artefacts, purposely shaped for particular uses, were tools, but at the same time never thought that clay pots made by the same people should also be called tools. The archaeologists with whom I was working didn't call them tools, either.

During the 1950's Raymond Dart (1957 and many separate papers) was trying to convince people that the broken bones being excavated from the australopithecine site of Makapan in South Africa had been purposely broken into particular shapes for particular jobs, and so were tools. Since "Man Maketh Tools" (a popular expression of the time) the australopithecines would then necessarily be hominids and not apes, as many people then considered them. This line of reasoning was soon to be considered invalid.

In 1960 Jane Goodall observed that the chimpanzees she was studying constructed two kinds of tools; they crushed wads of fresh leaves lightly between their teeth to increase their absorbent quality and then dipped such a wad in water to use as a sponge, and, even more startling, they prepared slim sticks or the stalks of coarse grasses and used these objects to probe into termites' nests. The termites would seize the intruding objects in their jaws and then be pulled out to be eaten by the chimpanzee. Here was a particular object prepared in a particular way for a particular purpose, and then used for that purpose. Further, remembering from past experience where a termite mound was, the chimpanzees would sometimes prepare their termite-sticks before they could see the mound, then carry the prepared objects to the nest and use them. Here was an example of "time-binding," also thought until then to be a unique human ability.

Louis Leakey, who had supported Goodall's research, said of these discoveries, "Scientists ..... must accept chimpanzee as man, they must redefine man, or they must redefine tools." (van Lawick-Goodall 1967, p. 32). Leakey's influence, both among scientists and the lay public, was important in publicizing the new discoveries, but actually no spectacular new definitions were made; instead the word 'tool' was accepted to mean simply an object purposely made and used in a certain way, and not defined in terms of who made it. Last ditch efforts to preserve human uniqueness were made by the claim that man was the skilled tool maker (Oakley 1969) and the statement by Gruber (1969) that a true tool must have been made by the use of another object. These attempts met the same failure as had Owen's effort a century earlier to distinguish apes from humans by a supposedly distinctive character in the latter's hippocampus. Generally accepted, thus, was the realization that at least one kind of ape made tools, forcing abandonment of a basic belief held for thousands of years.

Then some people began to wonder: if a chimpanzee's termite-stick is a tool, why not a chimpanzee's nest (Lancaster 1968)? Is it not a purposefully made structure for a particular purpose? Most people retorted automatically that a nest wasn't a tool, and besides if one regards a

chimpanzee's nest as a tool, why not a bird's
nest, or an alligator's 'nest' for her eggs, or a
termites' mound? One has to have a boundary
somewhere, you know! Beck (1980) has considered
these and other problems of "What is a tool?" more
thoroughly than others and concluded (p. 10) that
to be a tool an object must be free of any fixed
connection with the substrate, must be outside the
user's body at the time of use, and the user must
hold or carry the tool during or just prior to
use. (Beck's discussion is more complex and
thorough than my abstract of it, above, and should
be read in its entirety.)

However, before Beck's conclusions were
published, I had begun to ponder the same problem
of finding that elusive boundary between tool and
non-tool, and thought too of that earlier question
"Is a domestic animal a tool?" If true, must we
also admit that an ant's domestic aphids are
tools? One can use a finger as a lever; is a
finger, then, a tool or is one merely using it as
a tool? Is fire, if controlled, also a tool, a
chemical tool? Where was a boundary? Then I
realised, as Beck was doing about the same time,
that neither I nor anyone else had ever actually
decided what a tool really was. At that point I
began to think: What factor in common did all of
these possible tools have?

A tool is an object or process (if fire is
possibly a tool) which allows the user to accomplish work or to accomplish it more efficiently, or allows the user to conserve energy more
efficiently, than could be done without the use of
that tool. To put the matter more fundamentally,
more energy can be extracted from a given environment, or the same energy extracted more quickly,
or the available energy can be better conserved,
than if the tool were not used. Quite as in the
history of the study of thermodynamics, energy and
its utilization were the key factors. Who made an
object or controlled a process, or how particular
objects were made or used were secondary considerations. By contemplating man and his works as
primary we had been starting at the wrong end of

biological evolution, whereas by contrast energy
and its utilization were factors that permeated
that evolution. So I turned the problem around
and started at the beginning.

Without energy there would never have been
living cells. Whatever the process of the origin
of life may have been, the earliest living cells
functioned by themselves as gatherers and conserv-
ers of energy; I call them primary energy-traps,
and any living protoplasmic entity remains today
a primary energy-trap, whether you or I or a
bacterium or an oak tree.

I. PRIMARY ENERGY-TRAPS

The cellular, or protoplasmic part, of any
living being. All the living cells, one or many,
plant or animal, are the primary energy-trap of
that individual. Individuals that are solely
protoplasmic, solely primary energy-traps, whether
one-celled or multi-celled, are necessarily
relatively simple in construction, for as cells
aggregate and their numbers increase, non-living
structures must be added to them. Such additions
may be organic or non-organic, mono-molecular or
multi-molecular, inside a body or outside of it;
the basic requirements of these secondary energy
traps are: they aid the individual primary
energy-trap in acquiring or conserving energy, and
they are not part of the living protoplasm.

II. SECONDARY ENERGY-TRAPS

A. Protoplasmic Secretions: (Fluids or other
materials produced intracellularly and then secreted through the plasma membrane; their use as
energy-traps is extra-cellular): 1) Such secondary
energy-traps used within the body: tissue-fluids,
plasma, extra-cellular antibodies and hormones;
neurohumors; 2) Such secondary energy-traps used
outside the body: digestive enzymes, as secreted
by parasitic amoebae and by the digestive
epithelia of Metazoa; buccal and other external

secretions and their sometimes poisonous
derivatives of many animals; pheromones; mucus;
animal silk and its derivatives (threads, webs,
cocoons); waxy secretions on the surfaces of
some leaves; poisonous secretions of some plants;
etc.; 3) Supportive secondary energy-traps inside
and/or outside the body: Collagenous (animal) and
woody (plant) tissues; non-cellular parts of
cartilage and bone; dentin; enamel; skeletons of
strontium sulphate in some Radiolaria and of
silicious compounds in others and in basket
sponges; external calcareous skeletons of
Foraminifera, corals, many molluscs, and internal
skeletons of echinoderms; etc.

B. Features of the habitat used temporarily
by an animal or animals to gain or conserve energy
without disruption of the environment: 1) Escape
of predatee: Natural objects used for hiding or
escape, preserving the life and thus conserving
and continuing the energy of the predatee; grass,
thickets, trees, cliffs, ponds, etc.; success of
predator; use of objects in stalking, thus aiding
the acquiring of energy by the predator; grass,
thickets, trees, boulders, even automobiles
occasionally. For the animals the object used
is a secondary energy-trap, although that object
plays a completely passive part in the succesful
escape or hunt; 2) Life-preserving and thus energy
preserving escape from extremes of environment:
shade (a space of lowered temperature) for protection by many animals from the sun ("mad dogs ....
...... go out in the noonday sun."), thickets,
hollows, caves, pools, etc.; the same often serve
as secondary energy-traps for protection against
cold, wind, rain, and snow; some Japanese macaques
in winter use hot-springs as secondary energy
traps.

C. Symbiosis: one or more other organisms
used as a secondary energy-trap; 1) Intra-
specific:Social cooperation, from that of a pair
to that between millions of the same species; by
cooperation each individual and/or the young of
the particular pair or group is/are provided with
greater energy resources or with more conservation

of energy than any individual could provide
alone. In this sense, a well functioning human
spouse is a secondary energy-trap. 2)
Interspecific (Mutualism): The individual of each
species in such a pair-bonding is a secondary
energy-trap for the other member; examples are: a)
complete anatomical mixture of the two kinds of
individuals (lichen: fungus and algae; green
hydra: a coelenterate and algae); b) one kind of
species living in cavities in the other
(cellulose-digesting protozoans and/or bacteria
living in digestive tracts of termites or some
kinds of ungulates); c) two kinds of animals or
plants physically separate from each other, but at
least one of the two completely dependent upon the
other for survival (several kinds of small
crustaceans or fish which groom larger fish); most
of the kinds of animals (not always insects) which
cross-pollinate flowers; small fish which live
among the tentacles of sea-anemones; etc.); d) two
kinds of animals which act as secondary energy-
traps for each other but presumably could survive
alone (the honey badger and the honey guide; the
ox-pecker and its ungulate roosts); e) two kinds
of animals or plants which live together for mut-
ual benefit, but one is dominant, the other is
more like a servant (humans and their domestic
animals and plants are in this category), and so my
original question, the one that started me
thinking about energy-traps, "Are domestic animals
human tools?" is answered NO!

   D. Tools: A tool is a particular kind of
secondary energy-trap, an object or a controlled
chemical process (fire for example) in the
production of which the environment is modified.
The object or controlled chemical process is or
was used to acquire or to conserve energy more
efficiently than could the individual or social
group without the use of that object or process.

   This definition of tool reaches far beyond
that of any previous one, and in some of its
consequences stands in marked contrast to the
definition of Beck (1980), but emerges automatically from the concept of different kinds of

energy-traps in the biological world. Some individuals may be a bit dazed by such a new and revolutionary consideration of the meaning of 'tool,' so discussions of some examples are necessary.

The simplest tool is a natural object removed, perhaps only temporarily, from its place in the environment by an animal (I know of no plants that make or use tools), used for a specific purpose, and then replaced or merely dropped. Homo has often been, or is, that animal but such use of a tool is no different from the use of a smoothing stone by the North American burrowing wasp, Ammophilia, or use of a stone by a sea otter as an anvil against which to smash shell fish (Oakley 1954). But that is tool-use, some people will say; certainly no such 'lower' animals make tools. Those people are wrong.

The most primitive animals that make tools, insofar as I know, are members of the protozoan family Difflugiidae, each of which amoeboid uni-cellular individuals manufactures for itself an enveloping 'test,' a shell of sandgrains, bits of floating detritus, or other materials from the water in which the animal lives. That round or flask-shaped test, with one opening for the pseudopods, is a tool. Yes, one group of one-celled animals makes tools, and so do thousands of other kinds of animals; all nests are tools, and so are all burrows produced by physical removal of substrate (alteration of the environment). Lists of all the kinds of animals that make nests or burrows cannot be presented here, but they would be impressive! All animals on those lists are tool-makers. The nests of ants, below ground, and the more imposing 'nests' of termites, whether subterranean, above ground, or in trees, are all tools. The holes in timbers of molluscan 'shipworms,' crayfishes' burrows and beavers' dams, probing spines and twigs of the Galapagos finch Cactospiza and the artifical structures of the bower birds, etc. & etc.---: all are tools.

Tool-making is and has been of two kinds,

instinctive and cultural. The latter, the learning anew of tool-making by each individual of each generation, represents the activities of only a few kinds of primates and perhaps elephants. Otherwise all tool-making is instinctive, even the seemingly purposive and complex hive-building by some bees and wasps and dam-building by beavers.

Tool-making has no over-all continuity in a phylogenetic pattern, but the practice evolved independently here and there in many phyla. Particular tool-making primates represent only one or a clustered few of the twigs on the Tree of Life. That Tree is a vast one, with millions of terminal twigs, and we cannot focus on one twig, such as Homo, to dictate laws to which all other living twigs must conform. Man, only one twig among so many, is not necessarily the measure of all twigs.

REFERENCES

Beck BB (1980). "Animal Tool Behavior". New York: Garland STPM Press.

Dart RA (1957). "The osteodontokeratic culture of Australopithecus prometheus". Pretoria: Transvaal Mus Mem 10.

Gruber A (1969). A functional definition of primate tool-making. Man 4:573.

Lancaster JB (1968). On the evolution of tool making behavior. Amer Anth 70:56.

Oakley KP (1949). "Man the Tool-Maker". Chicago: University of Chicago Press.

Oakley KP (1969). Man the skilled tool-maker. Antiquity 43:222.

van Lawick-Goodall J (1967). "My Friends the Wild Chimpanzees". Washington, D.C.: National Geographic Society.

PART II
THE VOICE OF TODAY

Ancestors of the Ancestors

Hominid Evolution: Past, Present and Future, pages 101–106
© *1985 Alan R. Liss, Inc.*

AFRICAN ORIGIN, CHARACTERISTICS AND CONTEXT OF EARLIEST
HIGHER PRIMATES

Elwyn L. Simons

Director, Duke Primate Center
James B. Duke Professor of Anthropology
3705 Erwin Road, Durham, NC, USA 27705

The Oligocene continental deposits of the Fayum
Province of Egypt yield the oldest well-documented series
of terrestrial vertebrates in Africa. The land vertebrate
fauna has been most recently listed in Bown et al., 1982.
Among the numerous vertebrates are species of seven genera
of primates: Apidium (Osborn 1908), Propliopithecus and
Parapithecus (Schlosser 1911), Oligopithecus (Simons 1962),
Aegyptopithecus (Simons 1965), Qatrania (Simons and Kay
1983) and a new genus and species of prosimian (Simons and
Bown in press, 1985).

Of the six described genera two, Propliopithecus and
Aegyptopithecus, can be assigned to the family Propliopithecidae and a third, Oligopithecus may provisionally be
associated with them. The remaining three genera, Apidium,
Parapithecus, and Qatrania all belong in another family
Parapithecidae (Simons and Kay 1983). The best known of
these primates are Apidium and Aegyptopithecus. Species
of each of these two primates are documented not only from
large series of upper and lower jaws, but from skulls or
skull parts and many postcranial bones as well. These have
served as the basis for very specific analysis of the
probable adaptations and locomotor function of Aegyptopithecus (Fleagle, Simons and Conroy 1975; Fleagle and
Simons 1978; Fleagle and Simons 1982). Studies of the
adaptations and postcranial anatomy of Apidium have
been published (Fleagle and Simons 1978; Fleagle and
Simons 1983). Details of the cranial anatomy of
Aegyptopithecus have also been reported on in several
publications (Simons 1967; Simons 1972; Fleagle and Kay 1983).

The ecology of the Fayum primates has been analyzed (Kay and Simons 1980) and their degree of sexual dimorphism interpreted (Fleagle, Kay and Simons 1980). A complete revision of the African Oligocene apes has recently been published (Kay, Fleagle and Simons 1981) and Kay and Simons are currently preparing a similar monograph on the parapithecids, Apidium, Qatrania and Parapithecus.

Of all the many and varied African Oligocene primates only one, Aegyptopithecus zeuxis, appears to have given rise to later descendants. This species makes a good structural ancestor to species of the Eastern African Miocene genus Proconsul. The Proconsul group in turn, it is generally agreed, gave rise to the lineage which has produced the great apes and humans. Even though one stage, the late Miocene African hominoids, is poorly understood it would appear that the course of evolution of our lineage of higher primates, in the thirty million years between Aegyptopithecus and Australopithecus, took place exclusively in Africa. What had happened in the evolutionary development of Anthropoidea in Africa before Fayumian times is speculative. The age of the capping basalt overlying the Fayum sedimentary series has recently been redetermined (Fleagle, Bown, Obradovich and Simons, in press 1985). This change is due to the discovery that several successive basalts overlie the Jebel Qatrani Formation. The lowest of these lies on an erosion surface and gives a date of 31.0 $\pm$ 1 million. This means that the upper fossil wood zone primates, by simple averaging of the sediment thickness between the top of the Eocene below and the basalt above, would be about 33 million years old. Likewise, the primates from the lower zone, Qatrania and Oligopithecus, would be about 35 million years old. The only record of primates in Africa older than those of the Fayum is Azibius from the Eocene of Algeria (Gevin, et al. 1975). This animal was tentatively referred to as an archaic prosimian group and bears no apparent relationship to the Fayum primates. Because of a few broad similarities in the overall plan of molars and premolars it is possible to speculate that Propliopithecidae and Parapithecidae are derived from a common ancestor that lived perhaps 5 to 10 million years earlier, most probably in Africa. Thus, it seems reasonable to conclude that the higher primates or Anthropoidea, just like the much more recent lineage that gave rise to Homo, had their origin in Africa. The time concerned was about or prior to 45 million years ago. We are not certain as

to the group from which this stock could have been derived.
Origins among both Adapidae and Omomyidae have been suggested.
The molar anatomy seen in Cantius or Pelycodus makes a good
structural model from which to derive that of Aegyptopithecus.

The total number of primate specimens now known from
the Fayum probably stands close to 1,000 if all isolated
teeth and questionably referred postcranial elements are
included. Of these close to 100 are propliopithecids, the
remainder parapithecids. Of the many postcranial elements
recovered we know best the scapula, pelvis and limbs of
Apidium (Conroy 1976, Fleagle and Simons 1979b, Anapol 1983,
Fleagle and Simons 1983). These bones show that Apidium
was a small arboreal quadruped adept at leaping and springing.
Forelimbs are well adapted for clinging, climbing and
suspension during landing. A partial association of hind-
limb bones of Apidium was found in 1980. This shows an
unusual distal apposition of tibia and fibula. Although
not fused this apposition would stabilize the limb distally,
much as in Tarsius. Associated with tibia and fibula in
this specimen are talus and calcaneum. Distally both tibia
and fibula articulate with spool-like, prominent ridges that
would largely prevent rotational movements at the tibio-
talar joint. These adaptations enhance leaping capacities.
Apidium, the most abundant Fayum mammal, can be pictured as
inhabiting a densely forested, riverine environment, leaping
among tangled vines, various hardwoods and mangrove trees.
Dental anatomy of Apidium and Parapithecus suggests that
these animals were frugivorous-folivorous. Qatrania the
smallest known member of Old World Anthropoidea with a body
weight estimate of about 300 grams was only a little larger
than a mouse lemur. Its molars, only weakly crested,
suggest that it may have been fruit or gum eating (Simons
and Kay 1983).

Postcranial anatomy of Aegyptopithecus strongly suggests
that like Apidium it was an arboreal quadruped. There is
no feature of any known limb or foot bone in either of these
two primates that would suggest terrestrial adaptations
(Fleagle, in Bown et al. 1982). Both ulna and humerus of
Aegyptopithecus show that it was a robust, and perhaps
slow-moving arboreal quadruped with an over-all locomotor
resemblance to the South American monkey Alouatta (Fleagle
and Simons 1982 and op. cit.). The large number of
mandibles now known for Aegyptopithecus, Propliopithecus
and Apidium allowed Fleagle, Kay and Simons (1980) to

demonstrate that these Fayum anthropoids had distinctly
different sized sexes. This difference is marked in
Aegyptopithecus. It strongly suggests that A. zeuxis lived
in multiple male - multiple female groups, thus having a
social organization more complex than typifies most prosimians.
Analysis of the teeth of Aegyptopithecus suggests that it
may have been frugivorous (Kay and Simons 1980). Although
a stocky, arboreal animal Aegyptopithecus foreshadows
Proconsul postcranially as well as dentally. The upper
molars in the two have a similar overall plan including
beaded lingual cingula. The broadened humeral trochlea
and details of the condyles resemble especially the Miocene
apes and not more primitive primates. The hallucial
metatarsal, talus and calcaneum of this Fayum ape are all
quite similar to those of Proconsul. In turn some of the
new discoveries about the foot and hind limb of Proconsul
show that in certain ways the latter resembles Pan.
Anatomy of earliest Australopithecus suggests derivation
from a Pan-like animal (Walker and Pickford 1983).

The traditional scientific account of the rise from ape
to man starts with early apes living in the trees. To
this hypothesis all that we now know of Fayum primates gives
support. One publication (Kortlandt 1980) posits erroneously
that in Oligocene times the Fayum region was a dry grassland and that the Fayum primates were terrestrial animals.
There is no substance to his theory which is contradicted
by the whole broad range of new evidence now available
(Bown, et al. 1982). Not only are the postcranial remains
of both Apidium and Aegyptopithecus conclusive on the point
of their arboreality but the evidence from Fayum plants,
trace fossils, other vertebrates and sedimentology all
converge to suggest a monsoonal rain-forest.

Particularly interesting is the evidence from bird
fossils which are remarkably abundant in the same quarries
that have yielded the bulk of Oligocene primates (Rassmussen,
Olson and Simons in the press, 1985). Five families of
birds appear to have their earliest known occurrence in
the Fayum Oligocene. These are the jacanids (jacanas or
lilly trotters), the pandionids (ospreys), the musophagids
(touracos), the podicipedids (grebes), and balaenicipitids
(whaleheads or shoebill storks). Together with these are
four other families: acciptrids (hawks and eagles),
rallids (rails), ciconiids (storks) and ardeids (herons).
The acciptrid resembles Haliaetus, the fish eagle, and the

ardeid will be assigned to genus <u>Nycticorax</u>, the night heron.
Nearly all these birds live and feed in swamps or near running water. All but one of the African turacos are forest-
dwelling.

Fayum plants are known from leaf-yielding quarries and
sites where silicified fruits, nuts and seed pods occur.
The plants include mangroves, water lilies, aquatic ferns
(family Salvinaceae), Nelumbonaceae, figs, palms, cinnamon,
a variety of hardwoods, Juglandaceae and Araceae (genus
<u>Epiprimnum</u>). These plants have tropical Indo-Malaysian
affinities and suggest that the flora of the Fayum was
part of a larger, widespread Tethyian flora, not closely
related to floras of the African interior (Wing and
Tiffany in Bown <u>et al</u>. 1982).

The ichnofossils of the Fayum are the most abundant,
best preserved and most diverse of any trace fossil
assemblage found in river lain deposits anywhere in the
world. The clearest environmental indicator is the termite
trace fossil <u>Termitichnus</u> - the nest of a subterranean termite. Such termites occur only in tropical regions (Bown
1982). Sedimentological evidence suggests a monsoonal
climate. Fayum paleosols have a lower horizon which
indicates periodic wetting and drying. These soils also
contain clay enriched pans that suggest heavy absolute
rainfall at intervals. Thus, it is safe to conclude that
earliest known ancestors of man lived in groups as frugivores
in a tropical rain-forest.

Anapol F (1983). Scapula of <u>Apidium phiomense</u>; a small
    anthropoid from the Oligocene of Egypt. Folia primat
    40:11.
Bown TM (1982). Ichnofossils and rhizoliths of the nearshore
    fluvial Jebel Qatrani Formation (Oligocene), Fayum Province,
    Egypt. Paleogeog Paleoclimat Paleoecol 40:255
Bown TM, Kraus MF, Wing SL, Fleagle JG, Tiffney BH, Simons
    EL, Vondra CF (1982). The Fayum primate forest revisited.
    J Human Evol 11:603.
Conroy G (1976). Primate postcranial remains from the
    Oligocene of Egypt. Contrib to Primat 8:1.
Fleagle JG, Bown TM, Obradovich JD, Simons EL (1985). How
    old are the Fayum Primates? Proc IX Cong Int Primat Soc,
    in press.
Fleagle JG, Kay RF (1983). New interpretations of the
    phyletic position of Oligocene hominoids. In Ciochon RL,

Corruccini RS (eds): "New Interpretations of Ape and Human Ancestry" Plenum Press, p 181.

Fleagle JG, Kay RF, Simons EL (1980). Sexual dimorphism in early anthropoids. Nature 287:328.

Fleagle JG, Simons EL, Conroy G (1975). Ape limb-bone from the Oligocene of Egypt. Science 189:135.

Fleagle JG, Simons EL (1978). Humeral morphology of the earliest apes. Nature 276:705.

Fleagle JG, Simons EL, (1979). Anatomy of the bony pelvis in parapithecid primates. Folia primat 31:176.

Fleagle JG, Simons EL (1982). The humerus of Aegyptopithecus zeuxis a primitive anthropoid. Amer J Phys Anthropol 59:175.

Fleagle JG, Simons EL (1983). The tibio-fibular articulation in Apidium phiomense, an Oligocene anthropoid. Nature 301:238.

Gevin P, Lavocat R, Mongereau N, Sudre J (1975). Découverte de Mammifères dans la moitie inferiéure del'Eocène continental du Nord-Ouest du Sahara. C R Acad Sci Paris 280:967.

Kay RF, Fleagle JG, Simons El (1981). A revision of the Oligocene apes of the Fayum Province, Egypt. Amer J Phys Anthropol 55:293.

Kay RF, Simons EL (1980). The ecology of Oligocene African Anthropoidea. Int J Primatol 1:21.

Kortlandt A (1980). The Fayum primate forest: did it exist? J Human Evol 9:277.

Osborn HF (1908). New fossil mammals from the Fayum Oligocene, Egypt. Bull Amer Mus Nat Hist 24:265.

Rasmussen DT, Olson S, Simons EL (1985). The avifauna of the Oligocene Jebel Qatrani Formation, Fayum, Egypt. J vert paleont, in press.

Schlosser M (1911). Beitrage zur Kenntnis der Oligozänen Landsäugetiere aus dem Fayum (Ägypten). Beitr Pal Geo Ost-Ungarn 24:51.

Simons EL (1962). Two new primate species from the African Oligocene. Postilla 64:1.

Simons EL (1965). New fossil apes from Egypt and the initial differentiation of Hominoidea. Nature 205:135.

Simons EL (1967). The earliest apes. Sci Amer 217:28.

Simons EL, Kay RF (1983). Qatrania, new basal anthropoid primate from the Fayum, Oligocene of Egypt. Nature 304:624.

Walker,AC, Pickford M (1983). New postcranial fossils of Proconsul africanus and Proconsul nyanzae. In Ciochon RL, Corruccini RS (eds): "New Interpretations of Ape and Human Ancestry" Plenum Press, p 325.

Hominid Evolution: Past, Present and Future, pages 107–112
© *1985 Alan R. Liss, Inc.*

KENYAPITHECUS : A REVIEW OF ITS STATUS BASED ON NEWLY
DISCOVERED FOSSILS FROM KENYA

M. PICKFORD, Ph.D.

Institut de Paléontologie
8 Rue Buffon, Paris 75005

Kenyapithecus, frequently referred to the genus Ramapithecus, has for more than twenty years formed the basis
for much publicised claims for the existence of the family
Hominidae in the middle Miocene period. A lively debate,
totalling over fifty papers, followed Leakey's (1962) announcement of the discovery of Kenyapithecus wickeri at Fort
Ternan, Kenya (Pickford 1985). Subsequently dated at 14 m.y.,
the Fort Ternan find was often quoted in support of early
divergence of hominids from the great apes. The bulk of the
papers dealing with Kenyapithecus wickeri and subsequently
Kenyapithecus africanus (Le Gros Clark and Leakey, 1951)
(Leakey, 1967) dealt not so much with the hominid status of
the fossils, about which the main proponents in the debate
were generally agreed, but with the question of synonymy or
otherwise of Kenyapithecus with Ramapithecus.

Some twenty morphological characters were said to
support the inclusion of Kenyapithecus or Ramapithecus in
the Hominidae (Ciochon, 1983, Table 1), prominent among which
were its supposed parabolic dental arcade, small, sexually
monomorphic canines, zygomatic process set well forward,
small upper and lower incisors relative to cheek tooth size
and lower molar buccal cingula poorly developed or absent.
Several of these characters could not be observed in the
specimens of Kenyapithecus available up to 1979, but were
inferred from reconstructions based on fragmentary material.
These reconstructions were themselves the subject of debate,
but were nevertheless frequently used as evidence in support
of the hominid status of Ramapithecus, and have been widely
publicised, particularly in textbooks. In view of the very

fragmentary nature of the fossil evidence, the reconstructions probably reveal more about the scientists who made
them than they do about the species they purport to represent.

Recently, the fabric of the Ramapithecus debate has
shifted dramatically (Wolpoff, 1983), originally as a result
of the increased acceptance of molecular evidence concerning
the interrelationship between extant higher primates, followed closely by the impact of cladism on the systematics
of hominoids. A little later, comparative morphological re-
examination of Pongo and Indian Ramapithecus and Sivapithecus revealed the existence in all three genera of what is
generally accepted as being derived morphology of the subnasal alveolar process, morphology quite different from that
of African apes (both extant and extinct), Australopithecus
and man (Ward and Pilbeam, 1983). The result of the new
interpretations is that it seems likely that Ramapithecus
from the Indian sub-continent is not a hominid but a pongine
in the strict sense of the word (i.e. the subfamily which
includes Pongo, but excludes Pan and Gorilla). Where does
this leave Kenyapithecus? Should it also be rejected from
human ancestry on the grounds of its claimed synonymy with
Indian Ramapithecus? Could it not after all, be different
from Ramapithecus at generic and subfamily levels, and
therefore still be eligible for human ancestry? With the
nine specimen hypodigm that was available until 1979, there
was simply no way of making a convincing argument (Pickford,
1985) either in support of, or against its hominid status,
nor of investigating further its synonymy with Ramapithecus.
Since 1979, however, a greatly improved sample of Kenyapithecus fossils has become available from two areas in
Kenya. The first of these samples came from the middle
Miocene Maboko Formation in the Winam Gulf, Lake Victoria,
and consists of about 80 isolated teeth, a mandible, and
parts of the post-cranial skeleton of a young individual
(Pickford, 1985). The second, and more significant sample
comes from the Nachola area near Baragoi in Samburu District,
Kenya, and consists of some 150 specimens, including associated mandibles and maxillae, many isolated teeth and some postcranial remains. Both new collections are about 15 or 16 m.y.
old.

The Maboko sample which was collected by the author
from 1979 to 1984 revealed significant details concerning
the morphology and variability of Kenyapithecus africanus,

but left several important questions unanswered. For instance, it showed that simple molar cingula were relatively
common in Kenyapithecus, and that there seemed to be two
size groups of Kenyapithecus, possibly representing males
and females of a single dimorphic taxon. It revealed also
that the mandibles of the smaller group of Kenyapithecus
africanus differed from those of Ramapithecus in several
respects, especially in the morphology of the symphyseal
area. The arcade shape of the mandible, although somewhat
distorted due to post-mortem crushing, was not parabolic,
but more nearly U-shaped.

The hominid status of K. africanus was thus weakened,
especially if marked sexual dimorphism was a feature of the
genus. Its synonymy with Ramapithecus also seemed less
likely than before. On their own, the Maboko discoveries
revealed little about the status of Kenyapithecus wickeri,
the Fort Ternan species with which K. africanus was sometimes synonymised. It did show, however, that there were
significant differences between the samples from the two
sites and that they probably represent two separate taxa.

While re-assessing the entire hominoid collection from
Fort Ternan, the author found that specimens hitherto excluded from K. wickeri because they exhibited non-hominid
morphology, were rather similar to specimens from Maboko,
and quite different from Proconsul nyanzae, in which they
had previously been classified. If my reassessment of these
specimens is correct, then Kenyapithecus wickeri is characterized by marked sexual dimorphism, both in dental size and
canine morphology, has large third molars and relatively
large central incisors. It therefore does not display
special similarity to any Plio-Pleistocene and Recent hominids. Although it was possible to show that K. africanus
was significantly different from Ramapithecus, it is still
not possible to demonstrate conclusively that K. wickeri
differs markedly from Sivapithecus indicus (or Ramapithecus)
from the Indian sub-continent, except in the morphology of
the mandible. Until the Fort Ternan sample improves, these
uncertainties will probably persist.

In 1982 and 1984, the author, working with the Japan/
Kenya Palaeontological Expedition, found several early middle
Miocene localities which have yielded the fossilized remains
of Kenyapithecus africanus in much greater quantities than at
Moboko, and generally more complete and in better condition.

Over 150 specimens have been recovered, and the sites show
little sign of exhaustion. Particularly significant discoveries include associated mandibles, maxillae and postcranial remains.

The mandibles tend to confirm inferences based on the
Maboko specimens, that K. africanus was dimorphic, that
small individuals possessed small (female) canines, and
large specimens possessed large (male) canines, and that the
morphology of the symphysis was not particularly similar to
that of Sivapithecus from India. Of the greatest significance, however, was the discovery of maxillae complete
enough to exhibit the morphology of the subnasal alveolar
process. Preliminary assessment of this part of the maxilla
shows that K. africanus possessed the plesiomorphic dryo-
morph condition (Ward and Pilbeam, 1983), and none of the
derived ramamorph morphology expressed by Sivapithecus,
Ramapithecus and Pongo. The existence of deeply scooped
canine fossae bordered caudally by the vertical anterior
wall of the zygomatic process of the maxilla is confirmed,
but its conformation and position are not particularly Homo-
like. It does, however, recall to some extent the morphology
seen in Pongo and some of the Hadar hominids. In addition,
two maxillae from Nachola, which are otherwise compatible in
morphology, contain canines of markedly different size, which
is suggestive of significant sexual dimorphism in the species.

In summary :

(a) Kenyapithecus africanus can now be demonstrated to differ in several important respects from Ramapithecus and Sivapithecus from the Indian sub-continent. Its continued synonymy with Sivapithecus or Ramapithecus is, I believe, not
warranted.

(b) K. africanus differs in several respects from the younger species, K. wickeri, but the extent and significance of
the differences are difficult to assess in light of the expanded, but still inadequate, hypodigm of the Fort Ternan
species.

(c) Both K. africanus and K. wickeri appear to be sexually
dimorphic.

(d) K. africanus possesses none of the apomorphic characters
which define the Hominidae. It is therefore unlikely to

belong to that family, but it is not removed from a possible rôle in the eventual ancestry of hominids.

(e) Similarly Kenyapithecus africanus is sufficiently gen-eralised that it is plausible to consider that it might eventually have given rise to pongines. If this and point d) above can be substantiated, then K. africanus can be vis-ualised as predating the split between pongines and the African apes and man.

(f) Several points of similarity between K. africanus and Proconsul, when examined in the light of their chronological order, suggest the possibility that the two are phylogenetic-ally linked.

(g) The position of K. wickeri will remain problematic until better fossil material is found.

(h) Acceptance of the genus Kenyapithecus must lead to a re-examination of Middle Miocene Eurasian hominoids with thick enamelled molars currently identified as Sivapithecus darwini. Their continued classification as Sivapithecus is based on symplesiomorphic characters, and has chronological and zoogeographic implications which may not be warranted. It has yet to be demonstrated that any of the Middle Miocene Eurasian hominoids possess any apomorphic characters found in Sivapithecus.

ACKNOWLEDGEMENTS

I wish to thank the Director of the National Museum of Kenya, The L.S.B. Leakey Trust (London), the L.S.B. Leakey Foundation, and the Boise Fund. I particularly wish to thank Dr. H. Ishida for inviting me to participate in the Japan/Kenya Expedition, and Mr. K. Cheboi, my field assist-ant. Finally, I am indebted to Professor Phillip Tobias for the invitation to speak at the Taung 60 Symposium.

REFERENCES

Ciochon R (1983). Hominid cladistics and the ancestry of modern apes and humans. In Ciochon R, Corruccini R. (eds): "New Interpretations of Ape and Human Ancestry", New York: Plenum, p.783.
Leakey LSB (1962). A new lower Pliocene fossil primate from Kenya. Ann Mag Nat History 4: 689.

Leakey LSB (1967).   An early Miocene member of Hominidae.
Nature 213:155.

Pickford M (1985).   A new look at Kenyapithecus based on
recent discoveries in Western Kenya.   J Human Evol 15
(in press).

Ward S, Pilbeam D (1983).   Maxillofacial morphology of Miocene
hominoids from Africa and Indo-Pakistan.   In Ciochon R,
Corruccini R (eds): "New Interpretations of Ape and Human
Ancestry," New York: Plenum, p 211.

Wolpoff M (1983).   Ramapithecus and human origins : an
anthropologist's perspective of changing interpretations.
In Ciochon R, Corruccini R (eds): "New Interpretations of
Ape and Human Ancestry", New York: Plenum, p 651.

Rise of the Bipedalists

Hominid Evolution: Past, Present and Future, pages 115–127
© *1985 Alan R. Liss, Inc.*

HOMINID LOCOMOTION - FROM TAUNG TO THE LAETOLI FOOTPRINTS

Michael H. Day, MB BS, DSc, PhD, LRCP, MRCS

Head of the Division of Anatomy
United Medical & Dental Schools of Guy's
and St Thomas's Hospitals, London, England

The recovery and recognition of the Taung skull 60 years ago marked the beginning of an era in hominid studies not only in dental and cranial terms but, perhaps surprisingly, also in postural and locomotor terms.

The crucial observation that was made by Dart (1925) was that the position of the foramen magnum was tucked well beneath the skull when it was orientated in the Frankfurt horizontal. This clear indication of head poise, of upright posture and its possible association with a bipedal gait was not lost on Dart, neither was the profoundly shocking realisation that the Taung infant was a small-brained biped. This fact he knew would totally contradict the views of Elliot Smith his former teacher.
He and others including Keith believed that human evolution was "brain-led" out of some ancestral ape-like stage, glimmerings of which had come from the discoveries of Dubois (1894) in Java. The supposedly associated Pithecanthropus I skull cap and femur fitted this theory since they apparently provided evidence of increased brain size coupled with bipedal adaptation.

The Taung infant possessed an unlikely combination of features and, in view of the Javan finds, it originated from an unlikely continent, Africa, the origin of man being firmly believed at that time to have been most likely to have occurred in the Far East.

It was against this background of belief that Dart's claim for the hominid nature of his find was judged. Of

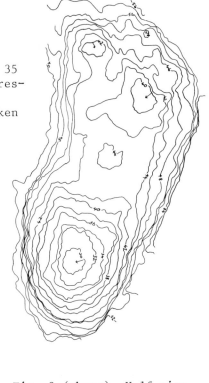

Fig. 1 (below) Half size

Laetoli hominid footprint G1 - 35
(left) showing a deep heel impression, a varus hallux position,
impressions of short toes, broken
ground behind the toe row, a
"barefoot gap" between hallux
and toes and hallucial drag at
"toe-off" as the foot began
the swing phase of walking.

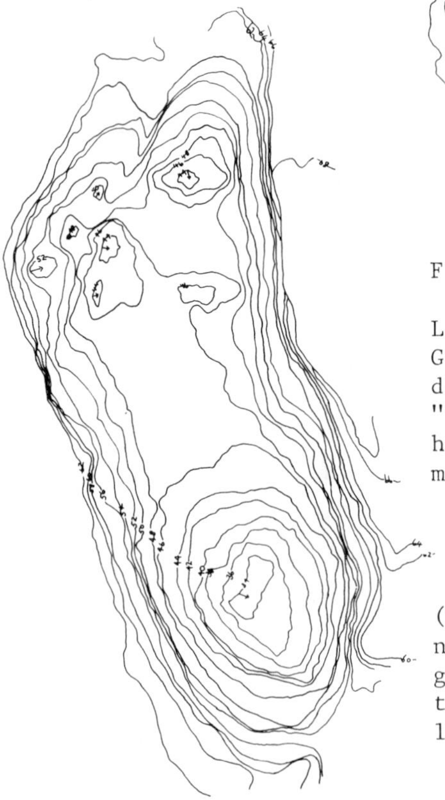

Fig. 2 (above) Half size

Laetoli hominid footprint
G1 - 34 (right) showing a
deep heel impression, a
"barefoot gap", a varus
hallux position and a
medial arch.

(Neither the pace length
nor the dynamic angle of
gait can be inferred from
the position of Figures
1 - 6).

all the doubts expressed by Dart's opponents those concerning the upright stance and bipedal gait of the australopithecines seem to have the longest history (Zuckerman 1966, 1967). The discovery by Dr M.D. Leakey in 1978 of bipedal hominid footprints in 3.6 million year old solidified volcanic ashes at Laetoli in Tanzania silenced the doubters and provided dramatic evidence of early hominid bipedalism (Leakey, Hay 1979), evidence indeed that is free of the locomotor interpretations of fossil postcranial bones and free of the difficulties encountered in the multivariate statistical analysis of morphometric data. This footprint evidence is a direct record of hominid locomotor activity contemporaneous with extinct animal tracks in datable rock. It is also one of the rare, if not unique, records of early hominid soft tissue anatomy.

The Laetoli footprint trails consist of two sequential sets of tracks of three individuals of differing foot sizes. On the left there is a single trail and on the right there are two overprinted trails. There are a number of methods of analysis that are applicable to a record of this kind, but an obvious, previously untried, method of analysis of soft substrate footprints was photogrammetry (Day, Wickens 1980). This is a stereoscopic photographic method that can produce contoured plots of very high accuracy - a method not to be confused with cartographic contouring that seeks to reduce landscape to the printed page and results in the swirling contours familiar from maps. Such plots are identifiable by the emphasis placed on each fifth contour, a mapmaker's convention.

A specially constructed camera used in medical photogrammetry, was transported to Tanzania and, with Dr Leakey's consent, the individual prints were recorded from casts made of the original trails. These photogrammetric plots of the Laetoli footprints were initially compared with experimental soft substrate human footprints made in moist sand. Characteristic human walking prints were established and their similarities noted (Day, Wickens 1980). Six of the best plots have been selected for further analysis, four from the G-1 individual (G-1 - 35 Left, G-1 - 34 Right, G-1 - 33 Left, G-1 - 27 Right) and two from the overprinted tracks of individuals G2 and G3 (G-2/3 - 27, G-2/3 - 26) (Figs 1-6).

Fig. 3 (below) Half size

Laetoli hominid footprint Gl – 33
(left) showing a deep heel impression, a deep and oblique ball
impression, a varus hallux position, a medial arch, short toe
impression, and a "barefoot gap".
There is also evidence of broken
ground behind the toe row as the
toes dug in on "toe-off".

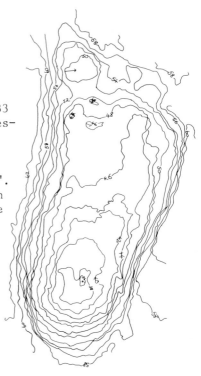

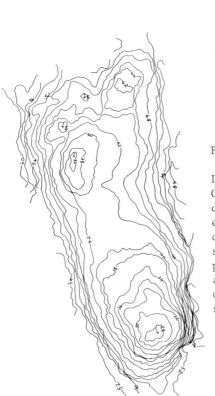

Fig. 4 (above) Half size

Laetoli hominid footprints
Gl – 27 (right) showing a
deep heel impression, lateral weight transference,
deep hallucial ball impression, a varus hallucial
position, a "barefoot gap"
and hallucial drag at "toe-
off" at the beginning of the
swing phase of walking.

Perhaps the most striking features of these footprints, apart from the human like mechanism of weight and force transfer through the foot reported earlier, are the consistent presence of a varus position of the great toe in all individuals, as well as a "barefoot gap" particularly well displayed in G1 and G3. [The position of the hallux is termed varus when it deviates towards the midline and valgus when it deviates away from the midline. This is distinct from the tilt of the terminal phalanx which is always valgus in man].

In a study of 327 male and 325 female footprints of Nigerians, unshod (or lightly shod), it was found that the mean hallux angle was 0° and that the "barefoot gap" was present. This is described as a divergence between toes I and toes II - V indicating a spread of the unconfined forefoot. Boot-wearing Europeans in the same study did not possess a "barefoot gap" yet did possess valgus toe positions that varied from $+ 6.9^o - + 11^o$. The difference between the unshod Nigerians and the shod Europeans was highly significant in terms of toe position (Barnicot, Hardy 1955).

Radiological evidence of hallux position in European infants of 4 - 6 years of age showed a very small hallux angle (Hardy, Clapham 1952), while a study of 38 examples of New Guinea unshod showed a range of -8° (Varus) - $+16^o$ (Valgus) with the majority at 0°. These results were based on footprints, radiographs and soft tissue examination (Barnett 1962).

In view of these results the presence of a "barefoot gap" and a marked varus position of the hallux in the Laetoli hominids takes on added significance. The mean varus angle of each of the Laetoli hominids can only be determined from all the available footprints of each individual, but from inspection of plots published here it would seem that the angle is high by comparison with the modern human unshod.

Other features of the plots to which attention should be drawn are the toe impressions of G1 - 35 and G1 - 33 disclosing short toes; the broken ground behind the toe row in G1 -35 and G1 - 34, the medial arch conformation as disclosed by G1 - 34, G1 - 33 and the great toe drag shown by G1 - 27 and G1 - 35. The "barefoot gap" shows itself

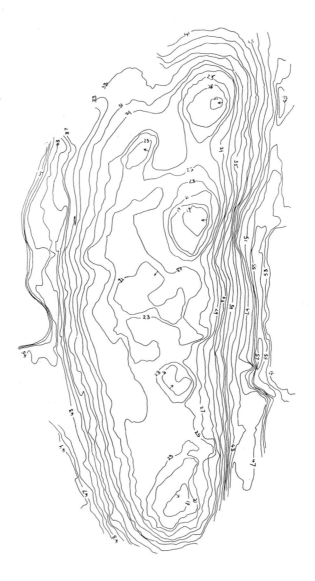

Fig. 5 (Half size)

Laetoli hominid footprints G2/3 - 26 (left overprinted)
showing a deep combined heel impression, hallucial impressions of both G3 (rear) and G2 (leading), varus hallucial
position in both, some short toe impressions in G2 and a
ball impression for G3.

as a deep indentation in the contours between the hallux and the rest of the toes in G1 - 35, G1 - 33, also to a lesser extent in G1 - 34 and G1 - 27. The overprinted contours of G2/3 are also revealing in that the "barefoot gap" is well shown in G3 - 27 as well as varus deviation of the hallux, a pronounced medial arch, a deep heel impression and an oblique ball impression (Fig. 6). G2/3 - 26 displays clear and deep hallux impressions for both prints and some toe impressions for G2 (Fig. 5).

What these footprints, and their photogrammetric analysis, shows is that bipedalism of an apparently human kind was established 3.6 million years ago. The mechanism of weight and force transmission through the foot is extraordinarily close to that of modern man. The spread of the forefoot as shown by the "barefoot gap", and the varus position of the hallux and the presence of short toes, have close similarities with the anatomy of the feet of the modern human habitually unshod; arguably the normal human condition.

Raymond Dart's faith in his morphological deduction of upright stance and bipedalism from the position of the Taung foramen magnum has been fully justified although it has taken nearly 60 years to prove.

But what has happened in the 60 years in terms of fossil finds and where have these finds led us in our search for answers to locomotor questions? I believe that they reveal two locomotor morphologies (other than that of Homo sapiens) within the family Hominidae that can be characterised by groups or constellations of features that, when fully developed, may well prove to be diagnostic of the genus Australopithecus or the species Homo erectus. These features can be grouped as femoro-pelvic, tibio-femoral, and pedal in terms of their functional anatomy.

Broom's and Robinson's recovery of the Sts 14 pelvis and partial skeleton of a gracile australopithecine from Sterkfontein provided the first evidence of the australopithecine femoro-pelvic complex and revealed such bipedal features as a broad backwardly angulated S-shaped iliac blade, a small acetabulum, the forward rake of the anterior superior iliac spine and a short ischium - despite the crushed and damaged nature of the specimen.

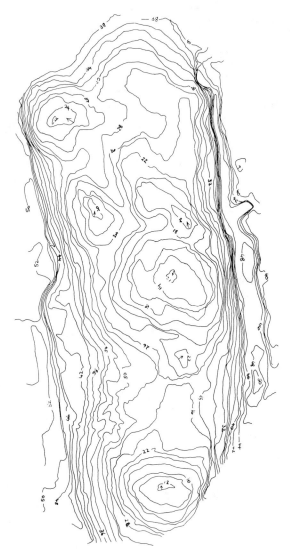

Fig. 6 (Half size)

Laetoli hominid footprint G2/3 - 27 (right) Overprinted contours showing a deep combined heel impression, deep hallucial
impressions for both G3 (rear) and G2 (leading), varus hallucial position in both, a "barefoot gap" in both, a deep
oblique ball impression in G3 as well as a pronounced medial
arch and evidence of lateral weight transfer.

Dart's finds from Makapansgat of 2 juvenile ilia and an ischium confirmed the general pattern (Dart 1949, 1958).

Le Gros Clark (1955) and Napier (1964) believed the gracile australopithecines to be bipedal on the basis of this evidence. Zuckerman (1966) believed this conclusion premature. It was not until 1973, after extensive but inconclusive metrical analysis, that Zuckerman et al. (1973) accepted at the Elliot Smith Symposium the possibility that the overall locomotor pattern of Australopithecus included "in addition to bipedalism components yet undefined". The recovery by Johanson of the AL 288 "Lucy" skeleton provided a dramatic confirmation of the form of the small australopithecine pelvis. The "Lucy" pelvis shows remarkable similarities to the Sterkfontein example in reconstructions by Robinson and Walker (in Shipman 1981) including the shape of the ilium, the small acetabulum, the short ischium and the position of the auricular surface. It seems clear that the pelvic morphological pattern characterised by the distinctive group of features shown by Sts 14, as well as by other specimens such as Sts 65 and SK 3155(B), has been repeated in AL 288 from a site 2,000 miles away and from deposits perhaps as much as 1,000,000 years older.

Let us now consider the upper end of the femur. The recognition by Napier (1964) of the hominid status of two proximal femoral fragments from Swartkrans, SK 82 and 97, was of crucial importance. It was Napier, in this seminal paper on bipedalism, who drew attention to their distinctive combination of anatomical features. He pointed out that these femora possessed; small heads, relatively long and flat necks, they lacked greater trochanter flare, they had a posteriorly placed lesser trochanter and little or no intertrochanteric line.

Some five years later (Day 1969), this group of features was increased in an account of Mary Leakey's find of OH 20 from Olduvai. In addition to the majority of features noted by Napier, I was able to add to the complex the presence of an obturator externus groove on the back of the neck and later a psoas major groove on the front. These features confirmed that the australopithecine hip was habitually held in an extended position during upright stance and bipedal gait. Since 1969, at a conservative estimate, there are now about 14 femoral necks from East and South Africa that show this group of features. The latest is from the Middle Awash (White 1984), dated at 4 my BP. This must make this specimen the oldest fossil evidence of the australopithecine lower limb locomotor skeleton from anywhere in the world.

These pelvic and femoral finds, and the analyses performed on them, including some that are metrical, have shown the essential similarity of the morphology of the australopithecine pelves and femoral necks; but what of the supposed morphological and locomotor differences between the so-called robust and gracile australopithecines suggested by Napier (1964) and Robinson (1972)? As long ago as 1971 grave doubts were expressed about the supposed locomotor difference between these two groups (Zihlman, 1971). I can see nothing in the material that has been found since then to assuage those doubts. The only so-called robust or <u>Paranthropus</u> pelvis known is the notorious SK 50 which is cracked, flattened and distorted. But what features there are seem reminiscent of the gracile pelves from elsewhere. Whatever the cranial and dental evidence may be of differences between gracile and robust australopithecines (at either generic or specific level), it is my view that the femoro-pelvic complex has nothing to contribute either morphologically or functionally to the debate. I see no reason to believe, on this evidence alone, that the larger form was distinct from or more arboreally adapted than the smaller.

If we turn to the knee, we will find that it remains (as any medical student will tell you) one of the most complex and difficult joints in the body. Early on Broom recovered TM 1513 and Sts 34 whose femoro-condylar angles and intercondylar grooves caused debate. Estimates of femur length have ranged from 250 mm - 360 mm on the basis of triangulation methods (Lovejoy, Heiple 1970) or reconstruction (Walker 1973).

The Hadar knee and others from Koobi Fora confirm the general pattern. A group of femora from Koobi Fora treated statistically by McHenry and Corrucini (1978) demonstrated two patterns of morphology, one australopithecine and one hominine. The Olduvai tibia and fibula (OH 35) lack upper ends but match, in size at least, the more gracile tibial shafts from other sites, and tibial plateaux of all the Koobi Fora hominids (741, 1500,

1810) reveal basically the same pattern with clear evidence of menisci and locking mechanisms.

Australopithecine foot material from South Africa is almost unknown and scanty from elsewhere. The Kromdraai broken talus (TM 1517) is part of the hypodigm of Paranthropus robustus, although the case for this status seems to rest almost entirely on association - a case weakened by recent research on the other postcranial elements in the assemblage (Day, Thornton in prep.).

The Olduvai foot is still the most convincing osteological evidence of bipedalism in the whole of the early hominid fossil record. When it was first described its very obvious bipedal adaptation (in conjunction with other features) was said to have convinced Phillip Tobias finally (pers. comm.) to agree to the erection of Homo habilis as a new species of man. At that time so perfectly adapted a bipedal foot was not anticipated for australopithecines (Day, Napier 1964). Perhaps we should bear in mind that the foot is fully adult (indeed arthritic); it is clearly not associated with the juvenile hand material. Secondly, the Laetoli footprints predate the Olduvai foot by 1.85 my. Thus it seems difficult to say now that the morphology of this foot is itself generically diagnostic of Homo alone. Perhaps the best answer for an unashamed gradist is that this "transitional morphology' is just what we should expect.

There is now no serious dispute as to the upright stance and bipedal gait of the australopithecines or of Homo erectus, even though the evidence produced by Dubois for this capability seems ironically to relate to Homo sapiens in the Trinil I femur and not to Homo erectus at all (Day, Molleson 1973; Day 1984). The early opinions of Dart, Broom, Le Gros Clark and Napier with regard to the bipedalism of the australopithecines were a triumph of anatomical reasoning that has been fully vindicated by subsequent fossil finds and finally by the dramatic and unassailable evidence of the Laetoli footprints. Weidenreich too, must take his share of credit - his impeccable observations on the Peking femora outlined a group of characters of the Homo erectus femoral shaft that have appeared again both in Africa and Europe (Day 1971; Sigmon 1982). In exactly the same way Napier's recognition of the group of femoral neck characters of

Australopithecus was a key observation that has been repeated
from many sites in Africa. In both cases almost all individual features in each group fall within modern human or other
comparative ranges. Taken together they form a clearly re-
cognisable pattern of morphology not repeated as a group in
any comparative series. I venture that the same thing obtains
for the femoro-pelvic complexes of Australopithecus and Homo
erectus.

   There are of course occasions when metrical and statistical analysis add weight to morphological arguments, but
anatomical analysis and comparison must come first. Here I
echo the plea Phillip Tobias made at the "Ancestors" conference in New York, but I would add that combinations of features have far more to tell, functionally and taxonomically,
than single features whose appearance in the fossil record
may seem to warrant a new junction on a cladogram.

   Two patterns of morphology in the lower limb have been
characterised from fossil material, much of it from East and
South Africa, the best evidence from the upper femur and pelvis. What do these differences mean in functional and taxonomic terms? Postcranial studies can provide two forms of inference - locomotor and taxonomic. Locomotor inference arises
from the study of fossil bones and more rarely footprints;
taxonomic inference can come from definitions of taxa that include locomotor criteria. If all bipeds were hominid, were
all hominids bipeds? At the moment the answer seems to be
yes, since evidence of bipedalism predates evidence in terms
of skulls and teeth. What preceded bipedalism or was pre-
adaptive for it? That is another story for another time.

Barnett CH (1962). The normal orientation of the human hal-
    lux and the effect of footwear. J Anat Lond 96(4):489.
Barnicot NA, Hardy RH (1955). The position of the hallux
    in West Africans. J Anat Lond 89:355.
Clark WE Le Gros (1955). "The Fossil Evidence for Human Evol-
    ution: an Introduction to the Study of Palaeo-anthropology":
    Chicago: Univ Press.
Dart RA (1925). Australopithecus africanus: The man-ape of
    South Africa. Nature 115:195.
Dart RA (1949). Innominate fragments of Australopithecus
    prometheus. Am J Phys Anthropol 7:301.
Dart RA (1958). A further adolescent australopithecine ilium
    from Makapansgat. Am J Phys Anthropol 16:473.
Dubois E (1894). Pithecanthropus erectus, eine menschen-

aehnliche Ubergangsform aus Java. Batavia: Landesdruckerei.

Day MH (1969). Femoral fragment of a robust australopithecine from Olduvai Gorge, Tanzania. Nature 221:230.

Day MH (1971). Postcranial remains of Homo erectus from Bed IV, Olduvai Gorge, Tanzania. Nature 232:383.

Day MH (1984). The postcranial remains of Homo erectus from Africa, Asia and possibly Europe. Cour Forsch Inst Senckenberg 69:113.

Day MH, Molleson TI (1973). The Trinil femora. In Day MH (ed): "Human Evolution," Symp SSHB 11 London: Taylor & Francis Ltd., pp 127-154.

Day MH, Napier JR (1964). Hominid fossils from Bed 1, Olduvai Gorge, Tanganyika: fossil foot bones. Nature 201:967.

Day MH, Wickens EH (1980). Laetoli Pliocene hominid footprints and bipedalism. Nature 286:385.

Hardy RH, Clapham JCR (1952). Hallux valgus: predisposing anatomical causes. Lancet 14 June:1180.

Leakey MD, Hay RL (1979). Pliocene footprints in the Laetolil Beds at Laetoli, Northern Tanzania. Nature 278:317.

Lovejoy CO, Heiple KG (1970). A reconstruction of the femur of Australopithecus africanus. Am J Phys Anthropol 32:33.

McHenry HM, Corruccini RS (1978). The femur in early human evolution. Am J Phys Anthropol 49:473.

Napier JR (1964). The evolution of bipedal walking in the hominids. Arch Biol Liège LXXV:673.

Robinson JT (1972). "Early Hominid Posture and Locomotion." Chicago and London: Univ Chicago Press, p 361.

Shipman P (1981). "Life History of a Fossil: An Introduction to Vertebrate Taphonomy and Paleoecology". Cambridge, Mass: Harvard Univ Press.

Sigmon BA (1982). Comparative morphology of the locomotor skeleton of Homo erectus and the other fossil hominids. Prem Cong Internat Paleo Hum Nice: Prétirage.

Walker A (1973). New Australopithecus femora from East Rudolf, Kenya. J Hum Evol 2:545.

White T D (1984). Pliocene hominids from the Middle Awash, Ethiopia. Cour Forsch Inst Senckenberg 69:57.

Zihlman AL (1971). The question of locomotor differences in Australopithecus. Proc 3rd Int Cong Primat Karger Basel 1:54.

Zuckerman S (1966). Myths and methods in anatomy. J Roy Coll Surg Ed 11:87.

Zuckerman S (1967). The functional significance of certain features of the innominate bone. J Anat Lond 101:608.

Zuckerman S, Ashton EH, Flinn RM, Oxnard CE, Spence TF (1973). Some locomotor features of the pelvic girdle in primates. Symp Zool Soc Lond 33:71.

Hominid Evolution: Past, Present and Future, pages 129–133
© *1985 Alan R. Liss, Inc.*

APE FOOTPRINTS AND LAETOLI IMPRESSIONS: A RESPONSE TO THE
SUNY CLAIMS

Russell H. Tuttle, Professor

Department of Anthropology,
Committee on Evolutionary Biology,
and The College, The University of Chicago
1126 East 59th Street
Chicago, Illinois 60637

and

Affiliate Scientist in Behavioral Biology
Yerkes Regional Primate Research Center

Between 1977 and 1979, at Laetoli in northern Tanzania,
M.D. Leakey and her associates uncovered trails that 4
bipedal individuals had made in volcanic ashfalls about 3.5
million years ago (Leakey 1978, 1979, 1981; Leakey and Hay
1979; Hay and Leakey 1982). These unique fossils are the
earliest evidence for hominid bipedalism and show that it
preceded stone tool making by 1.5 million years.

At site G in Laetolil tuff 7, there are 2 trails by
which 3 individuals are represented. One trail (G-1) was
made by a single individual (also designated (G-1). The
second trail (G2/3), running parallel with trail G-1, was
made by individuals G-2 and G-3. G-3 had overprinted and
partly obliterated the tracks of G-2. The fourth individual
(A) is represented by 5 prints at site A, 1.6 km from site
G.

The 3 individuals at site G were immediately recognized as members of the Hominidae. A is not so clearly
linked to the Hominidae. Indeed Tuttle's (1984) study on
bipedal bears indicates that it is ursid-like in a number
of features. Preliminary direct and interpretive studies,
using gross morphological examination (Leakey 1979, 1981;

Leakey and Hay 1979), photogrammetric contour diagrams (Day
and Wickens 1980) and comparisons with laboratory kinesiological data on Homo sapiens (Charteris et al. 1981, 1982)
indicated that the structure of the feet and inferred gaits
of G-1 and G-3 were very similar to those of Homo sapiens.

In 1981, Tuttle (in press) conducted a study on casts
of the southern (3.7 m) section of the G trails (19 prints)
and the entire A trail in order to infer the manner of
walking and taxonomic relationships of the Laetoli bipeds.
This was followed by studies on 29 casts of the best
individual prints from trails G-1 (n = 10) and G-2/3 (n =
19) at the Kenya National Museums and in Tuttle's laboratory
at the University of Chicago.

Tuttle (in press) concluded that in all discernible
morphological features, the feet of the individuals that
made the G trails are indistinguishable from those of
modern humans. The great toe was adducted to approximate
alignment with the lateral toes and it often left a deep
impression that is strongly reminiscent of the final toeing-
off prior to swing phase by human lower limbs during bipedal
locomotion. The lateral toes, which did not project beyond
the tip of the hallux, generally left barely discernible
impressions. Hominid G-1 had well developed medial
longitudinal pedal arches. The arches of G-3 are less
apparent, perhaps due to prior compaction of the substrate
by G-2. It appears that the stance phases of their progression began with substantial heel strikes. Weight was
borne initially by the heel and lateral sole. Then it
shifted medially onto the ball and hallux before toe-off
into the swing phase. Although the feet of G-1 and G-3
were relatively broad, their foot indices fall within a
large array of foot indices for modern Homo sapiens,
especially when one includes persons of relatively short
stature, who have never worn restraining footgear.

No hominid foot bones have been recovered from the
Laetolil Formation. Because of digital curvature and
elongation and other skeletal features that evidence
arboreal habits (Tuttle, 1981), Tuttle (in press) noted
that it is unlikely that Australopithecus afarensis from
Hadar, Ethiopia, could make footprints like those at
Laetoli.

Per contra, while championing a single morphotype for
eastern African Pliocene Hominidae, Stern and Susman (1983)
claimed that the morphology of the Hadar foot bones represents the morphology of feet that made the bipedal footprints
at Laetoli site G. To them, Australopithecus afarensis is
a veritable "missing link" which was still sufficiently
apelike that features of its footprints could be replicated
by modern chimpanzees. This stands starkly in contrast to
conclusions of all previous students of the prints who
noted that the Laetoli G impressions are remarkably human.

If the conclusion of Stern and Susman were based solely
on their limited access to and cursory examination of the
fossil prints, we could register it with other cases of
haste making paleoanthropological waste (not to be confused
with débitage) and focus on another topic. However, the
empirical record must be set straight anent the footprints
of humans and chimpanzees, which they have characterized
inadequately and erroneously. This is curious since humans
are not in short supply, Stern assisted me to film chimpanzee bipedalism in 1965 and subsequently borrowed the film
for detailed study, and Susman (1984) observed (but did not
describe) bonobo footprints along stream beds and in channel
sands in Zaire.

The main source of their mistakes is overgeneralization
from an unspecified human sample and 3 young incarcerated
chimpanzees: 2 subadult males that left bipedal footprints
in wet sand (Stern and Susman, 1983) and a 5-year old
female (Elftman and Manter, 1935). It appears that the
chimpanzees walked on the lateral aspects of their soles,
with the lateral toes "curled" to some degree, including
fairly tight flexion, and the hallux adducted close to,
but not against, the medial aspect of the second toe.
Indeed Susman and Stern (1983) proclaim that chimpanzees
do not walk on the ground with a widely abducted hallux.

Based on the film of 6 adolescent and 1 adult male
bipedal Pan troglodytes and a detailed study of the bipedal
footprint trails left on pressure sensitive paper and
moist earth by 8 juvenile and 2 adult male Pan troglodytes
and 1 adult female and 1 juvenile male Pan paniscus at
Yerkes RPRC, I conclude that marked hallucal abduction is
characteristic of both species when they walk bipedally.
Further, they generally do not flex their lateral toes
prominently unless they are frightened, injured or have

foreign substances on their soles. The mean distance
between opposite surfaces of the distal pads of toes I and
II is 56 mm (r = 9-102 mm) for 35 prints from 8 juvenile
Pan troglodytes that had foot lengths near or less than
the mean foot length of Laetoli G-1 (180 mm). The mean
distance for the young Pan paniscus is 62 mm (r = 47-81 mm,
n = 7 prints).

   While the heel and lateral sole are the most prominent
pressure points in prints of the subjects, often the medial
arching is as much due to bracing by the abducted hallux
as to a disinclination to place the sole flat on the substrate. In deformable substrates, they commonly leave
broad flat-footed impressions.

   If extended troglodytian toes had characterized the
Laetoli G hominids, one would expect the tips to have been
impressed at a notable distance from the distal ball of the
sole. And if they had flexed long toes, even deeper
knuckle impressions would be seen as evidenced by one
terrified Yerkes subject that took several steps with one
of its feet tightly flexed. Neither pattern exists at
Laetoli G so we can infer that their lateral toes were
quite short.

   The right foot of G-1 sometimes left peculiar marks
distal to the toe tips. These are best explained by the
greater wetness of the substrate in one spot (G-1/34) and
a tendency for G-1 to drag the foot on lift off probably
due to pathology of the lower limb (Tuttle, in press).

   Stern and Susman (1983) claimed that the Laetoli
prints are not humanoid because they lack deep broad
impressions from the post-hallucal ball and medial swelling at the base of the big toe which is supposedly
characteristic of human prints. But humans commonly leave
prints devoid of these features as may be seen in prints
on the beach. In slowly walking individuals the medial
ball need not leave marked impressions because, during the
double stance phase of a cycle, some weight has already
shifted to the contralateral foot before it is the chief
contact area of its foot. The hallucal pad, which has a
smaller surface area than the ball, commonly leaves a
deeper impression as in the Laetoli prints. Further, at
toe-off plastic substrates are likely to be pushed backward
into the area of the hallucal ball.

In brief, Stern and Susman (1983) raised no valid facts against the humanness of the Laetoli G-1 prints. Prince Charming was fortunate not to have had them as footmen; otherwise, he would have wed a stepsister instead of Cinderella.

Acknowledgements. I thank Mary Leakey for the opportunity to study the Laetoli prints and staff of Kenya National Museum and Yerkes Regional Primate Research Center for help. The research was supported by the Lichtstern Fund of the University of Chicago and NIH grant RR-00165 to YRPRC.

Charteris J, Wall JC, Nottrodt JW (1981). Functional reconstruction of gait from the Pliocene hominid footprints at Laetoli, northern Tanzania. Nature 290:496.

Charteris J, Wall JC, Nottrodt JW (1982). Pliocene hominid gait: new interpretations based on available footprint data from Laetoli. Am J Phys Anthropol 58:133.

Day MH, Wickens EH (1980). Laetoli Pliocene hominid footprints and bipedalism. Nature 286:385.

Elftman H, Manter J (1935). Chimpanzee and human feet in bipedal walking. Am J Phys Anthropol 20:69.

Hay RL, Leakey MD (1982). The fossil footprints of Laetoli. Sci Amer 246:50.

Leakey MD (1978). Pliocene footprints at Laetoli, northern Tanzania. Antiquity 52:133.

Leakey MD (1979). 3.6 million year old footprints in the ashes of time. Natl Geogr Mag 155:446.

Leakey MD (1981). Tracks and tools. Phil Trans R Soc Lond B-292:95.

Leakey MD, Hay RL (1979). Pliocene footprints in the Laetoli Beds at Laetoli, northern Tanzania. Nature 278:317.

Stern JT, Susman RL (1983). The locomotor anatomy of Australopithecus afarensis. Am J Phys Anthropol 60:279.

Susman RL (1984). The locomotor behavior of Pan paniscus in the Lomako Forest. In Susman RL (ed): "The Pygmy Chimpanzee," New York, Plenum, p 369.

Tuttle RH (1981). Evolution of hominid bipedalism and prehensile capabilities. Phil Trans R Soc Lond B-292:89.

Tuttle RH (1984). Bear facts and Laetoli impressions. Am J Phys Anthropol 63:230.

Tuttle RH (in press). Laetoli hominid footprints: kinesiological inferences and evolutionary implications. In Leakey MD, Harris JM (eds): "The Pliocene Site of Laetoli, Northern Tanzania," Oxford, Oxford Univ. Press.

Hominid Evolution: Past, Present and Future, pages 135–141
© *1985 Alan R. Liss, Inc.*

ANATOMICAL COMPARISON BETWEEN ENTIRE FEET OF CHIMPANZEE
GORILLA AND MAN

Patrice P. Le Floch-Prigent, Yvette Deloison
Laboratoire d'Anatomie, U.E.R. Biomédicale des
Saints Pères, Faculté de Médecine Paris-Ouest,
and Centre National de la Recherche Scientifique.
45 RUE DES SAINTS PÈRES, 75270 PARIS 6ème FRANCE

A morphological study of gorilla, chimpanzee and human
feet was undertaken in order to improve the interpretation
of hominid fossil remains. The specimens were C.T.-scanned
(Le Floch 1981 ; Deloison, Le Floch-Prigent 1982) and X-ra-
yed before dissection. This study complements previous studies on the anatomy of the foot which remain fundamental in
studies on man (Ledouble 1897 ; Pales, Chippaux 1953), gorilla (Straus 1930 ; Raven 1950), chimpanzee (Champneys 1872)
and comparative species (Testut 1884 ; Raven 1936 ; Sokoloff
1972). The ensuing section describes chimpanzee and gorilla
feet.

In dorsal view. The main characteristic is the large separation between the first ray and the other toes ; the *m.
extensor digitorum longus* is held by the fundiform ligament
which is attached to the sheath surrounding the *m. peroneus*
in the gorilla. On the medial side there is in Pan and Gorilla a *m. abductor hallucis longus* which does not exist in man
(Figs 1, 2).

Concerning the intrinsic muscles, the *m. extensor digitorum brevis* arises from the dorsal region of the *calcaneus*
near the tarsal sinus and, in Gorilla, on the sheath of the
m. peroneus where it forms a sling. *A*) In Pan, this muscle
is composed of two muscular bellies : one medial head going
to the hallux, and one lateral head composed of three bundles
for the second, third and fourth toes. *B*) In Gorilla, the
first bundle is isolated but the three other bundles (attached on the cuboid, presenting clearly distinct origins) quickly become one strip of muscle.

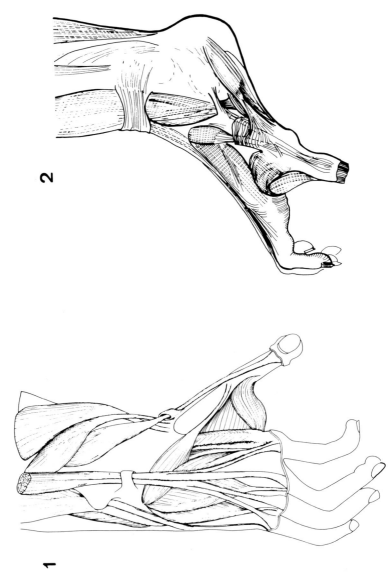

Fig. 1. Dorsal aspect of the foot in *Pan troglodytes* (from authors' dissection).
Fig. 2. Dorsomedial aspect of the foot in *Gorilla* (from Raven 1950).

The three tendons of the *m. extensor hallucis longus*, the *m. tibialis anterior* (which ends on the navicular bone) and of the *m. abductor hallucis longus* (which is attached to the base of the first metatarsal), run under the naviculo-metatarsal ligament. The *m. abductor hallucis longus* makes possible the abduction of the hallux from the axis of the foot and contributes to the mobility of the first toe.

In plantar view. Three muscular groups exist in Pan and Gorilla as in man : medial, intermediate and lateral. The intermediate group presents the *m. flexor digitorum brevis* which arises on the medial process and on the medial face of the *calcaneus*, and inserts as a tendon into the overlapping aponeurosis and into the bony attachment of the aponeurosis. This arrangement, similar to that of man, is formed in Pan by three bundles of muscles joined together and provides the tendons for the second, third and fourth toes; in Gorilla, it is round, is formed by two joined bundles and provides two tendons for the second and third toes (Figs 3, 4).

In the extrinsic muscles, the *m. tibialis posterior* is thicker and rises higher than in man, and it has fleshy fibers descending to the medial malleolus. It ends, in Pan, on the posteromedial border of the tuberosity of the navicular bone by a thick tendon. This tuberosity of the *os naviculare* is very well developed in the chimpanzee. It ends in Gorilla, in a groove on the deep face of the navicular bone and attaches to the tubercle of the lateral cuneiform. The *m. tibialis posterior* is a powerful invertor. The *m. gastrocnemius* and *m. soleus* are bulky and fleshy to their termination on the *calcaneus*.

As far as the *m. flexor fibulae longus* and *m. flexor tibialis longus* are concerned, in Pan there is no *m. flexor hallucis longus* as in man. The tendon of the *m. flexor fibulae longus* is divided into three tendons for the first, third and fourth toes. The tendon of the *m. flexor tibialis longus* provides tendons for the second and fifth toes. These two last muscles are separate and distinct at their proximal attachment. A tendinous expansion exists between the two flexor muscles. In Gorilla, the *m. flexor tibialis* and *m. fibulae longus* are as in Pan. There is no *m. flexor hallucis longus*, but there is one slip of the tendon of the *m. flexor fibulae longus* which provides also the tendons for the third and fourth toes. As in the chimpanzee, the *m. flexor tibialis longus* supplies the tendons for the second and fifth toes, but in the gorilla it

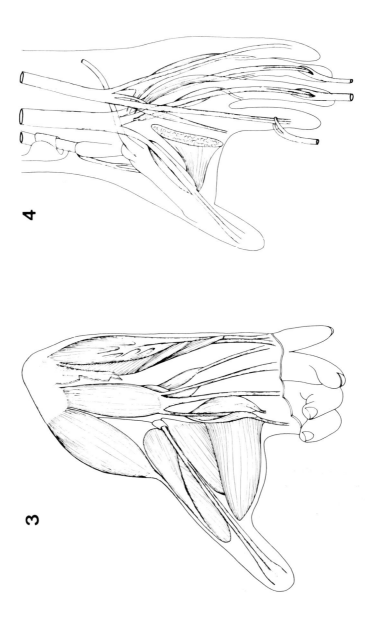

Fig. 3. Superficial plantar aspect of the muscular dissection of the foot in *Pan troglodytes*.
Fig. 4. Deep plantar aspect of the dissection of the foot in *Pan troglodytes*.

provides also the tendon for the fourth toe. After the bifur-
cation for the first toe, there is one tendinous expansion
between the tendon for toes I, III and IV, and the tendon for
toes II and V. There is also an expansion between the two
tendons for the second toe. As in Pan, there are special muscles around the last tendons of the *m. flexor longi* : the
co-contractors.

In the human foot, the *m. extensor digitorum brevis* is
a thin strip. The *m. abductor hallucis brevis* and the *m. abductor digiti quinti* are less developed than in Pan and Gorilla. There is no *m. abductor hallucis longus*, but a *m. flexor
hallucis longus* or *flexor fibulae longus*, and a *m. flexor digitorum longus* or *m. flexor tibialis longus* with its secondary
muscle, the *m. quadratus plantae* which is almost never found
in the apes. The *m. adductor hallucis* with its two heads,
transverse and oblique, is poorly developed in comparison
with the two African apes, in accordance with a different
mobility of the first metatarsal bone. In man, the hallux
converges with the other toes.

The foot of the chimpanzee and of the gorilla shows long
toes in comparison with those of Homo and a very divergent
hallux, this being the special muscular and architectural
development of their first ray. In man, the hallux lies parallel to the other toes. The comparison of the muscular anatomy
of the foot of these three genera highlights differences
in the arrangement of the muscles and tendons, in relation
to the bones upon which they operate.

For each, its morphology and its mode of locomotion influence the intensity and direction of movement, which determine in large part the form of the bones involved. The phalanges in apes are long in relation to the size of the foot ;
this feature and the large articular surfaces of the toes are
important for support in the trees. The cross-sections of the
shaft of the metatarsal bones, as seen in C.T.-scan, show a
circular morphology in Pan, whereas they are triangular in
man (Figs 5, 6).

Among the different genera of apes, the hand differs more
than the foot because its function is different for each ape.
The function of the foot is similar in each, becauses it serves to support the body. However the hands of the chimpanzee
and gorilla are more adapted to arboreal life than the hand
of man, but less so than the hand of Pongo. In the two Afri-

can apes, it supports in part the weight of the body in the usual ground locomotion by knuckle walking, a function which has totally disapeared in man.

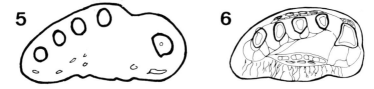

Fig. 5. Metatarsal cross-section in <u>Pan</u> (C.T.-scan).
Fig. 6. Metatarsal cross-section in <u>Homo</u> (anatomy).

The comparative anatomical study of the hand and of the foot can be useful in determining the form of locomotion which, in turn, can provide information on the place of a fossil species in the scale of evolution.

Despite their apparent differences, the anatomical homologies are very apparent between the three genera, <u>Pan</u>, <u>Goril</u>-la and <u>Homo</u>. This does not exclude the existence of very significant differences between <u>Pan</u> and <u>Homo</u>. There is true paralelism at the anatomical level, but the bones and the muscles have a particular arrangement which has a strong influence on the modes of locomotion : the false plantigrade locomotion of <u>Pan</u> and <u>Gorilla</u>, and the true plantigrade locomotion characteristic of <u>Homo</u>, with three supporting points, two supporting arches (lateral and anterior), and a suspended medial arch, on and over the ground, respectively.

REFERENCES

Champneys F (1872). On the muscles and nerves of a chimpanzee (*Troglodytes niger*) and a *Cynocephalus anubis*. J Anat Physiol 6:176.
Deloison Y, Le Floch-Prigent PP (1982). Étude tomodensito-métrique d'un pied entier chez *Pan* : comparaison avec des coupes anatomiques et tomodensitométriques chez *Homo*. C R Acad Sc Paris 295:239.
Ledouble AF (1897). "Traité des variations du système musculaire de l'homme et de leur signification au point de vue de l'anthropologie zoologique." Paris: Schleicher.

Le Floch P (1981). "Coupes horizontales sériées des membres, anatomie et tomodensitométrie ; coupes anatomiques du nouveau-né." Paris, Thèse de Doctorat d'Etat en Biologie Humaine, Université René Descartes, Paris V.

Pales L, Chippaux C (1953). Myologie comparative du pied. Cinquante dissections de colorés. Bull Soc Anthrop Paris 3:284.

Raven HC (1950). "The Anatomy of the Gorilla." Raven Memorial Volume. New York: Columbia University Press.

Sokoloff S (1972). The muscular anatomy of the chimpanzee foot Gegensbaurs Morphol Jb Leipzig 11:86.

Straus WL (1930). The foot musculature of the highland gorilla (*G. beringei*). Quart Rev Biol 5:261.

Testut L (1884). "Les anomalies musculaires chez l'homme expliquées par l'anatomie comparée." Paris: Masson.

Hominid Evolution: Past, Present and Future, pages 143–147
© *1985 Alan R. Liss, Inc.*

COMPARATIVE STUDY OF CALCANEI OF PRIMATES AND PAN-
AUSTRALOPITHECUS-HOMO RELATIONSHIP

Yvette Deloison

Chargé de Recherche CNRS
Laboratoire d'Anthropologie
Museum National d'Histoire Naturelle
Musée de L'Homme Place du Trocadero - 75116 PARIS

The fossil bones concerned are casts of the calcanei of
australopithecines, the originals of which were found in 1975
by Professors Y.Coppens, D. Johanson and M. Taieb in the
Awash Valley in the Afar region of Ethiopia (Latimer et al
1982). They date to around three million years. I compared
them with the calcanei of modern humans and chimpanzees.

The three fossil bones do not differ significantly in
size from the average human calcanei. In this paper, we consider them in four views: superior, inferior, lateral and
posterior. Like most fossils, they are incomplete and some
faces are more instructive than others.

Figure 1

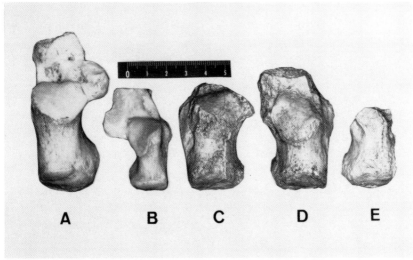

Figure 1 shows in superior view : A - left calcaneus of Homo; B - right calcaneus of Pan; C - AL-333 -55 left calcaneus of australopithecine; D - AL-333-8 right calcaneus of australopithecine; E - AL-333-37 right calcaneus of australopithecine.

The calcanei of Homo and Pan differ : 1 - in size; 2 - in shape : the human calcaneus shows a longitudinal and straight axis while that of the chimpanzee has an angulated axis; 3 - the apex of this angle is formed by the strongly developed peroneal trochlea; 4 - Pan differs also by the medio-lateral narrowness of the tuberosity in comparison with the human bone and 5 - by the angle of the curvature of the posterior articular surface for the talus.

I calculated an index of curvature for this articular surface in my human and chimpanzee populations. These two groups are clearly different: Homo 13,5% and Pan 23,4%. The curvature index of the posterior articular surface for the talus of AL-333-8 gives a value of 22%, intermediate between the means of Pan and Homo, but this value falls within the normal variation of the chimpanzee. The range of variation of Pan does not overlap with that of the human sample.

The three fossil bones show a mediolateral thickness of the tuberosity closer to that of Homo than that of Pan, but in its angulation the longitudinal axis is closer to that of Pan.

Figure 2

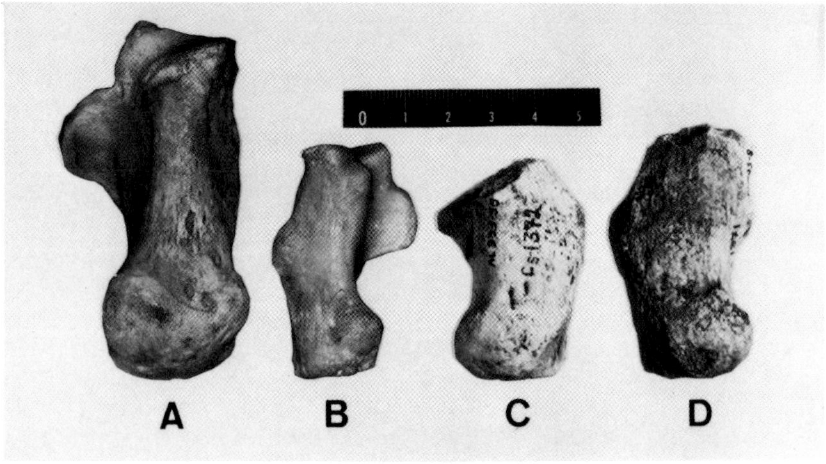

Figure 2 in inferior view (A - <u>Homo</u>: B - <u>Pan</u>; C - AL-333-55; D - AL-333-8) shows the axis of orientation of the bones which is straight in man, whereas in the chimpanzee it is curved, as in the fossil bones. In each case, the medial process is prolonged on the inferior or plantar face; the lateral process is well developed in man but it does not exist in the chimpanzee, other apes or the fossil bones. We remark the concavity of the medial face and the well developed peroneal trochlea which continues in the fossil bones, as in the chimpanzee, to give overall shape to the bone. That is to say, it marks the angle of the axis of the bone.

<u>Figure 3</u>

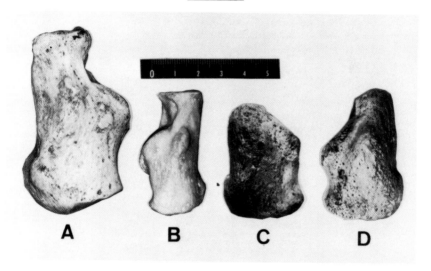

Figure 3 shows in lateral view A - <u>Homo</u>; B - <u>Pan</u>; C - AL-333-55; D - AL-333-8. In humans, this face is fairly flat, which is not at all the case in <u>Pan</u>. The chimpanzee has a peroneal trochlea which seems to be pinched between the tendons of the muscles peroneus brevis and peroneus longus. Usually it is difficult to distinguish this feature in the human calcaneus; when it exists to a very marked degree, it seems to be superimposed on this face and in no way modifies the relatively flat appearance of this surface.

The fossils differ from <u>Homo</u> in the presence of a large peroneal trochlea which alters this surface, making it raised as in the chimpanzee. In AL-333-55, this process is broken, but it is evident that if it were present, it would be even more developed than on the last fossil.

On AL-333-8 the groove left by the tendon of the muscle peroneus brevis should be noted: the question arises as to why this process is so important. It would seem that the shape of the bone was influenced during the growth of the child by the relative importance of certain muscles and that australopithecine locomotion was different from that of modern human beings.

Figure 4

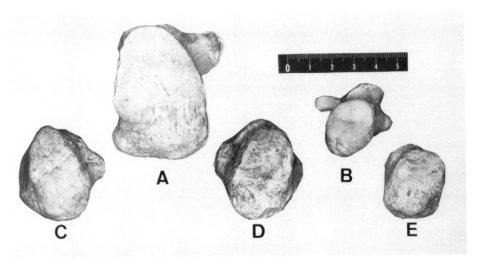

In figure 4 the calcanei are seen in posterior view : A – Homo; B – Pan; C – AL-333-55; D – AL-333-8 and E – AL-333-37.

In Homo, the form of the posterior face is subtriangular with an inferior base bordered by the medial and lateral processes (Forster 1928). The main axis of this face is almost vertical and tilted towards the medial side, in most human calcanei.

In Pan, this posterior face is clearly oval with the main axis tilted about thirty degrees towards the lateral side. This fact is not unusual since we know that apes walk on the lateral side of the foot, that is, they walk "in varus". In the chimpanzee there is no lateral process; the heel of the foot "rolls" on the ground towards the lateral side.

The posterior face of the fossil bones shows a form clearly ovoid with no lateral process. The main axis of this face is inclined towards the lateral side, less than in Pan but more than in Homo.

CONCLUSION

The human calcaneus, in its normal position on the ground, has the base of its posterior face with the two lateral and medial processes on a horizontal plane (Laidlaw 1904, 1905). It contrasts with the position in the australopithecines. It would seem that Australopithecus could not rest its foot on the ground and walk as modern man does. The calcaneus of Pan with its oval posterior face, with the curvature of its longitudinal axis and with the pronounced concavity of its medial face, is the result of truly non-plantigrade locomotion.

The apes' foot is in accord with its function, especially its role in climbing; the foot of man, in contrast, is adapted for plantigrade locomotion, with support points on the plantar tripod, and for perfect bipedal locomotion. The study of the fossil calcanei shows that the feet of these australopithecines seem to point to plantigrade locomotion intermediate between that of pongids and that of Homo. If they used bipedal locomotion, they would certainly not have had such good stability as in modern man.

The possibility is not excluded that the Laetoli footprints were left by this kind of foot.

Forster A (1929). Le calcaneum dans la série des mammifères et chez l'Homme. Étude de la statique et de la dynamique du pied. Arch Anat Hist Embyrol, Strasbourg 10:54.
Laidlaw PP (1904). The variety of the os calcis. J Anat Lond 38:10.
Laidlaw PP (1905). The os calcis. Part II : the processus posterior. J Anat Physiol 39:17.
Latimer BM, Lovejoy CO, Johanson DC, Coppens Y (1982). Hominid tarsal, metatarsal, and phalangeal bones recovered from the Hadar Formation : 1974-1977 collections. Am J Phys Anthropol 57:18.

Southern Sites

Hominid Evolution: Past, Present and Future, pages 151–164
© *1985 Alan R. Liss, Inc.*

RECENT GEOLOGICAL, STRATIGRAPHIC AND PALAEONTOLOGICAL STUDIES
AT MAKAPANSGAT LIMEWORKS

Judy M. Maguire, Ph.D.

Bernard Price Institute for Palaeontological
Research, University of the Witwatersrand
1 Jan Smuts Avenue, Johannesburg, 2001, RSA

The Makapansgat Limeworks cavern deposit is important
as a major source of Plio-Pleistocene fossil vertebrates
which include remains of the hominid *Australopithecus
africanus*, the phylogenetic placement of which is controversial (Johanson, White 1979 ; Tobias 1980; White et al 1981,
1982; Partridge 1982d). A central issue in the phylogenetic
interpretation of the South African gracile australopithecines is their chronological placement. However, controversy
still surrounds the origins, stratigraphic interpretation and
dating of the Makapansgat Limeworks sediments, due largely to
a lack of adequate sedimentological studies (Turner 1980).
Recent investigations have been aimed at a better understanding of the deposit, particularly the 'red muds' (Wells,
Cooke 1956; Brain 1958) or 'Member 2' (Partridge 1979).
Studies have included observation of sedimentation in analogous situations in recent dolomite caves in the Makapansgat
valley, a study of the configuration of the travertine floor
at Limeworks and its effects on localized sedimentation within
the cave, field studies of contact relationships between
members and palaeontological and chemical analyses of selected
sediments.

REVIEW OF STRATIGRAPHIC INTERPRETATIONS AND DATING

By arranging the geological and stratigraphic papers on
the Makapansgat Limeworks chronologically, it has been
possible to trace the conceptual evolution and emergence of
the current stratigraphic framework, i.e. that of Partridge
(1979), as used by Tobias (1980), Brock et al (1977), McFadden
(1980) and others.

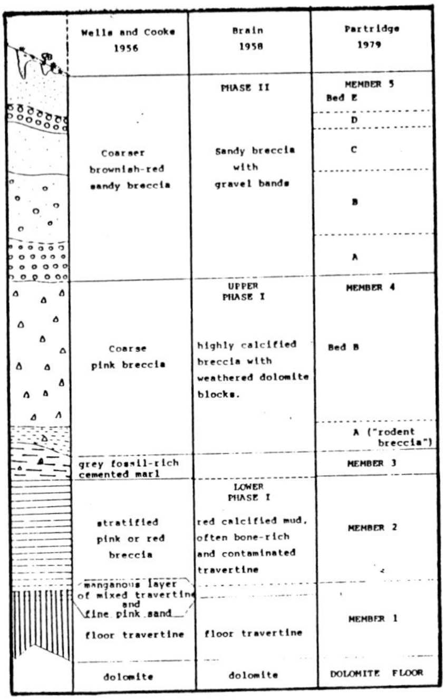

Wells and Cooke 1956	Brain 1958	Partridge 1979
	PHASE II	MEMBER 5 / Bed E
		D
Coarser brownish-red sandy breccia	Sandy breccia with gravel bands	C
		B
		A
Coarse pink breccia	UPPER PHASE I / highly calcified breccia with weathered dolomite blocks.	MEMBER 4 / Bed B
		A ("rodent breccia")
grey fossil-rich cemented marl		MEMBER 3
stratified pink or red breccia	LOWER PHASE I / red calcified mud, often bone-rich and contaminated travertine	MEMBER 2
manganous layer and fine pink sand		
floor travertine	floor travertine	MEMBER 1
dolomite	dolomite	DOLOMITE FLOOR

Fig. 1. Various stratigraphic subdivisions of the Makapansgat Limeworks deposit.

Geological and stratigraphic research falls into two periods, the first culminating in the important stratigraphic subdivision of Wells and Cooke (1956) and the pioneering geological interpretation of Brain (1958). These two essentially similar stratigraphic subdivisions (Fig.1) are in close agreement with that announced earlier by Dart (1952). The second period, following an hiatus of 20 years, resulted in the lithostratigraphic subdivision currently in use. Chronological arrangement of publications demonstrated that successive interpretations represent little more than successive refinements and renaming of the same basic stratigraphic concept.

Embodied in the stratigraphic columns shown is the idea of 'a stratigraphic sequence decreasing in age from bottom to top', a concept specifically stated by Wells and Cooke (1956) and implying that a simple superposition of units could be applied to the Limeworks. So pervasive has been the idea of a 'sedimentary succession' (perhaps reinforced by the series of diagrams used to illustrate the general model for a 'deposition cycle' in dolomite caverns proposed by Brain 1958, Fig. 13) that Partridge's lithostratigraphy (1979) comprises a numbered sequence of discrete Members, from Member 1 (the

travertine floor at the base) to Member 5 at the top -
although in reality, nowhere is the complete set of Members
present as a vertical sequence. The profiles shown in Fig.1
are thus "composite". Similar numbered and chronologically
based layer-cake stratigraphies have been proposed for the
Sterkfontein (Partridge 1978) and Kromdraai (Partridge 1982a;
Vrba 1981; Vrba, Panagos 1982) hominid sites.

The Makapansgat sequence has recently been dated using
palaeomagnetic techniques (Brock et al 1977; McFadden 1980;
McFadden et al 1979), sampling for palaeomagnetic assessment
being undertaken on the basis of Partridge's (1979) lithostratigraphic subdivision. This assumes that all the red muds
belong to a single unit (Member 2), and that they are penecontemporaneous, although the red muds are present in various
isolated depositories separated by a high area in the travertine floor (Member 1) with no demonstrable lateral continuity
(Maguire et al 1980). These assumptions, as well as assumptions that deposition rates in the different depositories
were 'approximately equal' (McFadden et al 1979) and that
depth of sediment can be related to passage of time, are
highly questionable. The conclusion, based on a composite
palaeomagnetic sequence, is that Member 3, from which all but
one of the *A. africanus* specimens have come (Tobias 1980), is
either just over 2,9m.y. or just over 3,06m.y. (McFadden et al
1979; McFadden 1980). Because Member 3 itself was not sampled
on account of its poor mechanical properties, the ages cited
are based on the age matchings for the 'overlying' Member 4
and 'underlying' Member 2. Nowhere in the literature on the
palaeomagnetic dating or the stratigraphy of the Limeworks is
the possibility conceded that Member 3, which is almost
certainly the result of animal bone-accumulating agents
(Maguire et al 1980),may have accumulated *concurrently* with
at least part of Member 4, or even at the same time that
certain of the red muds (all referred to Member 2) were
accumulating elsewhere in the cavern system. It is highly
probable that normal cave sedimentation proceeded, whilst bone-
accumulating agents such as striped hyaenas, porcupines or even
hominids were gradually adding to the bone bed comprising
Member 3 (Maguire et al 1985). Yet it is always shown in
stratigraphic columns and palaeomagnetic sequences as being
sandwiched between Members 2 and 4, and has thus been assumed
to occupy a position intermediate in age. That it is conceived of as such is demonstrated by the following statement
'.... if the less likely possibility is accepted that the
Mammoth event occurred in the *interval* represented by the thin
Member 3 ....' (Partridge 1982d).

The palaeomagnetic dates for the Makapansgat sequence, described as providing a 'firm age' for *A. africanus* (Partridge 1982c), are of primary consideration in the ongoing controversy surrounding the phylogenetic placement of *A. africanus* and *A. afarensis*. Since even fractions of a million years appear to be of significance in the argument (2,6 m.y. of White et al 1981, 1982, as opposed to 3.0 m.y. of Tobias 1980, Partridge 1982c,d), the probability of concurrent accumulation of Members deserves serious consideration.

REPORT ON ONGOING RESEARCH

Studies of Sedimentation in Recent Dolomitic Caves

Important lessons in the interpretation of processes operating in ancient caves may be learnt from observation of sedimentation in modern analogous cave situations, such as Ficus Cave and Peppercorn's Cave (Fig.2) in the Makapan valley and caves elsewhere. Being a cavern which has developed in the flank of a steep-sided valley, it is highly

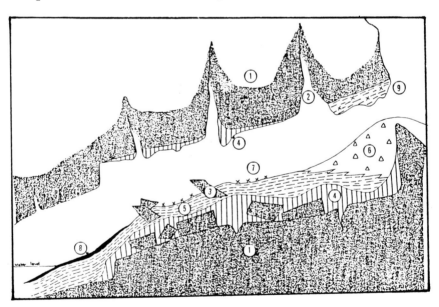

Fig. 2. Schematic cross-section through Peppercorn's Cave in the Makapansgat Valley.

probable that the ancient Limeworks cave was an elongate
lateral-entranced cavern (Brain 1958, Fig.6) for much of the
period that sedimentation was occurring. Partridge (1982b)
has suggested a lateral extent of about 250m. for the Limeworks cavern. This is in contrast to cup-like sedimentary
receptacles with a vertical or near-vertical, shaft-type
entrance, such as are characteristic of dolomitic caverns
developed in country of low relief (Brain Fig.6), although
initially, the entrance to the Limeworks was probably small
and shaft-like or fissure-like (Brain 1958; Maier 1973).

At Peppercorn's Cave, the general relationships between
the dolomite parent rock (Fig.2 (1)), travertine (4) and
sediments (5, 6, 7, 8) are similar to those which existed in
the original Limeworks cavern. Near the laterally-placed
entrance, a coarse stony breccia talus cone of predominantly
extra-cavernous material has developed (6) and is continuously
being added to. In deeper recesses, red muds (5) have been
and are being deposited contemporaneously. Continued addition of material is both by means of 'vertical' sedimentation
- the addition of insoluble residue of dolomite resulting
from episodes of active dissolution of walls and roof - and
'lateral' sedimentation by means of flushing of fines from
the talus cone. Also present in the deeper parts of the cave
are extensive deposits of bat guano (8), some dry and some
lying as a superficial colloidal suspension on the surface of
fine mud at the bottom of underground pools. These deposits
are continuously being added to because of the seasonal occupation of Peppercorn's Cave by vast hordes of colonial bats.
Present in the central floor area is a substantial accumulation of Iron Age archaeological relics including bone (7),
which is already being covered by collapsed blocks of roof
dolomite, red mud and travertine. In Peppercorn's Cave, a
transition between contemporaneous facies clearly occurs: one,
the stony breccia (6) being equivalent to Members 4/5 at the
Limeworks, the bone accumulation (7) approximating to Member
3 and the red muds (5) to some of the red muds representing
Member 2 at the Limeworks.

Also represented at Peppercorn's Cave is a unit of well-
stratified calcareous red mudstone adherent to the cave roof,
and containing fossil bone (9). The unit is clearly a much
older cave filling than the underlying talus cone material,
having been left as an eroded hanging remnant during the
formation of the present-day Peppercorn's Cave. Should continued talus cone sedimentation occur, the fossil unit would

ultimately come to lie on top of the talus cone. This
illustrates acutely the limitations of the use of the principle of superposition in caves. An additional complication
is that matrix and fossils collapsing off the lower surface
of the fossil unit are becoming incorporated in the much
younger talus cone below. Brain (1976, 1981, 1982a, 1982b)
has reported similar stratigraphic complexities for the
Swartkrans hominid site.

The Configuration of the Travertine Floor (Member 1):
the Problem of Separated Depositories

It has been assumed that units with a similar appearance
and lithology but separated spatially were deposited penecon-
temporaneously, e.g. the red muds at Makapansgat Limeworks
(Partridge 1979; McFadden et al 1979). Red muds were in fact
deposited in four different depositories, two to the west and
two to the east of a central floor high which isolates them
from one another   (Maguire et al 1980), Fig. 3. As no
spatial contact can today be demonstrated between the various
red mudstone units, any age relationship must remain inferential.

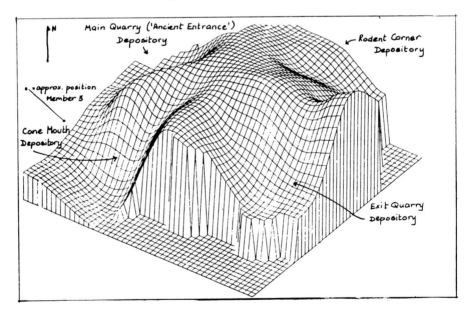

Fig. 3. 3-Dimensional reconstruction of the original travertine floor (Member 1) at Makapansgat Limeworks showing separated sedimentary depositories.

Contact relationships between certain red muds and other sediments, their differing fossil microfaunal content and chemistry suggest that the red muds may well differ in age. This is contrary to Partridge's statement (1979; McFadden et al 1979) that deposition in eastern and western deposi- tories was penecontemporaneous and that deposition rates were approximately equal - a point referred to in the latter publication in discussing the sampling and dating of Member 2. Even assuming an equal sediment supply, depositional *rate* will be controlled by the volume of the depository relative to the size (and number) of entrances to that depository. The volumes of the red mud depositories are unknown, and there is thus no meaningful way of relating depth of sediment to time lapse. Cross-correlation of absolute sediment depth from one depository to another must remain speculative. Similar problems of assuming stratigraphic equivalence for litbologi- cally discontinuous units have been noted at Sterkfontein (Wilkinson 1983).

Contact Relationships Between Members

An interesting contact relationship between Members 1, 2 and 4 is present on the south side of the 'rodent corner archway', against its western pillar: a unit of stony pink breccia (Member 4) has been eroded and some matrix material removed to leave a rounded boss. A layer of travertine (equivalent to Member 1?) has been precipitated over this surface and between fragments, apparently sealing and post- dating the Member 4. Subsequently, a unit of well stratified red mudstone (Member 2) containing cave pearls and numerous microfaunal fossils has been deposited in an onlapping fashion against the travertine surface, giving rise to a cave discon- formity. Clearly, in this area, the red mudstone (Member 2) would appear to be younger than either the pink breccia (Member 4) or the travertine. Further 'anomalous' contact relationships (in terms of the current lithostratigraphy) are discussed in Maguire et al (1985). Observations such as these demonstrate the stratigraphic complexity of the cave fill at Makapansgat, and a reality which cannot be accommo- dated by the existing conceptual framework.

Palaeontological Studies

a) <u>Microfaunal Analyses</u>. Pocock (1985a, b) has sampled
extensively from a red mudstone unit in the south-eastern
depository (the Exit Quarry Depository - Maguire et al 1985)
(see Fig. 3) termed the Exit Quarry red muds (EXQRM). Also
sampled were red mudstone units (Member 2) on the south and
north faces of the 'Rodent Corner Archway', as well as a unit
of pink, highly consolidated breccia (Bed A of Member 4) which,
on the north side of the archway at least, overlies the red
mudstone. This area lies in the Rodent Corner depository
(Fig. 3). A striking difference in the fossil microfauna from
the two depositories is the abundant presence of the extinct
dwarf cricetine rodent *Mystromys darti* Lavocat in the Rodent
Corner depository red muds (Member 2) *and* pink breccia (Member
4). Over 246 individuals of this species were recovered from
22.4kg of breccia and it is the second commonest species
present, being exceeded only by the vlei rat *Otomys cf.
gracilis* (272 individuals) (Pocock 1985b). However, *Mystromys
darti* is entirely absent from the EXQRM, despite the fact that
over seven times the volume of sediment has been processed -
over 150kg. This leaves no doubt that the difference is statistically valid and cannot be due to sampling error. Pocock
is therefore quite certain that the two assemblages (otherwise
essentially similar) cannot be contemporaneous, the simplest
explanation being that the Rodent Corner material (from both
Members 2 and 4) is older than the EXQRM, with *Mystromys darti*
(shortly to be renamed, Pocock 1985b, and transferred to the
genus *Stenodontomys*) declining from abundance to extinction in
the interval of time that elapsed between the deposition of
the Rodent Corner red muds and Member 4, Bed A, and the
deposition of the EXQRM. A further indication of the relatively greater age for the red mudstones and Member 4 at the
Rodent Corner is that *Stenodontomys* is known, in the form of a
slightly different species, only from the Langebaanweg deposits (probable age 5 m.y., Hendey 1981), but from no other
hominid-bearing site except the Limeworks, presently considered on other evidence to be the oldest South African hominid
locality (Vrba 1982).

Taphonomic differences between the Rodent Corner depository material and the EXQRM suggest that the former accumulated relatively closer to an entrance - an opinion expressed
also by Turner (1980) and indeed to be expected, as the
Rodent Corner area is downslope and therefore likely to be
first exposed by erosion of the hillside surface. A feature

of the Rodent Corner microfaunal assemblage and particularly of the Member 4, is an abundance of isolated teeth, such as elephant shrew incisors and molars of juvenile rodents. Such teeth are seldom retained, even in the mandibles found in owl pellets, and are liable to be lost or destroyed should owl pellet debris be subject to transportation. This implies *in situ* fossilization of Member 4, Bed A material, and, because owls by preference roost at no great distance from a cavern entrance, it suggests that this was close, probably immediately downslope.

Clear evidence of transportation of the EXQRM assemblage is the near absence of isolated teeth. The deposit itself shows much stratification - some more or less horizontal but elsewhere not at all so: the latter observation being seemingly incompatible with the suggestion that the red muds were deposited in standing water (Partridge 1979). One fossiliferous block was remarkably full of skulls, suggesting that these had accumulated by rolling down a slope, a phenomenon often seen at modern hillside roosts (Levinson 1982). However, evidence of periodic water transport does occur in the form of axial orientation of microfaunal long limb bones and sorting according to particle size - one mudstone block contained nothing but mandibles all in the same orientation. The EXQRM also contains a variety of other fossil material, such as seed husks of the grass *Setaria* and other plant seed remains, arthropod fragments (especially millipede rings), crab pinchers, insect carapaces, rat and pigeon coprolites (Pocock pers comm), none of which occurs at the rodent corner, and which are possibly of hillwash origin.

   b)   Macrofauna. Evidence is accumulating that the immensely fossil-rich Member 3 has been accumulated mainly by the fossil striped hyaena *Hyaena hyaena makapani* (Maguire et al 1980). Striped hyaenas are known to use the depths of caves as lairs, to scavenge from the carcase of almost any animal, and to be capable of cracking bones of camel-sized and even larger animals (Kruuk 1976; Maguire et al 1980; Skinner et al 1980). The feeding strategy of striped hyaenas is aimed at dismembering carcases and carrying away large parts (Skinner et al 1980). Skinner reports that a characteristic of all dens visited was one area used for feeding. Any cave that has been open for thousands of years may accumulate hundreds of thousands of bones. Some 30% of the Member 3 fossil bones show clear evidence of carnivore damage - a percentage apparently diminished by severe surficial weathering, to which many bones appear to have been subject.

The Member 4 assemblage contains a preponderance of monkey and baboon material (80%), highly suggestive of the leopard-based bone accumulations documented by Brain (1981). Previously, differences between the Member 3 and 4 faunal assemblages were attributed to a difference in age and/or palaeoenvironmental circumstances, whereas these may be due to no more than a difference in bone accumulating agent, with leopards periodically occupying situations closer to the cavern entrance, where their prey remains would add to the talus cave material (Eisenhart 1974), while striped hyaenas periodically occupied deeper recesses of the cavern. In fact, evidence is accumulating that the Member 3 and 4 faunas are rather more similar than previously thought, Vrba (1982), noting the similarity of the bovids and Eisenhart (1974), the almost 100% agreement to species level of primates from Members 3 and 4, primates being a group exploited by both leopards and hyaenas. *Gypsonhychus makapani*, the rare extinct giant rodent mole, is present only in Members 3 and 4 at Makapansgat; the only other site where the genus is represented is Taung. The suggestion that there is no great age difference between upper and lower levels at Makapansgat was made long ago by Cooke (1952), Oakley (1954a, b) and Brain et al (1955).

Chemical Analyses

Chemical analyses of various sedimentary units in the Limeworks have been facilitated by the availability of a modern Inductively Coupled Plasma Atomic Emission Spectrophotometer. Chemical analysis has confirmed that there is a fundamental difference between the Main Quarry ("Ancient Entrance") and EXQRM on the one hand and the Rodent Corner red muds on the other, a suggestion put forward by Turner (1980) on the basis of sedimentological differences.

Samples of red mudstone from the Exit Quarry chimney showed a relatively high phosphate content and appreciable quantities of zinc and copper. 'Hard' samples from this area showed even higher calcium phosphate percentages, suggesting that the cementing agent in this area is collophane - an amorphous form of calcium hydroxyphosphate, and not calcium carbonate. Two possible sources of phosphate are bones and bat guano: it seems plausible that vast hordes of bats, roosting on the cavern roof over pools of standing water for long periods, would provide sufficient guano to phosphatize pool sediments.

Rodent Corner red mudstones differ from all other red mudstones in that phosphate is only a minor constituent, and copper and zinc percentages are negligible. These mudstones have evidently been consolidated by the precipitation of calcium carbonate from bicarbonate-rich solutions probably of roof drip origin.

CRITIQUE OF THE PRESENT MAKAPANSGAT LITHOSTRATIGRAPHY

Use of the Principle of Superposition. The principle of superposition does not always apply in cavern depositories. Later cavern development may intersect or be contained within previous cave filling, resulting in older deposits overlying younger deposits, as occurs at Swartkrans (Brain 1976, 1981, 1982a,b).

Contemporaneous facies. The present lithostratigraphic framework does not appear to allow for the possibility that different lithologies (Members) represent contemporaneous facies deposited in different parts of the cave, e.g. part of Member 2, Member 3 and the lower part of Member 4. Such lateral facies equivalents would be particularly likely in an elongate cavern with a lateral entrance (Schmidt 1963) such as the Limeworks cavern probably was.

Lithostratigraphic nomenclature. Shortcomings of the present system have been referred to by Turner (1980), who was replied to by Partridge (1982b), although the present author cannot agree that "no particular difficulties are involved in numbering rather than naming members". The use of numbers implies an ordering which is unjustified, if the limitations on the law of superposition in cave fillings and the possibility of contemporaneous facies are recognized.

Travertines as "Member 1". Referring to cave travertine as "Member 1" has specific problems. It is true that a major phase of travertine precipitation occurred prior to significant sedimentation. In addition, however, considerable travertine precipitation occurred during sedimentation, and perhaps later. For example, the thick stalagmite layer overlying portion of Member 3, where this is preserved against the western wall of the cavern, is at least as thick as Member 3 itself. These various travertines do not necessarily connect either temporally or spatially with the earlier precipitation phase.

Sampling for dating and stratigraphic diachronism.
'Member 2' appears to incorporate in one unit deposits which
are separated spatially and probably temporally. Samples of
Member 2 for palaeomagnetic dating - all yielding normal
polarity data - have been taken from both eastern and western
depositories and incorporated into a composite 'vertical'
palaeomagnetic sequence. The EXQRM from the eastern depository, because they probably postdate the lower part of Member
4, could fit into the Gauss normal epoch postulated for this
member (McFadden 1980). If part of Member 1 was accumulating
simultaneously with part of Member 2,and Member 3 at the same
time as part of Member 4 (and Member 2), portions of units
currently referred to as Members 1, 2, 3 and 4 could in fact
be coeval. Since nowhere does a complete vertical 'sequence'
exist, polarity data for a laterally sampled composite vertical
sequence could fit almost any position in the palaeomagnetic
time scale.

REFERENCES

Brain CK, van Riet Lowe C, Dart RA (1955). Kafuan stone
    artefacts in the post australopithecine breccia at
    Makapansgat. Nature Lond 175:16.
Brain CK (1958). The Transvaal ape-man-bearing cave
    deposits. Mem Transv Mus 11:1.
Brain CK (1976). A re-interpretation of the Swartkrans
    site and its remains. S Afr J Sci 72:141.
Brain CK (1981). "The Hunters or the Hunted? An Introduc-
    tion to African Cave Taphonomy." Chicago, London:
    University of Chicago Press.
Brain CK (1982a). Cycles of deposition and erosion in the
    Swartkrans cave deposit. Palaeoecology of Africa 15:27
Brain CK (1982b). The Swartkrans site: stratigraphy of the
    fossil hominids and a reconstruction of the environment
    of early Homo. In 1er Congrès Internat Paléont Humaine:
    Nice: CNRS, Vol 2. p 676.
Brock A, McFadden PL, Partridge TC (1977). Preliminary
    palaeomagnetic results from Makapansgat and Swartkrans.
    Nature Lond 266:249.
Cooke HBS (1952). Quaternary events in South Africa. Proc
    1st Pan-Afr Cong Prehist 1947. Oxford: Blackwell, p 26.
Dart RA (1952). Faunal and climatic fluctuation in Makapans-
    gat Valley: their relation to the geological age and pro-
    methean status of Australopithecus. Proc 1st Pan-Afr Cong
    Prehist 1947. Oxford: Blackwell, p 96.

Eisenhart WL (1974). The fossil cercopithecoids of Makapans-
    gat and Sterkfontein. Unpubl BA thesis, Dept Anthropology,
    Harvard College, March 1974.
Hendey B (1981). Palaeoecology of the late Tertiary fossil
    occurrences in "E" Quarry, Langebaanweg, South Africa, and
    a reinterpretation of their geological context. Ann S Afr
    Mus 84(1):1.
Johanson DC, White TD (1979). A systematic assessment of
    early African hominids. Science 202:321.
Kruuk H (1976). Social behaviour and foraging of the striped
    hyaena (Hyaena vulgaris). E Afr Wildl J 14:91.
Levinson M (1962). Taphonomy of microvertebrates - from owl
    pellets to cave breccia. Ann Transv Mus 33(6):115.
Maguire JM, Pemberton D, Collett MH (1980). The Makapansgat
    Limeworks grey breccia: hominids, hyaenas, hystricids or
    hillwash? Palaeont afr 23:75.
Maguire JM, Schrenk F, Stanistreet IG (1985). The litho-
    stratigraphy of the Makapansgat Limeworks australopi-
    thecine site: some matters arising. Ann Geol Surv S Afr
    (in press).
Maier W (1973). Palaoökologie und zeitliche Einordnung der
    Südafrikanischen Australopithecinen. Z Morph Anthrop 67:70.
McFadden PL (1980). An overview of palaeomagnetic chronology
    with special reference to the South African hominid sites.
    Palaeont afr 23:35.
McFadden PL, Brock A, Partridge TC (1979). Palaeomagnetism
    and the age of the Makapansgat hominid site. Earth Planet
    Sci Lett 44:373.
Oakley KP (1954a). The dating of the Australopithecinae of
    Africa. Am J Phys Anthropol 12:9.
Oakley KP (1954b). Study tour of early hominid sites in
    southern Africa. S Afr Archaeol Bull 9:75.
Partridge TC (1978). Re-appraisal of lithostratigraphy of
    Sterkfontein hominid site. Nature Lond 275:282.
Partridge TC (1979). Reappraisal of lithostratigraphy of
    Makapansgat Limeworks hominid site. Nature Lond 279:484.
Partridge TC (1982a). Some preliminary observations on the
    stratigraphy and sedimentology of the Kromdraai B hominid
    site. Palaeoecology of Africa 15:3.
Partridge TC (1982b). Brief comments on "Sedimentological
    characteristics of the 'red muds' at Makapansgat Limeworks"
    by BR Turner. Palaeoecology of Africa 15:45.
Partridge TC (1982c). The chronological positions of the
    fossil hominids of southern Africa. In 1er Congrès
    Internat Paléont Humaine: Nice: CNRS, Vol 2. p 617.
Partridge TC (1982d). Dating of South African hominid sites.

S Afr J Sci 78:300.

Pocock TN (1985a). Plio-Pleistocene mammalian microfauna in southern Africa. Ann Geol Surv S Afr (in press).

Pocock TN (1985b). Plio-Pleistocene fossil mammalian micro-fauna of southern Africa - a preliminary report including description of two new fossil muroid genera. S Afr J Sci (in press).

Schmidt E (1963). Cave sediments and prehistory. In Brothwell D, Higgs E (eds): "Science in Archaeology." London: Thames and Hudson, p 123.

Skinner JD, Davis S, Ilani G (1980). Bone collecting by striped hyaenas, Hyaena hyaena, in Isreal. Palaeont afr 23:99.

Tobias PV (1980). "Australopithecus afarensis" and A. africanus: critique and an alternative hypothesis. Palaeont afr 23:1.

Turner BR (1980). Sedimentological characteristics of the "red muds" at Makapansgat Limeworks. Palaeont afr 23:51.

Vrba ES (1981). The Kromdraai australopithecine site re-visited in 1980: recent investigations and results. Ann Transv Mus 33(3):17.

Vrba ES (1982). Biostratigraphy and chronology, based par-ticularly on Bovidae, of southern African hominid-asso-ciated assemblages. In 1er Congrès Internat Paléont Humaine: Nice: CNRS, Vol 2. p 707.

Vrba ES, Panagos DC (1982). New perspectives on taphonomy, palaeoecology and chronology of the Kromdraai apeman. Palaeoecology of Africa 15:13.

Wells LH, Cooke HBS (1956). Fossil Bovidae from the Lime-works quarry, Makapansgat, Potgietersrust. Pal afr 4:1.

White TD, Johanson DC, Kimbel WH (1981). Australopithecus africanus: its phyletic position reconsidered. S Afr J Sci 77:445.

White TD, Johanson DC, Kimbel WH (1982). Dating of South African hominid sites: White, Johanson and Kimbel reply. S Afr J Sci 78:301.

Wilkinson MJ (1983). Geomorphic perspectives on the Sterk-fontein australopithecine breccias. J Archaeol Sci 10:515.

Hominid Evolution: Past, Present and Future, pages 165–170
© *1985 Alan R. Liss, Inc.*

LOWER-LYING AND POSSIBLY OLDER FOSSILIFEROUS DEPOSITS
AT STERKFONTEIN

M. Justin Wilkinson, M.A.

Geography and Environmental Studies
University of the Witwatersrand, Johannesburg
Republic of South Africa, 2001

Analysis of the internal morphology of the Sterkfontein
Cave system from detailed maps strongly suggests that consolidated, fossiliferous deposits at low levels are of significantly greater age than sediments at higher levels. It
is argued that geological controls of original cavern shape
have determined the location and conical mode of deposition
of several major cave fill masses.

The Sterkfontein Cave system, which lies 50km northwest of Johannesburg, comprises three voids. The largest is
the Tourist Cave, a complex of chambers measuring broadly
350m E-W and 120m N-S, with a known vertical extent of 55m.
The upper units of the Sterkfontein Formation breccias, which
have yielded australopithecines, are exposed at the surface
directly above the central sector of the Tourist Cave (Fig.1a).

FRACTURE CONTROL OF CAVERN DEVELOPMENT

Fractures within the host rock are the major controls
of original cavern form (Wilkinson 1973, 1983). Dolomite
along near-vertical fractures and fracture zones has been
preferentially dissolved to produce an angular pattern of
chambers (Fig.1a). Larger chambers occur where fracture
lines intersect, and the largest known void is located in
the center of the system where two sets of closely-spaced
(10-15m), parallel fractures cross (F2/F3 and F4/F5, Fig.1a).

Cavern form reflects these controls: single fracture
voids are relatively narrow, but elongated laterally (Fig.1a)

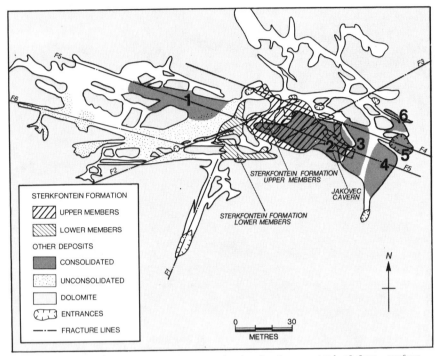

Fig.1a Simplified plan of Tourist Cave showing fracture control of form, surface and underground exposures of Sterkfontein Formation, and six other consolidated externally derived sediment bodies (after Wilkinson 1983).

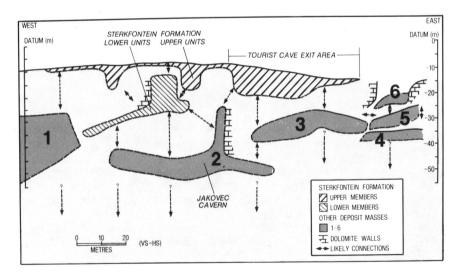

Fig.1b Schematic section through central deposits occupying fractures F4,F5 showing position and known thickness of six major deposits at and below level of the Sterkfontein Formation, and several likely connections between these.

and vertically (Fig.2a); multiple fracture cavities are
more varied since dolomite between parallel fractures and at
fracture intersections, may be entirely removed by phreatic
solution (Fig.2b), the dominant process of cavern erosion at
Sterkfontein. Impermeable and less soluble bands of shale
exert lesser control and form planar roofs in some larger
chambers.

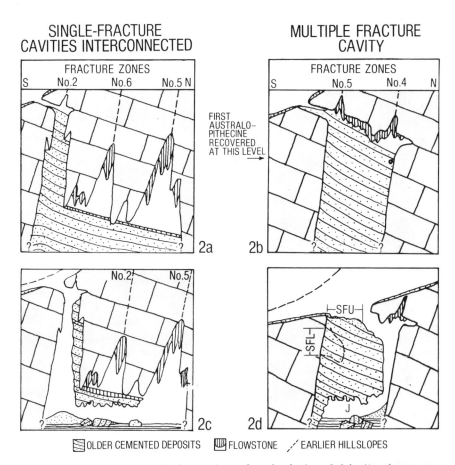

Fig.2a-d Model of chamber development by preferred solution of dolomite along near-vertical fracture planes. Fills accumulate from lowest levels of system upwards. Re-solution above and below results in removal and collapse of fills in places. Sterkfontein Formation upper units (SFU) exposed in surface excavations, lower units (SFL) exposed underground.

DEPOSITIONAL CONTROL BY VERTICAL CAVERN MORPHOLOGY

As a result of the pattern of strong <u>vertical</u> development of Tourist Cave chambers, soils and colluvia -- with
bone material -- enter the chambers and accumulate in simple
fashion by upward growth of conical masses, cone bases lying
at or below the lowest known parts of the system (Figs.2a,
2b). Such a depositional model appears realistic in the
absence of countervailing evidence.

Subsequent changes in the sediment masses are complex
however: (1) syn- or post-depositional cementation by $CaCO_3$
-charged percolating meteoric water, (2) subsequent re-solution of calcite cement at upper and lower levels (in the
course of surface lowering by erosion and by fluctuating
water levels respectively, Figs.2c, 2d). Similar processes
of sedimentation, cementation, and erosion have been documented at the Swartkrans Site 1km west of Sterkfontein Cave
(Brain 1976).

EXTENT OF THE STERKFONTEIN DEPOSITS

Externally derived materials are encountered throughout
the fracture-controlled sectors of the cave system, particularly in the Tourist Cave (Fig.1a), because fracture-related
cavities have opened to the hillside at various points
(Fig.2a, 2b). Precise stratigraphic connections are not known,
but considering that the mode of accumulation is often one of
interdigitating cones, relationships are probably complex. A
preliminary generalization can be offered, however: deposits
at different levels and locales <u>may be connected physically</u>
if they lie on the same fracture line, that is, within an
original fracture-controlled void.

Indeed several such vertical and horizontal connections
between specific deposits are considered likely. These are
represented in Fig.1b by arrows for the centrally located
deposits. It can be seen that the Sterkfontein Formation
which has long monopolized attention (more recently for
example, Brain 1958, Robinson 1962, Butzer 1971 and 1984,
Partridge 1978, Stiles and Partridge 1979), has likely
connections with three major deposits (Nos. 1-3, Fig.1b)
which lie directly beneath it along a distance of more than
100m, and with three to the east (Nos. 4-6, Fig.1b).

THE JAKOVEC DEPOSIT : LOCATION AND AGE

A connection of special interest concerns the large
Jakovec Cavern deposit (No.2, Figs.1a,1b), a mass at least
20m thick, 40m long and 20m wide. The deposit forms the
roof and three walls of the Jakovec Cavern, and lies as much
as 25m below Member 1 of the Sterkfontein Formation (Fig.2d).
A detailed examination of the three-dimensional relationships
of the central caverns (Wilkinson 1983) reveals that the
Jakovec deposit is most plausibly interpreted as a lower,
and hence older, unrecognised part of the Sterkfontein Formation cone. The reasons for this interpretation are (1) the
position of the deposit squarely within the zone of greatest
fracturing in the dolomite, (2) its location beneath the
Sterkfontein Formation, and (3) its consolidated nature,
suggesting that it is an older generation deposit.

The pronounced vertical development of the caverns
appears to preclude the existence of a dolomite bedrock base
to the Sterkfontein Formation at a level of 25-30m below
surface as suggested by Partridge (1978) and Stiles and Partridge (1979). Furthermore, because it occupies a position
directly beneath the inferred apex of the Sterkfontein Formation cone, the Jakovec deposit is probably not simply a
stratigraphic extension downslope of higher-lying Members.

Thus, the simplest explanation of the existence of the
Jakovec deposit is that it is an older basal mass underlying
and continuous with the Sterkfontein Formation (Fig.1b).

P.V.Tobias (1979) has stated that the lower Members of
the Sterkfontein Formation, situated 15-30m below surface
"may well contain earlier forms of Australopithecus, perhaps
comparable in age and morphology with those of Laetolil in
Tanzania" (p.164). The same claim can be made even more
justifiably for the Jakovec Cavern deposit at 40-50m below
surface, since it is fossiliferous and at least 20m thick in
a central exposure.

Tobias (1979) reasoned on biostratigraphic grounds that
Members 1-3 of the Sterkfontein Formation may be older than
3.0 m.y.BP, even suggesting ages as great as 3.6-3.8 m.y.
Considering the probable stratigraphic relationships, it is
not impossible that the Jakovec sediments are twice the age
of Mbs 1-3 since they lie at twice the depth. Certainly
they could be one, two or even three million years older.

In light of this crude but not necessarily incorrect perspective on the age of the Jakovec deposit, investigation of the sediments is important for a better understanding of the stratigraphy at Sterkfontein Cave, and perhaps for paleoanthropology in southern Africa.

ACKNOWLEDGEMENTS

I owe sincere thanks to members of the cave mapping teams, to Profs. KW Butzer and PV Tobias, to Dr BP Moon for reading an interim draft, to Mrs MD Whitehouse for typing the manuscript, and Mr P Stickler for preparing the figures.

REFERENCES

Brain CK (1958). "The Transvaal Ape-man-bearing Cave Deposits." Memoir, Transvaal Museum, No.11. Pretoria.
_____ (1976). A re-interpretation of the Swartkrans site and its remains. S Afr J Sci 72:141.
Butzer KW (1971). Another look at the australopithecine cave breccias of the Transvaal. Am Anthropol 73:1197.
_____ (1984). Archeogeology and Quaternary environment in the interior of southern Africa. In Klein RG (ed) "Southern African Prehistory and Paleoenvironments." Rotterdam, Boston : Balkema, pp 1-64.
Partridge TC (1978). Re-appraisal of lithostratigraphy of Sterkfontein hominid site. Nature 275:282.
Robinson JT (1962). Sterkfontein stratigraphy and the significance of the Extension Site. S A Archaeol Bull 17:87.
Stiles DN, Partridge TC (1979). Results of recent archaeological and palaeoenvironmental studies at the Sterkfontein Extension Site. S Afr J Sci 75:346.
Tobias PV (1979). The Silberberg Grotto, Sterkfontein, Transvaal and its importance in palaeo-anthropological researches. S Afr J Sci 75:161.
Wilkinson MJ (1973). "Sterkfontein Cave System : Evolution of a Karst Form." M.A.Thesis. Dept. of Geography and Environmental Studies, Univ. of the Witwatersrand, Johannesburg (Unpublished).
_____ (1983). Geomorphic perspectives on the Sterkfontein australopithecine breccias. J Archaeol Sci 10:515.

Hominid Evolution: Past, Present and Future, pages 171–187
© *1985 Alan R. Liss, Inc.*

SPRING FLOW AND TUFA ACCRETION AT TAUNG

T.C. Partridge, Ph.D.

University of the Witwatersrand,
Johannesburg.

INTRODUCTION

The calcareous tufas at Buxton on the Ghaap escarpment
were first quarried for lime early in 1916 (Gordon 1926). Primate (baboon) remains were recovered from 1919 onwards; the
vastly more important Taung hominid was retrieved five years
later in 1924, and described by Raymond Dart early the following year. The exact provenance of this find is obscure owing
to the failure of the quarry staff to make the necessary re--
cords or to take photographs at the time. It must, however,
be conceded that the significance of the specimen was appreciated only after Dart's removal of the face from the enclosing
breccia, a process which took fully four weeks. By that time
continued quarrying operations would have severely modified the
discovery site; its destruction was, in fact, said to be complete by the time that RB Young, Professor of Geology at the
Witwatersrand University, visited the site in February 1925
(Young 1926). Quarrying at Buxton was, from the outset, carried out on a large scale; explosives were used to advance an
excavation face some 30m high over a wide front, and it is unlikely that the debris produced by a single large blast could
have been related with any confidence, by the quarrymen who
lacked a formal geological training, to a precise locality on
the newly exposed face.

Despite these difficulties the approximate position of the
discovery can be reconstructed from contemporary accounts (Peabody 1954, Fig.3). All accounts agree that the specimen came
from a body of well cemented sandy and silty infill material
within the Thabaseek Tufa. The position of this tufa and its

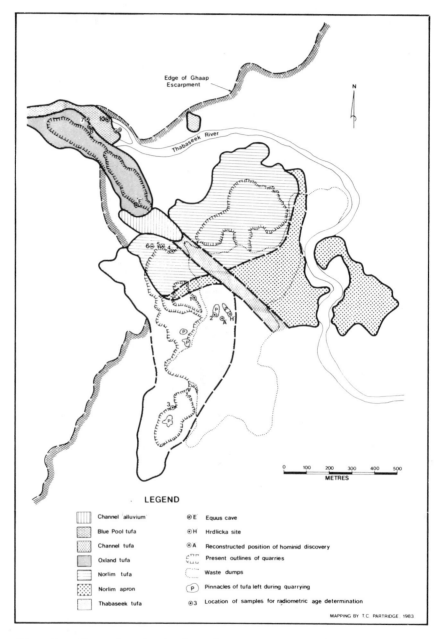

Fig. 1. Map of principal tufa carapaces at Buxton
limeworks, Taung.

relationships to the three succeeding carapaces at Buxton are
shown in Fig. 1.

Young (1926) and geologists who followed him in studies
of the site all agreed that these deposits were formed by local
spring activity. Indeed Butzer (1974) and Butzer et al (1978)
speculated on the role of past climates in the springflow regime, and proposed a series of spring cycles to account for the
sequence of depositional and erosional events which they have
identified in the tufas. In none of these studies, however,
has attention been given to the springs themselves or to the
evidence from relevant hydrological analyses. The present
paper examines this evidence and its significance for understanding the history of the tufas.

THE THABA SIKWA SPRINGS

The edge of the Ghaap escarpment is formed by the lowermost beds of the Precambrian Ghaap Plateau Dolomite Formation.
The tufa carapaces have been deposited in steplike succession
where the Thabaseek River cuts through the scarp face, which is
located chiefly in shales of the Schmidtsdrif Formation (Fig.2).
These latter, although they contain subordinate dolomite horizons, have a relatively low vertical permeability. The overlying Ghaap Plateau Dolomite Formation is characterized by
massive dolomite beds which are readily susceptible to karst
weathering, with the formation of an interconnected network of
subterranean solution channels along enlarged joints. The
typical cycle of karst development in southern Africa has been
documented by Brink and Partridge (1965). As the cycle pro-
ceeds surface drainage becomes progressively less well integrated as surface flow is diverted to subsurface paths. Springs
develop most commonly where incision has exposed an impermeable
chert or shale horizon within the dolomite, or has reached the
top of the Schmidtsdrif shales, flow emerging at the upper contact of the impermeable stratum.

The Thabaseek River is fed by two springs which emerge on
the escarpment near the village of Thaba Sikwa. These springs
are located some 2.25 and 3.5 km west of the uppermost (Blue
Pool) tufa and are, for practical purposes, perennial. To the
west of their emergences, the catchment of the Thabaseek River
extends for some 60 km across the Ghaap Plateau to a watershed
almost 20 km north-west of the town of Reivilo (Fig.3). However, a reasonably well integrated channel network is present

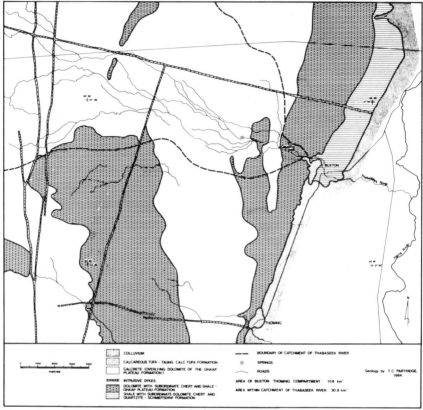

Fig. 2.   Geology in the vicinity of Buxton.

for no more than 35 km of this distance, giving an effective
catchment area above Buxton of c. 460 km^2. This definition of
the effective catchment accords well with that of Middleton et
al (1981). The runoff upstream of the Thaba Sikwa springs is
strictly episodic, in conformity with the local rainfall pattern, which is typical of the semi-arid western interior of
the southern African plateau. Most of the ground surface is
made up of rock or calcrete (caliche) outcrop, which leads to
rapid concentration of that part of the precipitation which
does not find its way into subsurface channels. Hence, while
regular spring flow characterizes the Thabaseek channel downstream of the Thaba Sikwa springs, the upper part of the catchment is subject only to intermittent flood discharges.

Like most dolomitic areas of southern Africa, the Ghaap

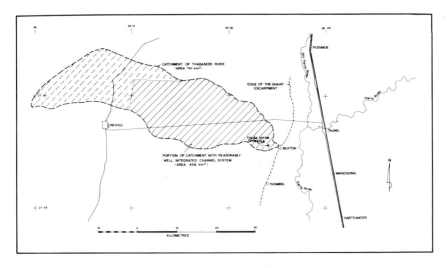

Fig. 3. Catchment of the Thabaseek River above Buxton.

Plateau is crossed by a network of intrusive dykes which intersect to form groundwater compartments. The generally impermeable dykes often form effective aquicludes so that individual
compartments operate as geohydrological units. Fig. 3 shows
the compartment which supplies the Thaba Sikwa springs. The
same compartment feeds also the springs at Thoming, 9 km south
of Buxton, where analogous tufa deposits have been produced.
However, this part of the compartment lies south of the
Thabaseek River catchment.

   Combined flow from the Thaba Sikwa springs was measured at
a gauging weir from 1960-1980; the results are presented as bar
charts in Fig. 4. If the mean flow over this period is plotted
against the rainfall which feeds that part of the compartment
lying within the catchment of the Thabaseek River, the point
falls precisely on the line reconstructed by Enslin (1970) for
groundwater recharge in summer rainfall areas of South Africa
(Fig. 5). Of the mean annual rainfall of 425mm, only 3.2% contributes to the spring flow. A plot of mean annual flow against mean annual rainfall shows some scatter over the 20-year
survey period (Fig. 6), but a linear relationship is nonetheless evident. The Thaba Sikwa springs are therefore typical,
hydrologically, of semi-arid dolomitic springs; this allows
greater reliance to be placed on the application of uniformi-
tarian principles in the assessment of the possible behaviour
of the springs under varying climatic circumstances in the past.

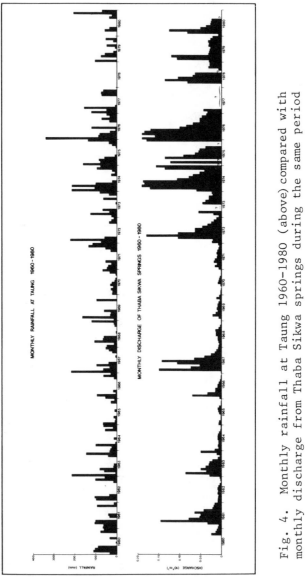

Fig. 4. Monthly rainfall at Taung 1960-1980 (above) compared with monthly discharge from Thaba Sikwa springs during the same period (below).

    When flow is compared with rainfall on a monthly basis (Fig. 4), a clear and direct response is evident. As is expected for a rocky catchment with little weathering and skeletal soil cover, the lag between rainy spells and increase in spring discharge is about six weeks, indicating a low storage volume.

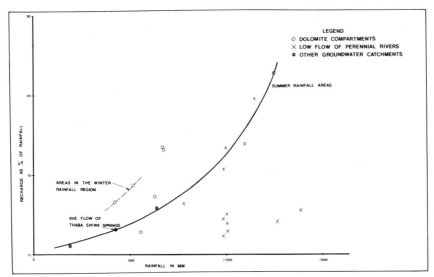

Fig. 5. Groundwater recharge as a percentage of rainfall
in South Africa (after JF Enslin 1970). Average flow of
Thaba Sikwa springs is plotted as filled circle.

Fig. 4 reveals that both rainfall and spring flow show distinct oscillations. For instance, the decade 1961-1970 was,
on average, significantly drier than that from 1971 to 1980, a
trend clearly reflected in flow records. But while rainfall
during the wetter decade was only about 1.5 times that of the
drier, spring flow increased 3.6 times, indicating marked sensitivity to increased rainfall in this geohydrological regime.
The short-term oscillations recorded in the Taung area over the
20-year period exactly parallel those that prevailed over much
wider areas of South Africa (Tyson et al 1975; Tyson 1983).
The rainfall at Buxton is somewhat lower than at the Taung
Police Station where the measurements were made, but after the
appropriate correction it is still somewhat surprising that
the maximum annual rainfall during the wetter decade was c.
800mm, nearly double the mean value.

Observations of the effects of spring discharge and surface storm runoff in the Thabaseek channel above the Buxton
tufas confirm the view of previous workers that tufa accretion
is produced almost exclusively by the former. Indeed, a new
tufa carapace is forming today upstream of the Blue Pool tufa,
where the spring water traverses a series of rapids. At this
point the flow divides into a series of shallow channels, each

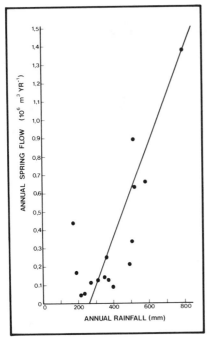

Fig. 6. Relationship between annual flow of Thaba Sikwa springs and annual rainfall.

of which feeds a succession of tufa-rimmed pools. The carbonate charged water spills from one pool to another. These shallow receptacles serve as evaporating pans so that, by the time the spring flow reaches the toe of the carapace, most has been lost to the atmosphere or has infiltrated into the banks of the channel. Algae are present, although not prolific, suggesting that some part of the carbonate precipitation is due to loss of carbon dioxide rather than to evaporation; however, a dense growth of trees and shrubs next to, and overshadowing, the growing tufa accounts for the profusion of plant macrofossils preserved here and in some earlier carapaces. The growing tufa body has a convex longitudinal profile: towards the back, successive depositional layers are inclined at a gentle angle downstream, while, towards the toe, these layers are almost vertically disposed as a series of foreset beds. As the tufa grows downstream the foreset beds are progressively buried under more gently inclined strata. In periods of low flow the surface of each layer dries out and forms a hard, impermeable crust, from the undulating surface of which succeeding flows are evaporated. The tufa body clearly operates as a system in equilibrium with the spring discharge, which is effectively dispersed across its surface in a manner with a high degree of evaporative loss. A sample from the surface of this tufa gave an ionium age of zero.

As indicated, spring flow varies directly and dramatically with fluctuations in rainfall; however, the total volumes involved are small in comparison with the surface flood discharges. Maximum spring flows exceed 190 000m^3 per month, while minimum flows fall to less than 1 000m^3 per month. The average over the 20-year period is 34 900m^3 per month. In contrast, the average annual surface runoff from the Thabaseek catchment is c. 1 600 000m^3, or an average of about 133 300m^3 per month (Middleton et al 1981). As is discussed below, runoff volumes many times greater than this average annual figure occur periodically, the volume being, on average, proportional to the length of the return period. Because of the rocky, poorly vegetated nature of the catchment, these surface discharges tend to concentrate rapidly, and flood peaks are high, but of short duration, with intervening periods of no surface flow (other than that from the springs). This pattern conforms to the pattern of precipitation in the area, which is characterized by episodic convectional showers of high intensity, mainly between November and April. Hence the pattern of surface runoff is scarcely conducive to the deposition of dissolved carbonate by the mechanisms of tufa accretion described above: flows are too violent and transient to permit dissipation and slow evaporation from the surface of the carapace; such evaporation and deposition occur with the modest, more regular, carbonate charged flows from the springs. Instead, such discharges might well inflict erosional damage on newly deposited carbonate layers, a process which has been observed at this locality by the writer during a heavy thunderstorm. It may, therefore, be concluded that little, if any tufa growth is due to runoff from surface sources. Indeed, tufas have formed along the Ghaap escarpment only where springs with a more or less perennial regime emanate from near the scarp crest.

Under these circumstances the chemistry of the spring waters is paramount to the rate of tufa deposition. In analyses carried out by the National Institute for Water Research, the calcium carbonate content of the water from the Thaba Sikwa springs varied from 368 to 445 mg/l (mean 407 mg/l). This accords closely with the average values obtained by Smit (1977), for groundwaters from various localities in the Ghaap Plateau area. From the above value and the present mean spring flow of 419 000m^3 per year, some 63m^3 of $CaCO_3$ could be deposited, on average, per year if evaporation were 100% effective. In order to assess the effectiveness of evaporation, the original volumes of the tufa carapaces at Buxton were calculated using data from the recent, highly accurate, 1 : 2500 scale contour

survey of the quarries and surrounding areas. The results are
as follows (volumes in m^3): Blue Pool tufa: 2.17 x 10^5; Oxland
tufa: 1.08 x 10^6; Norlim tufa: 4.40 x 10^6; Thabaseek tufa:
2.72 x 10^6. These volumes have been computed from the distribution of the remaining portions of the lower three carapaces,
and have been adjusted to allow for the volumes removed naturally by post-depositional incision. They are not, therefore,
precise, but are likely to represent reasonable estimates based
on controlled measurements. It is theoretically possible for
the Blue Pool tufa to have been deposited by spring flow in
c. 3500 years. Ionium dates are now available for the upper
and lower levels of this tufa (Vogel, Partridge 1984); these
indicate that the carapace, as preserved today, accumulated
over some 70 000 years. From this figure, it appears that only
about 5% of the calcium carbonate contained in the spring water
(on present flow volumes) was deposited in the course of tufa
accretion.* Dates for the Oxland tufa range from 230 000 to
343 000 yr (Vogel, Partridge ibid), giving a figure of about
15%. The available dates suggest also that about 130 000 years
intervened between the accumulation of the Oxland tufa and the
onset of the Blue Pool deposition. If similar rates of tufa
accretion prevailed during past cycles of deposition, and if
similar time intervals separated the older tufas, all four
Buxton tufas could have been deposited within the space of
somewhat under 1.4 m.y. Preliminary dating evidence now suggests that the upper levels of the Thabaseek tufa may be no
more than about 1 m.y. old (Vogel, Partridge 1984; Vogel this
volume); this, in turn, implies a marginally increased rate of
calcium carbonate deposition, amounting to about 16% of the
presently available $CaCO_3$ volume, during the accretion of the
Norlim tufa and the higher levels of the Thabaseek tufa.

THE ROLE OF SURFACE RUNOFF

Research in South Africa over the past 15 years has led to
the formulation of relationships linking climatic zones, rainfall events of specified magnitude, and catchment area, to
flood discharges of specified frequency. As observed above,
the mean annual discharge from surface runoff in the Thabaseek

*Analyses show that the Buxton tufas are composed almost
exclusively of calcium carbonate. Magnesium carbonate derived
from dissolution of dolomite is more than 100 times as soluble
as $CaCO_3$ and is thus seldom redeposited except when near total
evaporation, in fact, occurs.

River catchment above Buxton is approximately 1.6 million m^3 (Middleton et al 1981). It can be shown also that, if the annual rainfall were to increase to double that of today, i.e. to about 850mm, the mean annual discharge would be approximately 27.6 million m^3, or more than 17 times the present mean volume. In contrast, spring flow would increase by only about 3.6 times (Fig. 6). It is therefore clear that the onset of more humid climatic conditions would markedly decrease the significance of spring discharge in relation to that of surface runoff. Since, as argued previously, tufa accretion is overwhelmingly the product of spring discharge, while rapid surface runoff results in erosion of the carapaces so formed, it is clear that a change to mesic conditions would favour tufa incision at the expense of carbonate accretion. Butzer et al (1973) have presented compelling evidence from studies of the Alexandersfontein Pan near Kimberley that annual rainfall was approximately double the present 400mm average, during the latter part of the Last Glacial Period.* That this period was one of tufa destruction at Buxton seems to be confirmed by the failure of the recent dating programme for the Blue Pool tufa to reveal deposits younger than c. 30 000 yr (Vogel, Partridge 1984). The northern flank of the Blue Pool tufa has, in fact, been sectioned by erosion to a depth of some 20m. Subsequently, new tufa accretion commenced upstream of this area, and this process is continuing today. The implications of these findings are far-reaching: tufa accretion is occurring today and may even be favoured by semi-arid conditions as at present; a modest increase in rainfall, such as might have characterized the South African interior during Early Glacial times, would almost certainly have been accompanied by continued deposition, but the effect of increased spring volumes might well have been partly offset by lower temperatures, leading to reduced evaporation. More notable increases in precipitation are inferred for periods just before and after the Last Glacial Maximum (Deacon et al 1984); these would have produced incision, and widespread pitting and corrosion such as are evident in the surfaces of the carapaces. Gorges have been cut through both the Thabaseek and Norlim carapaces to depths of over 20m. In short, tufa deposition is likely to be at a maximum during semi-arid intervals and at the drier ends of the waxing and waning phases of humid cycles. Under extreme aridity spring activity is terminated and tufa deposition halted; significantly increased

*The peak annual rainfall of approximately 800mm recorded in 1976 approximates closely to the average inferred for the area during Late Glacial times.

precipitation during humid maxima would lead to the destruction
of newly-deposited tufa skins, and to pitting and gullying.
In the model proposed by Butzer et al (1978), both tufa accretion and gorge cutting are ascribed to a single humid cycle
associated with increased spring discharge; the evidence presented above runs counter to this general proposition. It is
at variance also with the conclusions of Marker (1971), who
suggested that tufas are formed preferentially in the waning
phases of pluvial periods. Better dating control for the
younger tufas may help to resolve these differences.

It may be argued that fluctuations in total rainfall are
less significant than infrequent, large flood discharges, for
fluvial erosion in semi-arid areas. To test this hypothesis
peak flood discharges and volumes were calculated for different return periods in the Thabaseek catchment, by use of the
relationships established by the Hydrological Research Unit
(1972). From experience in the interior of South Africa, a
4-hour storm duration was selected as likely to produce maximum
runoff. The results are presented in Figs. 7 and 8. If the
accumulated flood peaks are plotted against flood return period
(Fig. 9), it is apparent that the 5-year flood is of greatest
hydrological significance. Only when the environs of stream
channels consist of unconsolidated deposits (which is not the
case in the Thabaseek catchment), are larger but less frequent
events likely to have a greater local erosive effect. If
everything else is equal, both flood peaks and volumes vary
in direct proportion to rainfall (Pitman 1971); hence, if
rainfall doubled, flood peaks and volumes would likewise increase about twofold. Clearly, therefore, in the Thabaseek
catchment, frequent, moderate flood discharges have a greater
geomorphic significance than large infrequent discharges. But
doubling of the annual rainfall would, in turn, be more significant than the cumulative effect of successive 5-year floods
under a semi-arid regime, in producing long-term erosion.

What significance do these analyses have for the reconstruction of the likely environment of the Taung child? Since
much evidence has been destroyed, it is uncertain whether the
accumulation and cementation of the cave filling containing
the fossils was more-or-less contemporary with the formation
of the youngest part of the Thabaseek carapace, as was suggested by Peabody (1954) and is supported by the observations
of Brain (this volume), or whether the cavity was formed by
solution and filled after the end of this phase of tufa deposition, as Butzer (1974) contended. The writer has previously

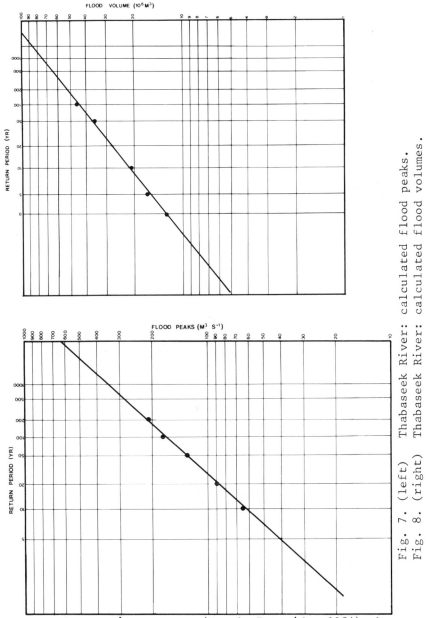

FLOOD VOLUME (10⁶ M³)

RETURN PERIOD (YR)

FLOOD PEAKS (M³ S⁻¹)

RETURN PERIOD (YR)

Fig. 7. (left) Thabaseek River: calculated flood peaks.
Fig. 8. (right) Thabaseek River: calculated flood volumes.

supported Butzer's argument (Vogel, Partridge 1984), but, after further review, concedes that there is a dearth of evidence on which to reject Peabody's view. Contemporaneous filling of the hominid cavern with the "dry phase" sands,

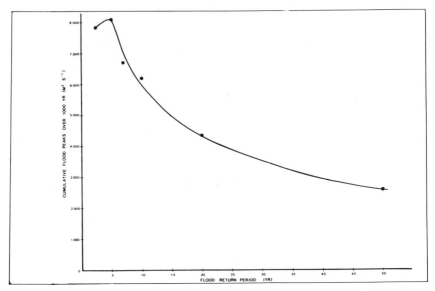

Fig. 9. Thabaseek River: cumulative discharge during
flood peaks of differing return period.

which, according to Peabody (1954), yielded most of the baboon
fauna, is certainly compatible with the model of tufa accretion under a semi-arid regime proposed above. Peabody concluded that the hominid deposit overlay the baboon sands; the
clayey matrix from which the Taung cranium was extracted, and
the frequent occurrence of calcite veins and void fillings
within it (Butzer 1974, 1980) do suggest a more humid environment during its accumulation. On the present model, this
phase of clastic deposition within the hominid cave could
well have occurred concurrently with the commencement of a
humid phase, which effectively ended the Thabaseek cycle of
tufa accretion.

SPRING FLOW AND THE AGE OF THE TAUNG CHILD

Various methods have been used in an attempt to date the
Taung skull (Partridge 1973, 1982; Butzer 1974, 1980; Vogel,
Partridge 1984; Vogel, this volume). All of these attempts
suggest a much younger age than had hitherto been considered
likely. The most recent radiometric analyses of JC Vogel,
using two independent methods, suggest that the specimen may

be no more than about 1 m.y. old. If these results are correct, and if the Taung child is a member of the same species of Australopithecus as is represented by the many individuals recovered from Sterkfontein Member 4 and Makapansgat Members 3 and 4, then the Taung child must have belonged to a relict population which had persisted with little change for more than 1 m.y. This would apply also to the baboon fauna, though on the analyses of Delson (1984), they are compatible with remains from East African sites dated to a little earlier than 2 m.y.

Much evidence from African sites suggests that pronounced aridification occurred over large areas some time after 2.5 m.y. The Ghaap Plateau, on the edge of which the Buxton quarry is located, forms a dolomite "island" some 12 000 km^2 in area, from around the northern and eastern periphery of which powerful springs emerge. The areas surrounding the limits of the dolomite are marginal to the Kalahari Desert, and are, in contrast, poorly watered; under marked aridification, they would have constituted effective barriers against the migration of populations into better watered areas to the north and east. It may well have been the presence of this line of springs around the Ghaap Plateau that enabled a relict population of hominids to persist for long in this remote enclave. A sample from fairly high in the tufa sequence at Ulco south of Buxton gave an infinite uranium series age (JC Vogel, pers. comm.), indicating that springs were active in that area earlier than those which deposited the oldest Buxton tufa.

ACKNOWLEDGEMENTS

The help of the Directorate of the Weather Bureau, Department of Transport Affairs, is acknowledged with gratitude. Flow records for the Thaba Sikwa springs were kindly supplied by the Director-General of Water Affairs. The writer owes a special debt of gratitude to Dr WV Pitman for advice on the computation of flood discharges and for providing runoff data. The chemical analyses were carried out by the National Institute for Water Research of the South African C.S.I.R. The diagrams were drawn by Mrs PJ Moon and the text typed by Mrs IM Partridge and Mrs JI Britt.

REFERENCES

Brink ABA, Partridge TC (1965). Transvaal karst: some consi-
    derations of morphology and development with special refer-
    ence to sinkholes and subsidences on the Far West Rand.
    S Afr Geog J 47:11.
Butzer KW, Fock GJ, Stuckenrath R, Zilch A (1973). Palaeo-
    hydrology of late Pleistocene lake, Alexandersfontein,
    Kimberley, South Africa. Nature 243:328.
Butzer KW (1974). Paleoecology of the South African austra-
    lopithecines: Taung revisited. Curr Anthrop 15:367.
Butzer KW, Stuckenrath R, Bruzewicz AJ, Helgren DM (1978).
    Late Cenozoic paleoclimates of the Ghaap Escarpment, Kala-
    hari margin, South Africa. Quat Res 10:310.
Butzer KW (1980). The Taung australopithecine: contextual
    evidence. Palaeont afr 23:59.
Deacon J, Lancaster N, Scott L (1984). Evidence for Late
    Quaternary climatic change in southern Africa. In Vogel JC
    (ed): "Late Cainozoic Palaeoclimates of the Southern Hemi-
    sphere", Rotterdam and Boston: AA Balkema, p391.
Delson E (1984). Cercopithecid biochronology of the African
    Plio-Pleistocene: correlation among eastern and southern
    hominid-bearing localities. In Andrews P, Franzen JL (eds):
    "The Early Evolution of Man", Frankfurt: Cour Forsch Inst
    Senckenberg 69:199.
Enslin JF (1970). Die grondwaterpotensiaal van Suid-Afrika.
    Waterjaar 1970: Convensie: Water vir die Toekoms, Dept
    Waterwese: 1.
Gordon GB (1926). Discussion of "The calcareous tufa depo-
    sits of the Campbell Rand, from Boetsap to Taungs Native
    Reserve" by RB Young. Proc Geol Soc S Afr 28:xxvii.
Hydrological Research Unit (1972). Design flood determina-
    tion in South Africa. Univ Witwatersrand HRU Report 1/72.
Marker ME (1971). Waterfall tufas: a facet of karst geomor-
    phology in South Africa. Z Geomorph NF Supp Bd 12:138.
Middleton BJ, Pitman WV, Midgeley DC (1981). Surface water
    resources of South Africa. Water Res Comm, Univ Witwaters-
    rand HRU Report 8/81, Vol II: Drainage Region C - The Vaal
    Basin.
Partridge TC (1973). Geomorphological dating of cave open-
    ing at Makapansgat, Sterkfontein, Swartkrans and Taung.
    Nature 246:75.
Partridge TC (1982). The chronological positions of the
    fossil hominids of southern Africa. Proc 1st Int Cong
    Human Palaeont (Nice), Paris: CNRS 2:617.
Peabody FE (1954). Travertines and cave deposits of the

Kaap Escarpment of South Africa and the type locality of Australopithecus africanus Dart. Bull Geol Soc Am 65:671.

Pitman WV (1971). Amendments to design flood manual HRU 4/69. Univ Witwatersrand HRU Report 1/71.

Smit PJ (1977). Die geohidrologie in die opvanggebied van die Moloporivier in die noordelike Kalahari. D Sc Thesis Dept Geol Univ OFS Bloemfontein.

Tyson PD, Dyer TGJ, Mametse MN (1975). Secular changes in South African rainfall: 1880 to 1972. Quart J R Met Soc 101:817.

Tyson PD (1983). The great drought. Leadership SA 2:49.

Vogel JC, Partridge TC (1984). Preliminary radiometric ages for the Taung tufas. In Vogel JC (ed): "Late Cainozoic Palaeoclimates of the Southern Hemisphere", Rotterdam and Boston: AA Balkema, p507.

Young RB (1926). The calcareous tufa deposits from Boetsap to Taungs Native Reserve. Trans Geol Soc S Afr 28 (1925): 55.

Hominid Evolution: Past, Present and Future, pages 189–194
© *1985 Alan R. Liss, Inc.*

FURTHER ATTEMPTS AT DATING THE TAUNG TUFAS

John C. Vogel

National Physical Research Laboratory
CSIR, Pretoria

The radiometric measurements on the Taung tufas which were reported recently (Vogel, Partridge 1984), suggested that the oldest carapace – The Thabaseek tufa – is only about 1 million years old. Using the conventional Th-230 (ionium) dating method, ages of 30 ka (1 ka = 1000 years) and 103 ka were obtained for the youngest Blue Pool carapace and of 230 ka for the Outer Oxland tufa. Data for these samples indicated furthermore that the U-234/U-238 activity ratio in the water of the dolomite spring which forms the source of the tufas has remained virtually constant over time. This observation enabled us to calculate ages of 764 ka and 942 ka from the respective U-234 contents of the Basal Norlim tufa, and of the Outer Thabaseek tufa.

From a radiochemical point of view the data all seem highly satisfactory. It is conceivable, however, that the travertine does not represent a completely closed system and that some uranium was added after formation. Under such circumstances the uranium isotope dates would underestimate the age of the tufa. If this be the case, there would be little purpose in processing further samples in the same manner. Instead thermoluminescence (TL) dating seemed to offer a more promising approach.

Thermoluminescence is the phenomenon by which a crystal or mineral that has been subjected to nuclear irradiation, emits light when heated to moderate temperatures. Since the TL signal is effectively destroyed by exposure to sunlight, it can be used to date sediments (Wintle, Huntley 1982): when sand, for instance, lies in the sun for any length of time, the "TL clock" is set to zero, and when such sand grains become buried in sediment, the TL signal starts building up again at a rate corresponding to the nuclear radiation that it receives from the environment.

Current opinion seems to be that the dating range is only a few hundred thousand years. This is based on experience with undifferentiated "sediment" (clay minerals) and, for instance, with feldspar. The reasons for the limiting range are saturation of the TL centres and/or natural fading of the signal. Analyses that we have made with buried desert sands in the Namib, suggested, however, that quartz grains possibly show a slower response and that their TL signal may be more resistive to fading.

To test the suitability of quartz over a longer time range some samples of breccia from the fossil site of Swartkrans cave near Krugersdorp were investigated. These breccias are approximately dated by the stone artefacts or the faunal assemblages they contain. Four samples were selected from deposits with estimated ages of 100 ± 20 ka (MSA), 500 ± 100 ka (ESA), 1.2 ± .1 m.y. (Brown Breccia) and 1.65 ± .15 m.y. (Orange Breccia) respectively.

The material was boiled in 10N HCℓ and the solubles used for determination of the uranium and thorium isotopes by α-particle spectrometry. Potassium is not expected to make any significant contribution to the radioactive dose of these samples and was not measured. The insoluble fraction was sieved and passed through a magnetic separator to produce a pure quartz grain fraction in the size range 75–150 μm. The thermoluminescence was measured on a "Toledo" TLD recorder in which a few milligrams of the sample are preheated at 120 °C for 30 sec. and the glow curve then recorded, using a ramping rate of 2.4 °C/sec. to a maximum of 383 °C. The glow curves of the quartz grains were very uniform with a double peak between 300 °C and 340 °C.

The results are listed in Table 1 and the normalized TL signals plotted against the estimated ages in Fig. 1. It is evident that the TL signal grows steadily up to the 1.65 m.y. level, although there is apparently some flattening of the curve beyond 0.5 m.y. The shape of the curve would be compatible with an average lifetime of the trapped electrons of 2 m.y., and if this be the case, then pure quartz could be used effectively for dating samples

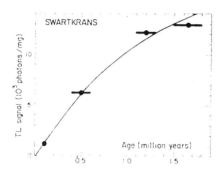

Fig. 1. Thermolumines-
cence signal of quartz
grains from four different
levels in Swartkrans cave
plotted against the esti-
mated ages of the depo-
sits.

throughout the Quaternary, provided the radiation dose rate
is low enough to avoid saturation.

The radioisotope analyses show that the radioactive
doses are at present similar to each other, but also that
the uranium-series isotopes of the older samples are not in
radioactive equilibrium and that additional uranium has
thus been introduced into the deposit in relatively recent
times. For this reason actual ages cannot be calculated
from the TL signals. In view of the radioisotope composi-
tions it seems remarkable that the relative intensities of
the TL signals are nevertheless compatible with the relative ages of the samples.

In the light of the encouraging results obtained on
the Swartkrans breccias, the tufas at the Buxton quarry at
Taung were scrutinized for interbedded sands and two such
layers were located: a sandy level near the tip of the Oxland carapace just below the Equus cave and a gravelly
layer in the Thabaseek tufa near the southern limit of the
carapace. The samples, together with some samples of breccia from different localities in the tufas, were analysed
in the same manner as those from Swartkrans (Table 1).

The two sandy samples contain very much less uranium
and thorium than do the samples from Swartkrans ($\frac{1}{2}$ % and
10 % respectively in terms of current radiation dose rate),
but the isotope measurements show that neither these nor
the breccias represent closed systems. The dose rates can
therefore not be used to calculate ages for the samples.
Comparison of the TL signal of the Thabaseek sand (U-339)
with those of the Thabaseek breccias does, however, indicate that the former has by no means reached saturation.
This result is encouraging and the selection of more suit-

Table 1.   Summary of measurements on samples from Swartkrans cave and from the Taung tufas.

Anal. no. Sample*	SWARTKRANS BRECCIAS				TAUNG SANDS AND BRECCIAS						TAUNG CALCITES	
	U-330 SK11 MSA	U-329 SK10 ESA	U-328 SK 8 Br.br	U-327 SK 2 Or.br	U-338 Oxl. sand	U-339 Thab. sand	U-337 Oxl.br MSA	U-332 Thab. br.	U-340 Thab. bab.br	U-331 Thab. Aus.br	U-166 Oxl. outer	U-271 Thab. outer
Estimated age (ka)	100	500	1200	1650	230		100				230	>>
TL signal (10³ photons/mg)	1.22	6.21	12.1	12.8	0.24	17.2	8.36	20.5	27.4	27.4	1.77	4.05
U-238 $\left(\frac{dpm}{kg}\right)$	254	183	798	372	1.61	24.2	123	35.2	321	2990	85.36	33.74
U-234	332	249	1053	498	3.22	34.0	218	43.5	403	3080	168.4	37.97
Th-230	216	144	591	272	1.49	30.3	109	48.6	271	1160	160.8	38.35
Th-232	307	271	322	337	0.50	38.2	48	191.4	94	102	2.21	8.79
U-234/U-238	1.31 ±.04	1.36 ±.03	1.32 ±.02	1.34 ±.06	2.00 ±.34	1.41 ±.06	1.77 ±.03	1.12 ±.06	1.25 ±.05	1.03 ±.03	1.97 ±.09	1.13 ±.03
Th-230/U-234	0.65 ±.01	0.58 ±.02	0.56 ±.02	0.55 ±.03	0.46 ±.06	0.89 ±.03	0.50 ±.01	1.12 ±.06	0.67 ±.03	0.38 ±.03	0.95 ±.04	1.05 ±.05
Current Dose rate (rad/ka)	73.0	57.6	138.6	87.7	0.31	9.27	24.0	49.2	55.0	244.3	23.2	6.77
Total Dose (k rad)											4.18	9.55

* U-332: Undefined piece of breccia from the Thabaseek carapace; U-340: breccia from the baboon-bearing breccia in Hrdlicka cave; U-331: breccia from block that contained the *Australopithecus africanus* skull.

able (compact) material may yet provide absolute ages for the tufas. The breccias, on the other hand, have too much recently introduced radioactivity and are less promising.

The following conclusions can be drawn from the above measurements:
1.   There is no indication of TL saturation of the quartz grains in the time range of 10^6 years with the radiation doses experienced by these samples.
2.   The Swartkrans samples may reflect natural fading, with an average lifetime of the trapped electrons of about 2 m.y. at ambient temperatures of ca 19 °C.
3.   The breccias do not represent closed systems as far as uranium is concerned and it will probably not be possible to calculate total radiation doses and thus TL-ages for such material.
4.   Sand grains interbedded in the tufas appear more promising, because of the low dose rates that they have experienced and better samples must be sought.

In the light of these partly negative results it was decided to investigate the thermoluminescence of the tufa calcite itself. It is already known that the Taung tufas represent, if not entirely, at least nearly closed systems (Vogel, Partridge 1984) and the radiation doses could be calculated from the uranium and thorium concentrations with some confidence. In doing this the gradual progression of the uranium series isotopes towards radioactive equilibrium needs to be taken into account. Using data given by Wintle (1978) and the measured α-particle activities of the samples, the total effective doses can be calculated as a function of age. In doing this the α-particle efficiency for producing thermoluminescence was taken as 40 % and the initial activity ratio (U-234/U-238) as 2.7 (Vogel, Partridge 1984).

The thermoluminescence of two calcite samples, one from the Outer Oxland tufa and one from the Outer Thabaseek tufa, was determined following the method proposed by Debenham and Aitken (1984), i.e. the 75 to 150 μm fraction of crushed calcite was treated with 1 % acetic acid and the glow curve measured using a Schott BG3 blue filter. The glow curves of the two samples were very similar, with overlapping maxima at 280 °C and 350 °C. For the present purpose it was thus immaterial whether only the 280 °C peak, or the total photon emission below 383 °C was used.

By accepting the ionium age of 230 ka for the Outer Oxland tufa, the age of the Outer Thabaseek tufa can be calculated using the simple relationship for the TL signal

$$(TL) = k. \text{ Dose.}$$

The age obtained is 1200 ka with an estimated uncertainty of 20 % or more. If the recorded non-linearity of the 350 °C peak reported by Debenham (1983) is taken into account, the age may increase by about 10 %.

As was the case with the dates derived directly from the uranium isotope activities, the TL glow curves suggest an age in the vicinity of 1 million years for the Outer Thabaseek tufa. These TL results should not at this stage, however, be considered as providing an absolute date for the tufa, but rather as a promising indication that the TL dating method can potentially be applied to these deposits.

ACKNOWLEDGEMENTS

Dr C K Brain and P B Beaumont are thanked for helping me collect samples from Swartkrans cave and Taung. A Hughes provided a piece of breccia from the original block that contained the A.africanus skull. The help of Dr E A Retief, E A Hecker, F T Wybenga and Miss M L du Preez with various aspects of the analyses (magnetic separation, blue glass filter, TL analyser) is gratefully acknowledged. I also thank Miss S Huber for preparing and analysing the TL samples, and A S Talma for extensive discussions and for writing the computer program for calculating the radiation doses of the samples.

REFERENCES

Debenham NC (1983). Reliability of the thermoluminescence dating of stalagmitic calcite. Nature 304:154.
Debenham NC, Aitken MJ (1984). Thermoluminescence dating of stalagmitic calcite. Archaeometry 26:155.
Vogel JC, Partridge TC (1984). Preliminary radiometric ages of the Taung tufas. In Vogel JC (ed.): "Late Cainozoic Palaeoclimates of the Southern Hemisphere", Rotterdam: Balkema, p 507.
Wintle AG (1978). A thermoluminescence dating study of some Quaternary calcite: potential and problems. Can J Earth Sci 15:1977.
Wintle AG, Huntley DJ (1982). Thermoluminescence dating of sediments. Quat Sci Rev 1:31.

Hominid Evolution: Past, Present and Future, pages 195–200
© *1985 Alan R. Liss, Inc.*

EARLY HOMINIDS IN SOUTHERN AFRICA: UPDATED OBSERVATIONS ON
CHRONOLOGICAL AND ECOLOGICAL BACKGROUND

Elisabeth S. Vrba

Transvaal Museum
P.O. Box 413, Pretoria 0001
South Africa

INTRODUCTION

How does hominid chronology compare with hypotheses of hominid phylogeny and evolution? Which environmental variables
were of major importance in the ecology of our early relatives? What were the physical and biotic changes that caused
hominid evolution? The hominid fossils on their own cannot
take us far towards answering these questions (Vrba 1985).
Instead, the changes in Man's family tree need to be studied
as an integral part of wider biotic, climatic and geological
rhythms. I have tried to do this by analysis, and comparisons
with other data, of the fossil record of African antelopes.
This is a summary of more extensive arguments that have been
presented previously (Vrba 1975, 1982, 1985; Greenacre, Vrba
1984).

CHRONOLOGY

I place Makapansgat Member 3 (MAK 3) as the oldest hominid-
associated assemblage in South Africa and close to, perhaps
a little later than, 3 million years (m.y.) in age. Evidence
for this (Vrba 1982) includes species shared with assemblages
from the Laetolil Beds (close to 3.7 m.y.), from the Hadar Sidi
Hakoma Member (about 3.12-3.32 m.y.), and from Shungura Member C (about 2.85-2.52 m.y., Brown et al. 1985).

Sterkfontein Member 4 (ST 4) and MAK 3 share not only
Australopithecus africanus, but also several antelope taxa
which are unknown after this time period. An example is *Maka-
pania* known only from MAK 3 and 4 and ST 2 and ST 4. Other
taxa are absent from the large MAK 3 sample but present in

the poor one of ST 4 and in all later Transvaal assemblages,
e.g. *Antidorcas*, the first record of which elsewhere is from
Shungura B Unit 11 (close to 2.9 m.y., Brown et al. 1985).
Based on such data I date ST 4 between 2.4 and 2.8 m.y.

On very slender evidence *A. africanus* from Taung and *A.
robustus* from Kromdraai B East Member 3 (KBE 3) may both belong between ST 4 and Swartkrans Member 1 (SK 1) in time, and
closer to SK 1 than to ST 4 (Vrba 1982).

Until ongoing research at Sterkfontein (e.g. Clarke 1985;
analyses of bovids by me) is further advanced, I prefer to
say little on the dating of materials that are currently referred to Sterkfontein Member 5 (ST 5). Among remains usually attributed to ST5 are the West Pit Extension Site assemblage (see
Vrba 1975), fossils assigned to *Homo habilis*, and recently
excavated stone tools and fauna (Clarke 1985). At least two
time periods are represented in this set of materials: one
that is earlier (close to 2 m.y.), indicating arid grassland
and probably associated with *H. habilis*; and another large,
late (Late Early Pleistocene-Middle Pleistocene?) sample
suggesting wetter conditions. ST 5 as currently conceived is
heterogeneous and the associations of fauna, hominids and
stone artefacts require further study.

The earliest Swartkrans bovid assemblage from SK 1 (including the Hanging Remnant and Lower Bank or Orange Unit,
Brain pers.comm.) is associated with early *Homo erectus*
(Clarke pers. comm.), *A. robustus* and stone artefacts. I
estimate its age as between 1.6-1.8 m.y. The earliest stratum
of what was previously called Swartkrans Member 2 (SK 2),
corresponding to the Brown Unit (Brain pers.comm.), also has
Homo, *A. robustus* and artefacts. It may belong after SK 1 but
previous to 1 m.y. Two additional chrono-assemblages, one
containing typical Middle Pleistocene and the other Middle
Stone Age faunal elements, are indicated by the Swartkrans
bovid evidence (Vrba 1982).

The differences between phylogenetic hypotheses for early
hominids chiefly concern the crucial lineage splitting event
that marked the origin of *Homo*. It is interesting to compare
alternative phylogenies with estimated dates for hominid
finds in southern and eastern Africa. Some time estimates for
South African hominids are shown in Fig. 1 together with dates
for earliest known *Homo* from the sub-KBS unit (Koobi Fora
Formation, East Turkana) near 2.1 m.y., for earliest evidence

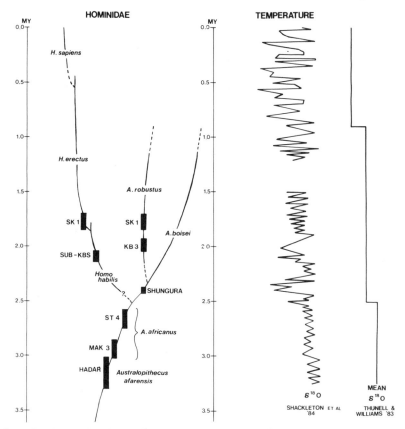

Fig. 1. Hominid tree (discussed in text) and two climatic curves. The curve adapted from Shackleton et al. (1985, Figs 3,4) refers to deep sea oxygen isotope data from a North Atlantic site. The curve adapted from Thunell and Williams (1983, Fig. 5) is a schematic summary of mean oxygen isotope values from sites in the Pacific, Caribbean, Mediterranean, off Northwest Africa.

of robust australopithecines from Shungura E close to 2.4 m.y. and for *Australopithecus* from Hadar from 2.9-3.3 m.y. (see solid vertical bars in Fig. 1; Howell 1978 for discussion of earliest fossils in the *Homo* and robust lineages; Brown et al. 1985 for East African dates). One view of hominid phylogeny (White et al. 1981) sees the Hadar hominids as belonging to the directly ancestral species of the crucial splitting event with *A. africanus* already diverged from the ancestry of *Homo*.

This implies that the *Homo* branch should emerge between Hadar and MAK 3, probably very close to 2.8 or 2.9 m.y. Other anthropologists, notably Tobias (1985), place *A. africanus* as the direct ancestor, which suggests a divergence of *Homo* and *robust* branches close to 2.5 m.y. (as in Fig. 1). If the Taung hominid is younger than previously supposed, as has been argued by some of us, then a persistent *A. africanus* stem (not shown in Fig. 1) may continue independently from the furcation point that gave rise to *Homo* and the robusts.

PALEOECOLOGY

Census data of African bovids show that today higher percentages of alcelaphines and antilopines (of all antelopes present in an area) are strongly associated with lower bush cover and with relatively more arid, open grassland, and vice versa (Greenacre, Vrba 1984). I argue that these habitat associations of bovid tribes have existed at least since the early Pliocene (Vrba 1985). The Transvaal evidence (Fig. 2) suggests major environmental change sometime between 2.5 and 2.0 m.y. ago, from the wetter and more bush-covered surroundings of MAK 3 and ST 4 *A. africanus* to more arid and open ones associated with early *Homo* and with the SK 1 robust apeman. Similar environmental and hominid changes occurred in East Africa. The Shungura sequence records significant turn-

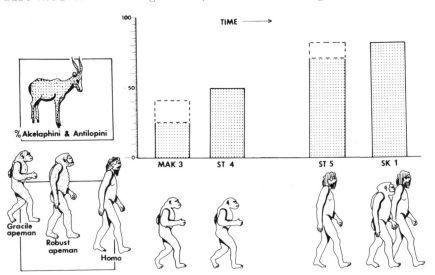

Fig. 2. Hominid and bovid tribal representation in four Transvaal limestone cave strata. See text for details.

over of micromammals (Wesselman 1985), pollen assemblages
(Bonnefille 1976) and bovids (Vrba 1985a), all indicating the
spread of arid open grassland, between the top of Member C
and Member F, i.e. from previous to 2.5 until about 2.4 m.y.
ago. During the same interval stone tools and robust australopithecines are first recorded (Howell 1978). It is likely
that the Transvaal changes reflect the same evolutionary pulse
at 2.4 m.y. On their own the paleontological data suggest that
wide-spread environmental changes, probably caused by global
cooling and associated alterations in rainfall, *caused* the
evolutionary changes (Vrba 1975, 1985). Fig. 1 compares one
alternative hominid phylogeny with recent evidence of the
first wide-spread North polar glaciation (Shackleton et al.
1984) and of a global 'mean cooling step' (Thunell, Williams
1983) near 2.4 m.y. Thus, independent analyses support the
proposal that global climatic change was the crucial driving
force that led to the origin of *Homo*. If the Taung child dates
to 2.3-2.0 m.y. (Delson pers.comm.) or later, then it may represent a rare population of *A. africanus* that survived unchanged through the 2.4 m.y. 'crunch'. Perhaps this latest
and southernmost record of *A. africanus* reflects a persistent,
wet and bush-covered refugium in the midst of Late Pliocene
changes.

I suggest that we may discover good evidence at other times
of coincident climatic, biotic and hominid evolutionary
events. A wave of bovid speciations and extinctions may have
coincided with the origin of *Homo sapiens* (Vrba 1985, Fig.2).
The global mean cooling step near 0.9 m.y. could be synchro-
nous with the extinction of robust australopithecines (Fig.1).
And late Miocene cooling may have coincided with the origin
not only of Hominidae (Brain 1981), but also of several en-
demic tribes of typical savanna antelopes (Vrba 1985, Fig.2).
Thus the Family of Man was probably a 'founder member' of the
African savanna biota. Perhaps hominids evolved then and since
in concert with other lineages, in response to common environmental causes.

Bonnefille R (1976). Palynological evidence for an important
    change in the vegetation of the Omo Basin between 2.5 and
    2 million years. In Coppens Y, Howell FC, Isaac GL, Leakey
    REF (eds): "Earliest Man and Environments in the Lake Rudolf
    Basin", Chicago: Univ. of Chicago Press, p 421
Brain CK (1981). The evolution of man in Africa: was it a con-
    sequence of Cainozoic cooling? Annex Transv Geol Soc S Afr
    84:1

Brown FH, McDougall I, Davies T, Maier R (1985). An integrated Plio-Pleistocene chronology for the Turkana Basin. In Delson E (ed): "Ancestors: The Hard Evidence", New York: Alan R Liss, in press.

Clarke RJ (1985). The ancient artefacts of Sterkfontein. In Tobias PV (ed): "Past, Present and Future of Hominid Evolution", New York: Alan R Liss, in press.

Greenacre MJ, Vrba ES (1984). A correspondence analysis of biological census data. Ecology 65:984.

Howell FC (1978). Hominidae. In Maglio VJ, Cooke HBS (eds): "Evolution of African Mammals", Cambridge, Massachusetts: Harvard University Press, p 154.

Shackleton NJ, Backman J, Zimmerman H, Kent DV, Hall, M, Roberts DG, Schnitker D, Baldauf JG, Desprairies A, Homrighausen R, Huddlestun P, Keene JB, Kaltenbach AJ, Krumsieck KAO, Morton AC, Murray JW, Westberg-Smith J (1984). Oxygen isotope calibration of the onset of icerafting and history of glaciation in the North Atlantic region. Nature 307:620.

Tobias PV (1985). Punctuation and phyletic evolution in the hominids. In Vrba ES (ed): "Species and Speciation", Pretoria: Transvaal Museum Monographs, in press.

Thunell RC, Williams DF (1983). The stepwise development of Pliocene-Pleistocene paleoclimate and paleoceanographic conditions in the Mediterranean: oxygen isotope studies of DSDP Sites 125, 132. In Meulenkamp J (ed): "Reconstruction of Marine Paleoenvironments", Utrecht Micropal Bull 30:111.

Vrba ES (1975). Some evidence of chronology and palaeoecology of Sterkfontein, Swartkrans and Kromdraai from the fossil Bovidae. Nature 254:301.

Vrba ES (1982). Biostratigraphy and chronology of southern African hominid-associated assemblages. In De Lumley H, de Lumley MA (eds): "Congres International de Paleontologie Humaine", Nice: UISPP, p 707.

Vrba ES (1985). Ecological and adaptive changes associated with early hominid evolution. In Delson E (ed): "Ancestors: The Hard Evidence", New York: Alan R Liss, in press.

Vrba ES (1985a). African Bovidae: evolutionary events since the Miocene. S Afr J Sci, in press.

Wesselman HB (1985). Fossil micromammals as indicators of climatic change about 2.4 million years ago in the Omo Valley, Ethiopia. S Afr J Sci, in press.

White TD, Johanson DC, Kimbel WH (1981). *Australopithecus africanus*: its phyletic position reconsidered. S Afr J Sci 77:445.

Australopithecus *in the Forest of Hominid Trees*

Hominid Evolution: Past, Present and Future, pages 203–212
© *1985 Alan R. Liss, Inc.*

THE MOST PRIMITIVE AUSTRALOPITHECUS

Donald C. Johanson Ph.D.
The Institute Of Human Origins
2453 Ridge Road
Berkeley, California 94709

INTRODUCTION

It is now more than a decade since Pliocene hominid fossils were first recovered in the Hadar Formation, Ethiopia. Since their initial discovery, the publication of these remains as <u>Australopithecus afarensis</u> (Johanson, White, Coppens 1978), has stirred controversy in paleoanthropology. Controversy is not rare in paleoanthropology. Indeed, this appears to be a normal state, caused by the fragmentary nature of the fossil evidence, the common lack of adequate geochronological control, and the likelihood that the fossils do not adequately sample the true range of variation of the paleospecies.

Disagreement concerning hominid paleontology has, however, stimulated scholars to reexamine their own views and to assess more thoroughly the strengths and shortcomings of their propositions. Such reassessment has certainly been the case in aspects of our work and it is our hope that continued scholarly exchange will further enhance understanding of the relevance of <u>A. afarensis</u> to hominid evolution.

At the present time we (Johanson and White, 1979; White <u>et al</u>., 1981) consider the fossil hominid sample currently known from the sites of Hadar and Laetoli to constitute a single species. This taxon is characterized by a suite of primitive dental, cranial and mandibular feat-

ures which make it distinct from other species of the genus
Australopithecus. This species, which was committed to
bipedal terrestrial locomotion, was "ecologically diverse,
geographically widespread and temporally enduring" (White
et al. 1981: 448). These conclusions are based on a
thorough study of the anatomy of the fossils as well as an
extensive comparative investigation of all Plio/Pleistocene
hominids found up to 1981. In our analyses we have heeded
the words of Tobias, "We must strive for a holistic
approach. While detailed studies on single characters,
single character-complexes and single regions are of
immense value, the broad systematic and evolutionary
inferences should be based on the totality of available
characters, complexes and regions" (in press). In such an
analysis, however, it is imperative to distinguish between
derived and primitive traits. A simple listing of traits
is not sufficient since it is the derived features that
distinguish the more evolved forms such as A. africanus and
Homo habilis from the less derived species A. afarensis.

In this paper it is not my intent to address all of the
critical commentary that our concept of A. afarensis has
received, for many of these have already been dealt with in
detail elsewhere (White et al. 1981; Kimbel 1984;
Kimbel et al. 1984, and in press; Lovejoy 1981;
Johanson, Taieb, Coppens, 1982; Rak 1983; Kimbel,
Rak in press; White in press). Rather, I intend to
review some of the most significant anatomical evidence for
the distinctiveness of A. afarensis from A. africanus; a
contention which has been criticized by a number of authors
(Boas 1979; Kennedy 1980; Tobias 1978, 1980). Emphasis
will be given to the primitive suite of traits which are
seen in the dental, mandibular, and cranial anatomy of A.
afarensis. As some of these features are shared with
Miocene fossils currently referred to as
Sivapithecus/Ramapithecus, as well as with extant great
apes, they characterize A. afarensis as a highly
generalized, basal hominid species. The species serves as
a reasonable ancestor to later hominids and in many
respects a good intermediate form between the Miocene
fossils and the more recent representatives of
Australopithecus and Homo.

DENTAL FEATURES

Although our earlier papers (Johanson 1980; Johanson and White 1979; White et al. 1981) listed many additional traits of the dentition, here I focus on those traits which have proven to be most diagnostic in differentiating A. afarensis from A. africanus.

1) A. afarensis exhibits a high frequency of mandibular C/P3 diastemata (9 out of 20); A. africanus exhibits only one out of 12 cases.

2) Upper and lower canines often exhibit pongid-like wear in A. afarensis, whereas canine wear is strictly apical in A. africanus.

3) Upper and lower anterior teeth in A. afarensis, compared with those in A. africanus, are relatively larger when compared to the posterior dental battery.

4) The lower third premolar in A. afarensis is usually an asymmetric oval and is unicuspid in males and females (6 out of 15 lack metaconids). In A. africanus the premolar is rounder in shape and biscupid.

5) In more advanced stages of cheek tooth wear the lower canines and third premolars of A. afarensis project above the tooth row, while in A. africanus all teeth are worn flat.

MANDIBULAR FEATURES

The following mandibular features are of significance in distinguishing A. afarensis from A. africanus (White et al. 1981):

1) The anterior corpus is receding and bulbous in A. afarensis; in A.africanus it is usually straighter and more vertical.

2) In A. afarensis the mental foramen is placed low and opens anterosuperiorly; in A. africanus it is at mid-corpus height and opens laterally or slightly anteriorly (STS 36).

3) In A. afarensis the lateral mandibular surface always bears a shallow depression, posterosuperior to the mental foramen; in A. africanus this region is usually swollen. There is some overlap (minor depressions in STS

36 and MLD 18), but no A. afarensis specimen exhibits the degree of mandibular swelling seen in A. africanus and no specimen of A. africanus shows the degree of hollowing seen in A. afarensis.

4) In A. afarensis there is a high frequency of straight or slightly laterally concave mandibular tooth rows; in A. africanus they are laterally convex.

5) In A. afarensis the superior transverse torus is weak to moderate; it is stronger in A. africanus.

CRANIAL FEATURES

It is unfortunate that A. afarensis is not represented by a complete cranium, but sufficient remains are known from Hadar which permit insight into important, diagnostic cranial anatomy. A composite reconstruction of an adult male skull (Kimbel et al. 1984) has provided valuable clues to A. afarensis cranial anatomy. A recently announced frontal fragment (Clark et al. 1983), dating to about 4.0 m.y., from Belohdelie, Ethiopia, only slightly alters the earlier reconstruction.

A. afarensis crania are distinct from A. africanus in the following features:

1) The "transverse buttress" on the face of A. afarensis is absent in A. africanus (Rak 1983).

2) The "asterionic notch" of A. afarensis, previously known only in apes, is absent in A. africanus (Kimbel and Rak 1985).

3) In A. afarensis the extensive pneumatization of the temporal squama is similar to the ape condition. In A. africanus temporal pneumatization is more limited to the mastoid region (Kimbel et al. 1984).

4) In A. afarensis males the canine fossa is so deep it gives a pinched appearance to the face; in A. africanus males it is usually reduced to a narrow groove or is absent (e.g. STS 71, TM 1511, but not TM 1514) (White et al. 1981; Rak 1983).

5) In A. afarensis the snout is characterized by a

semi-oval outline due in large part to the convex naso-alveolar clivus. In A. africanus the clivus is flatter and straighter (White et al. 1981; Rak 1983; Kimbel et al. 1984).

6) Male (A.L. 333-45, -84) and females (A.L. 162-28) A. afarensis crania possess compound temporal/nuchal crests similar to those seen in chimpanzees (with or without sagittal crests). This crest is absent in A.africanus (Kimbel et al., 1984).

7) A. afarensis has a flat, very shallow palate, while in A. africanus it is deeper with premaxillary shelving (White et al. 1981; Rak 1983).

8) The mastoid process is strongly inflected inferomedially in A. afarensis compared with A. africanus (Olson 1981; Kimbel et al. 1984).

9) In A. afarensis the mandibular fossa is shallow owing to a weak articular eminence. In A. africanus the eminence is moderate or strong and hence the fossa is deep (White et al. 1981; Kimbel 1985).

10) In basal view the tympanic in A. afarensis is tubular and horizontal. In A. africanus it is vertical with a strongly curved anterior face (White et al. 1981; Kimbel 1985).

11) A. afarensis has a high occipital scale ratio (nuchal plane dominance) in males and some females; a condition similar to apes. It is lower in A. africanus and only in males is it high (Kimbel et al. 1984).

12) A. afarensis has the steepest nuchal plane of any hominid, whereas in A. africanus it is more horizontal (Kimbel et al. 1984).

CONCLUSIONS

The list of 22 primitive morphological traits considered above forcefully argue for the specific distinctiveness of the fossils from Hadar and Laetoli within the genus Australopithecus. Following Le Gros Clark (1964), it is precisely this recognition of a total morphological pattern which substantiates the taxonomic position of the fossil samples considered in this presentation. To be sure, individually the traits oulined above do not always provide absolute, clear-cut criteria to differentiate between A. afarensis and other species of this genus, but it is the constellation of features which is most revealing.

In fact we recently wrote that the mere listing of morphological traits has its shortcomings. We stressed that it is important to remember:

First, such atomization of morphology is artificial. Most craniodental traits do not exist in isolation, but are constituent parts of larger morphological units, functional complexes. Thus, across the taxonomic spectrum of hominids...the polarity of most traits shifts congruently. Secondly, most characters of the skull and dentition are not invariant within a given hominid species. It is at the very best insufficient to reduce morphology to trait lists that are held to characterize a taxon in its supposed 'typical' state (Kimbel et al. 1984: 373-374).

Hence, in any analysis which entails individual trait comparisons, caution must be excercised so that the broader morphological (taxonomic) picture can be properly evaluated.

In our phylogenetic assessment of the placement of A. afarensis,we have proposed that a plausible interpretation places this species as an ancestor to all later hominids. One branch is viewed as ancestral to A. africanus which led to A. robustus/boisei and another as leading to Homo. On the basis of shared derived characters, we do not consider A. africanus as an appropriate ancestor to Homo, but view it as a precursor solely to the robust australopithecines (Johanson and White 1979; Rak 1983; White et al. 1981; Kimbel et al. 1984). Many of these characters are associated with a specialized masticatory apparatus. Considering A. africanus as an ancestor of Homo as well as of the later "robusts" requires a loss of these characters and hence an evolutionary reversal. We wrote, "It is possible that this was the case. It seems to us, however, more parsimonious to use the characters phyletically to link A. africanus exclusively with A. robustus + A. boisei" (White et al. 1981: 467).

Wolpoff (1983), McHenry (1984 and in press) and Skelton et al. (1984) believe that there is a sufficiently long list of derived features in Homo which are shared with A. africanus, A. robustus and boisei, to claim A. africanus as the last common ancestor to later "robusts" and Homo. If this was not the case they argue that, the shared traits would "have evolved independently in the transition between A. afarensis to H. habilis and from A. afarensis to A. africanus and A. robustus + boisei" (McHenry 1984: 302). They believe that because of the number of traits involved in such independent evolution such a scenario is unlikely.

We believe that some of the traits shared by an alleged-A. africanus-"robust" Australopithecus sister group arose in parallel. That is to say that while these characters are shared, they are not necessarily the result of common evolutionary history (see Kimbel et al. 1984 for a more detailed discussion).

Of all of the competing phylogenetic hypotheses that have been advocated about A. afarensis, we consider McHenry's and Wolpoff's to be the most plausible. In fact we (White et al. 1981) pointed out in a publication prior to those of McHenry and Wolpoff that an alternative hypothesis to the one we favored was one which sees A. africanus as the common ancestor. We believe, however, that there are very good morphological reasons which lead us to consider the probability of their hypothesis lower than the one which we have proposed. Our hypothesis reflects what we believe to have been the selective pressures which operated on hominids to produce evolutionary change in directions that we know occurred. Given the nature of the evolutionary trajectories in the two later known hominid lineages, if we assess A.africanus interms of these two trajectories, it fits best on the one that culminated in an extreme specialization in the masticatory apparatus.

The test of these hypotheses resides in the gap between 2.0 and 3.0 m.y. in East Africa.
Our hypothesis (Johanson and White 1979; White et al. 1981; Kimbel et al. 1984) predicts that when the fossil ancestor intermediate between A. afarensis and Homo habilis, is discovered it will not show the morphology of A. africanus. Let us hope that future fossil hunting in East Africa will locate this ancestor to assist us in illuminating further these alternative hypotheses.

ACKNOWLEDGMENTS

It gives me great pleasure to have participated in this salute to Professor Raymond Dart. Professor Dart has served as an inspiration to all students in the field of paleoanthropology. His fortitude and dedication to providing a better and fuller uderstanding of human origins have profoundly enriched the knowledge of our past.

I would especially like to express appreciation to Professor Phillip V. Tobias for inviting me to present this paper at the historic "Taung Diamond Jubilee International Symposium". It proved to be a superb forum in which to assess our current understanding of mankind's origins in the precise setting where it all began with the Taung discovery just over 60 years ago.

For their assistance and encouragement during the preparation of this paper I would like to thank William H. Kimbel, Gerald G. Eck and Tim White.

This paper is decicated to the memory of my mentor, Professor Paul Leser, who passed away shortly before Christmas.

REFERENCES

Boaz NT (1979). Hominid evolution in eastern Africa during the Pliocene and early Pleistocene. Ann Rev Anthropol 8: 71.
Clark WE Le Gros (1964). "The Fossil Evidence for Human Evolution." Chicago: University of Chicago Press.
Clark JD, Asfew B, Assefa G, Harris JWK, Kurashina H, Walter RC, White TD, Williams MAJ (1983). Paleoanthropological discoveries in the Middle Awash Valley, Ethiopia. Nature 307: 423.
Johanson DC (1980). Early African hominid phylogenesis: a re-evaluation. In Konigsson L-K (ed): "Current Argument on Early Man," Oxford: Pergamon Press, p 31.
Johanson DC, Taieb M, Coppens Y (1982). Pliocene hominids from the Hadar Formation, Ethiopia (1973-1977): Stratigraphic, chronologic, and paleoenvironmental contexts, with notes on hominid morphology and systematics. Am J Phys Anthropol 57: 373.
Johanson DC, White TD (1979). A systematic assessment of early African hominids. Science 203: 321.

Johanson DC, White TD, Coppens Y (1978). A new species of the genus Australopithecus (Primates: Hominidae) from the Pliocene of eastern Africa. Kirtlandia 28: 1.

Kennedy GE (1980). "Paleoanthropology." New York: McGraw Hill.

Kimbel WH (1984). Variation in the pattern of cranial venous sinuses and hominid phylogeny. Am J Phys Anthropol 63: 243.

Kimbel WH (1985). Ph.D. Dissertation, Kent State University.

Kimbel WH, Rak Y (in press). Functional morphology of the asterionic region in extant hominoids and fossil hominids. Am J Phys Anthropol.

Kimbel WH, White TD, Johanson DC (1984). Cranial morphology of Australopithecus afarensis: a comparative study based on a composite reconstruction of the adult skull. Am J Phys Anthropol 64: 337.

Kimbel WH, White TD, Johanson DC (in press). Craniodental morphology of the hominids from Hadar to Laetoli: Evidence of Paranthropus and Homo in the mid-Pliocene of eastern Africa? In Delson E, Tattersall I (eds): "Paleoanthropology: The Hard Evidence," New York: Alan R. Liss.

Lovejoy CO (1981). The origin of man. Science 211: 341.

McHenry HM (1984). Relative cheek-tooth size in Australopithecus. Am J Phys Anthropol 64: 297.

McHenry HM (in press). Implications of postcanine megadontia to the origin of Homo. In Delson E, Tattersall I (eds): "Paleoanthropology: The Hard Evidence," New York: Alan R. Liss.

Olson TR (1981). Basicranial morphology of the extant hominoids and Pliocene hominids: The new material from the Hadar Formation, Ethiopia and its significance in early human evolution and taxonomy. In Stringer CB (ed): "Aspects of Human Evolution," London: Taylor and Francis, p. 99.

Rak Y (1983). "The Australopithecine Face." New York: Academic Press.

Skelton RR, McHenry HM, Drawhorn GM (1984). Phylogenetic analysis of Plio-Pleistocene hominids. Am J Phys Anthropol 63: 219.

Tobias PV (1978). Position et rôle des australopithécinés dans la phylogenèse humaine, avec étude particulière de Homo habilis et des théories controversées avancées a propos des premiers hominidés fossiles de Hadar et de Laetolil. In: "Les Orgines Humaines et les Époques de l' Intelligence," Paris: Masson for Fondation Singer-Polignac, p. 38.

Tobias PV (1979). The Silberberg Grotto, Sterkfontein, Transvaal, and its importance in palaeo-anthropological researches. S Afr J Sci 75: 161.

Tobias PV (1980). "Australopithecus afarensis" and A. africanus: Critique and an alternative hypothesis. Palaeont Afr 23: 1.

Tobias PV (in press). Single characters and the total morphological pattern re-defined. The sorting effected by a selection of morphological features of the early hominids. In Delson E, Tattersall I (eds): "Paleoanthropology: The Hard Evidence," New York: Alan R. Liss.

White TD (in press). The hominids of Hadar and Laetoli: an element-by-element comparison of the dental sample. In Delson E, Tattersall I (eds): "Paleoanthropology: The Hard Evidence," New York: Alan R. Liss.

White TD, Johanson DC, Kimbel WH (1981). Australopithecus afarensis: its phyletic position reconsidered. S Afr J Sci 77: 445.

Wolpoff MH (1983). Australopithecines: The unwanted ancestors. In Reichs K (ed): "Hominid Origins," Washington, D.C: University Press, p. 109.

Hominid Evolution: Past, Present and Future, pages 213–220
© *1985 Alan R. Liss, Inc.*

AUSTRALOPITHECUS AFARENSIS: TWO SEXES OR TWO SPECIES?

Adrienne L. Zihlman, Ph.D.

Professor of Anthropology
University of California
Santa Cruz, California   95064

The early hominid fossils from Hadar, Ethiopia and
Laetoli, Tanzania, supposedly represent a new species
"Australopithecus afarensis".   According to Johanson and
White, "A. afarensis" is the oldest and most primitive
hominid species, and is ancestral to all other hominids.
The marked size differences observed in several parts of
the postcranial skeleton (e.g., distal femora, AL 333-4 and
AL 129-1a) reflect sexual dimorphism (Johanson,   White
1979; Johanson 1980).   It is questionable which, if any, of
these statements are valid.

The issue of primitive features of "A. afarensis" is
not straightforward: The dentition may show primitive
features, as argued by Johanson (this volume); however, the
postcranial fossils from the Hadar Formation can be matched
in the South African material.   For example, the AL 288
pelvis (Ethiopia) shows marked morphological similarity to
the Sterkfortein 14 pelvis and is nearly identical in
almost all measurements (Suzman 1982; Schmid 1983; Berge
1984).   Furthermore, Schmid (1983) argues that the AL 288-1
cranium does not differ from skulls recognized as
Australopithecus africanus.

Dating is obviously central in sorting out the phyletic
position of "A. afarensis".   The age of the fossil material
has been controversial, but most researchers place the
hominid fossils from the Hadar Formation between 2.8 and 3.3
m.y.BP (Brown 1982; Curtis 1981; Kalb et al. 1984).   These
relatively young dates may overlap the time range of
hominids in South Africa; Makapansgat may be as old as

3.1 my and Sterkfontein may date between 2.5 and 3.0 my (Tobias 1978). This overlap in the time range of supposed ancestral and descendant lines raises the question of how "A. afarensis" could be a stem ancestor to all other hominids.

Sexual dimorphism in the hominid fossil record has been a topic of interest for two reasons. First, it contributes to observed morphological variability and second, it has implications for behavioral correlates. The idea that the varieties of australopithecine fossils can be encompassed within one highly variable and dimorphic species is not new. For example, Brace (1973) and Wolpoff (1971) suggested that the robust and gracile South African australopithecines could be included within a single, highly dimorphic species. Their discussion centered mainly on cranial and dental anatomy rather than on the postcranial skeleton and the viewpoint did not receive wide support. Particularly convincing were the arguments that sexual dimorphism could not explain the variety of dental measurements (Pilbeam,   Zwell 1973; Greene 1973).

Now, sexual dimorphism has again been suggested to account for the observed variation within the postcranial skeleton of the Hadar hominids. The interpretation of extreme sexual dimorphism for these fossils has been a mere assertion from the beginning (e.g., Johanson et al. 1978) and has continued to be so (Johanson, White 1979; Johanson 1980). No measurements and little comparative data for the postcranial fossils have been given which would enable other researchers to assess the degree of dimorphism. There are at least two key questions: What is the size range of the Hadar limb bones? And, does the variability exist in size only, or is variability also observed in morphology of the the postcranial and cranial skeleton?

One problem in using sex differences to explain variation lies in the simplistic view of sexual dimorphism, which is usually presented as a shorthand way for referring to male-female differences in body size and weight. However, sexual dimorphism is not a unidimensional feature. It is most obviously expressed in body weight and canine tooth size, which do correlate with one another among anthropoids (Leutenegger, Kelly 1977). However, sexual dimorphism is also expressed in cranial capacity, pelvic measurements, and limb bone dimensions, such as long bone lengths and joint size, and these parameters do not always correlate with body weight and canine size (Zihlman 1976).

Thus among primates, sexual dimorphism is a mosaic of features, each of which is more or less dimorphic. The overall pattern appears to be unique for each species, as it certainly is among the African hominoids--gorillas, pygmy chimps, common chimps, and modern and extinct hominids (Zihlman 1976, 1983).

A comparison of pygmy (Pan paniscus) and common chimpanzees (Pan troglodytes) illustrates this mosaic pattern of dimorphism. In both species female body weight is 80-84 percent of male body weight. In other features the two species exhibit different patterns of dimorphism--cranial capacity (none in pygmy chimps; moderate in common chimps); canine tooth size (more marked in common chimps); limb bone dimensions and joint sizes (no differences in female and male pygmy chimps (Zihlman 1976; Cramer, Zihlman 1978). Thus, the overall pattern for each species is unique.

Given the variety of patterns among extant hominoids, it is probable that early hominids exhibited yet another pattern of dimorphism, as I suggested in 1976 prior to the discovery of most of the Hadar hominid material. The early hominid fossil record is limited in the information it yields on sexual dimorphism. Certainly within each of the South African robust and gracile groups, there is little evidence for much variation in cranial capacity, and there are only minimal canine tooth size differences. Indications of body weight are better reflected in joint sizes than in dentition and these parts of the anatomy are much less frequently preserved. However, a range in size of limb bones exists and suggests that at least for body weight there might have been moderate sex differences within each of the two South African species.

The Hadar hominid fossil sample consists of limb bones which exhibit a size range that is greater than that found in the two australopithecine species from South Africa. But this sample has been difficult to assess from the point of view of sexual dimorphism in the postcranial skeleton, because few measurements have been published which can be compared with hominoid species whose sexual dimorphism is known.

Table I shows relative dimorphism in three joints (humerus, femur, and tibia) for modern humans, gorillas, and orangutans. The latter two species are selected because they exhibit extreme sexual dimorphism in body size and weight. Comparable measurements for the Hadar fossils from different localities are then given for comparison. The largest and smallest examples of the sample are chosen, and the assumption is made that these fossils represent individuals around the mean rather than population extremes.

	Humerus	Femur	Tibia
Homo sapiens n=21	89%	90%	96%
Gorilla n=12	77%	81%	78%
Pongo n=10	77%	79%	77%
Hadar Fossils smallest/ largest	*69% or **74% (AL 288/ AL 333.107)	72.6% (AL 288/ AL 333.3)	77% 78% (AL 129 AL 288/ AL 333.26

Table I. Percentage of female/male means for living hominoids in humeral and femoral head diameters, and tibia breadth. For the fossils, measurements are taken from Senut (1981), Tardieu (1983), Johanson et al. (1982), Lovejoy et al. (1982a, 1982b).
*measurement from Lovejoy et al. (1982a).
**measurement from Senut (1981).

The Hadar fossils suggest even greater dimorphism than exists in orangutans, a species where males may be more than three times the body weight of females. This means that "A. afarensis" is more sexually dimorphic than any living hominoid. Thus, from the point of view of size, more than one species is strongly implied. Size alone cannot be conclusive, however, because size variation does exist in specimens from the same locality (e.g., AL 333). Furthermore, until the species question is settled, it is not possible to assess the degree of sexual dimorphism among the Hadar fossils.

The Hadar postcranial fossils support the proposal that the extreme size variation is better interpreted as more than one species. This possibility takes on added significance in light of accumulating evidence that demonstrates two morphological patterns in postcranial and cranial anatomy. Features which show two different patterns cannot easily be accommodated within one species. Furthermore, in some instances ancestral relationship between the Hadar hominids and all others is directly contradicted.

The upper limb material shows that, in addition to size variation between, as well as within, localities, two morphological patterns exist (Senut 1981; Senut, Tardieu 1985). Similarly, for the lower limb, there are two morphological patterns in the distal femur (Tardieu 1983; Senut, Tardieu 1985).

Other researchers have commented on other aspects of the postcranial fossils. Comparing the footprints from Laetoli, which are included in the "A. afarensis" designation, with the footbones from Hadar, Tuttle (1981; this volume) maintains that the footprints could not have been made by the Hadar footbones. Studying all the locomotor skeletal material, still other researchers who treat the material as one species conclude that the degree of the supposed sexual difference in locomotor behavior was greater than that found in any living ape (Stern, Susman 1983).

The basicranium and nasal regions may also exhibit two patterns; specimens from one site have one pattern and specimens from other sites show another (Olson 1981; 1985). Olson argues that these features do not change during ontogeny and are useful for taxonomic purposes. Therefore, the specimens lumped as a single species from Hadar in fact include two phyletic lines. Similarly, the venous drainage pattern in all available specimens from Hadar conform to the pattern recognized for the robust australopithecines from East and South Africa (Falk, Conroy 1983).

One final issue must be considered here. Frequently, it is argued that the prehominid ancestor was highly dimorphic, a pattern that continued in "A. afarensis." Recent evidence demonstrates that chimpanzees and modern humans are most closely related, with gorillas as the "outgroup" (Hasegawa et al. 1984; Sibley, Ahlquist 1984). These findings strongly suggest that the common ancestor, and therefore the hominid precursor, were very likely NOT highly dimorphic. Sexual dimorphism can, however, evolve quickly. The Hadar fossil groups may turn out to show a high degree of dimorphism, but that conclusion cannot be assumed or asserted at this time.

There is only one evolutionary story so that all lines of evidence must point in the same direction. In demonstrating the presence of more than one species among the Hadar Formation hominids, no line of evidence alone is compelling. That evidence--basicranium, venous drainage pattern, limb bone size and morphology--strongly questions the existence of a single species at Hadar as well as its validity as a new species.

Sherwood Washburn has pointed out that evolutionary studies are like a game where you play the odds. On the basis of these lines of evidence, I suggest that the odds are that the proposed "A. afarensis" consists of more than one species, and that the split between the robust and gracile lines is more ancient than now recognized at 2.0-2.5 m.y. BP.

For comments and assistance I thank B. Senut, C. Tardieu and C. Borchert. For financial support, I thank the Faculty Research Committee, University of California, Santa Cruz.

Berge TC (1984). Multivariate analysis of the pelvis for
hominids and other extant primates. J Hum Evol 13:555.

Brace LC (1973). Sexual dimorphism in human evolution. Yrb
Phys Anthropol 16:31.

Brown FH (1982). Tulu bor Tuff at Koobi Fora correlated with
the Sidi Hakoma Tuff at Hadar. Nature 300:631.

Cramer DL, Zihlman AL (1978). Sexual dimorphism in the pgymy
chimpanzee, Pan paniscus. In Chivers DJ, Joysey KA (eds)
"Recent Advances in Primatology", Vol. 3, London: Academic
Press, p.487

Curtis G (1981). Establishing a relevant time scale in
anthropological and archaeological research. Philos Trans R
Soc Lond B 292:7.

Falk D, Conroy G (1983). The cranial venous sinus system in
Australopithecus afarensis. Science 306:779.

Greene DL (1973). Gorilla dental sexual dimorphism and early
hominid taxonomy. In Zingeser M (ed) "Cranial Facial
Biology of Primates", Basel: Karger, p. 82.

Hasegawa M, Yano T, Kishino H (1984). A new molecular clock
of mitochondrial DNA and the evolution of hominoids. Proc
Japan Acad 60:95.

Johanson DC (1980). Early African hominid phylogenesis: a
re-evaluation. In Konigsson L-K (ed): "Current Argument on
Early Man", Oxford: Pergamon Press, p. 31.

Johanson DC et al. (1982). Morphology of the Pliocene
partial hominid skeleton (AL 288-1) from the Hadar
Formation, Ethiopia. Amer J Phys Anthropol 57:403.

Johanson DC, White TD, Coppens Y (1978). A new species of
the genus Australopithecus (Primates: Hominidae) from the
Pliocene of Eastern Africa. Kirtlandia 28:1.

Johanson DC, White T (1979). A systematic assessment of
early African hominids. Science 202:321.

Kalb JE, Jolly CJ, Oswald EB, Whitehead PF (1984). Early
hominid habitation in Ethiopia. Amer Sci 72:168.

Leutenegger W, Kelly JT (1977). Relationship of sexual
dimorphism in canine size and body size to social,
behavioral and ecological correlates in anthropoid
primates. Primates 18:117.

Lovejoy CO, Johanson DC, Coppens Y (1982a). Hominid upper
limb bones recovered from the Hadar Formation: 1974-1977
Collections Amer J Phys Anthropol 57:637

Lovejoy CO, Johanson DC, Coppens Y (1982b). Hominid lower
limb bones recovered from the Hadar Formation: 1974-1977
Collections Amer J Phys Anthropol 57:679.

Olson T (1981). Basicranial morphology of the extant
hominoids and Pliocene hominids: the new material from

the Hadar Formation, Ethiopia, and its significance in early human evolution and taxonomy. In Stringer CB (ed): "Aspects of Human Evolution", London: Taylor and Francis, p. 99.

Olson T (1985). Cranial morphology and systematics of the Hadar formation hominids and "Australopithecus" africanus. In Delson E (ed) "Ancestors: The Hard Evidence", New York: Alan Liss, in press.

Pilbeam DR, Zwell M (1973). The single species hypothesis, sexual dimorphism and variability in early hominids. Yrb Phys Anthropol 16:69.

Schmid P (1983). Eine rekonstruktion des Skelettes von A.L. 288-1 (Hadar) und deren Konsequenzen. Folia Primat 40(4): 283.

Senut B (1981). "L'humerus et ses articulations chez les Hominides Plio-Pleistocene", Cahiers de Paleontologie Paris: CNRS.

Senut B, Tardieu C (1985). Functional aspects of Plio-Pleistocene hominid limb bones: implications for taxonomy and phylogeny. In Delson E (ed): "Ancestors: the Hard Evidence", New York: Alan Liss, in press.

Sibley CG, Ahlquist JE (1984). The phylogeny of the hominoid primates as indicated by DNA-DNA hybridization. J Mol Evol 20:2.

Stern J, Susman RL (1983). The locomotor anatomy of Australopithecus afarensis. Am J Phys Anthropol 60(3):279.

Suzman, IM (1982). A comparative study of the Hadar and Sterkfontein australopithecine innominates. Am J Phys Anthropol 57(2):235.

Tardieu C (1983). "L'Articulation de genou: Analyse morpho-fonctionnelle chez les primates et les hominides fossiles", Cahiers de Paleoanthropologie, Paris: CNRS.

Tobias PV (1978). The earliest Transvaal members of the genus Homo with another look at some problems of hominid taxonomy and systematics. Z Morph Anthrop 69(3):225.

Tuttle RH (1981). Evolution of hominid bipedalism and prehensile capabilities. Phil Trans R Soc Lond B 292:89.

Wolpoff M (1971). Competitive exclusion among lower Pleistocene hominids: the single species hypothesis. Man 6(4):601.

Zihlman AL (1976). Sexual dimorphism and its behavioral implications in early hominids. In Tobias PV, Coppens Y (ed): "Les Plus Anciens Hominides", Paris: CNRS. p. 268.

Zihlman AL (1983). Sexual dimorphism in Homo erectus. In "L'Homo erectus et la place de l'Homme de Tautavel parmi les hominides fossiles", tome 2:949. Paris: CNRS.

Hominid Evolution: Past, Present and Future, pages 221–226
© *1985 Alan R. Liss, Inc.*

IS AURALOPITHECUS AFRICANUS ANCESTRAL TO HOMO?

Henry M. McHenry and Randall R. Skelton

Department of Anthropology
University of California
Davis, California   95616   U.S.A.

We reconstruct some aspects of the immediate ancestor of Homo habilis by assuming that it would possess those traits which Australopithecus robustus+boisei and H. habilis share. This assumption is based on two widely accepted propositions: (1) that these hominids are   contemporary species that must have shared a common ancestor and (2) unless proven to be cases of homoplasy, traits shared by these species must be present in their common ancestor. We identify 33 cranial and dental characteristics which are similar in H. habilis and A. robustus+boisei, but which at least one other hominid species does not share. Our list derives from several sources especially White et al. (1981), Johanson and White (1979), Kimbel et al. (1984), Ward and Kimbel (1983), Rak (1983), Tobias (1980a, 1980b,), and our own observations. We use the taxonomy presented by White et al. (1981).

Following these guidelines, the immediate ancestor of H. habilis would have a strong and cylindrical articular eminence, a deep mandibular fossa, a postglenoid process which merges inferiorly with the tympanic plate, a nearly vertical tympanic plate, a cone-shaped external auditory meatus, a reduced inflection of the mastoids under the cranial base, strong flexion of the cranial base, weakly developed posterior part of the M. temporalis, the posterior temporal lines would diverge above the lambdoidal suture, a medium to large planum occipitale, an absence of the compound temporonuchal crest in females, no asterionic notch, the nuchal plane would have no transverse concave curvature and would not be steeply inclined, there would be no basin-like canine fossa, reduced subnasal prognathism, the roots of the upper lateral incisors would lie medial to the level of the nasal apertures, non-projecting upper canine, no mesial or distal wear on the upper canine, distal accessory cusps would be common on the upper third molar, a divergent upper tooth row, a deep palate, lower central incisors would have 3 mammelons, the lower canine would have no lingual ridge and its mesial occlusal edge would be horizontal, curved, and relatively short, the lower first premolar would have a round occlusal outline, strong metaconid, and wear equal to the posterior teeth, a lower second molar with equally broad talonid and trigonid, the widest part of the extramolar sulcus would be at M3, a vertical mandibular symphysis, and a low, wide, and blunt deciduous canine.

In all of these traits of the immediate ancestor of Homo, A. africanus is a closer match than is A. afarensis. In nine other characteristics A. africanus is more similar to H. habilis than is either A. afarensis or A. robustus+boisei: length of the incisal edge of the upper central incisor, height of the lower first premolar talonid, occlusal outline of the upper premolars, approximation of the upper second premolar cusp apices, wear disparity between buccal and lingual molar cusps, degree of lower molar cusp swelling, position of the mental foramen, frequency of the mandibular diastema, and in the absence of a transverse buttress from the canine juga to the zygomatic arch.

These characteristics imply that, of the known species of Hominidae, A. africanus appears to be most like the immediate ancestor of the genus Homo. However, there are 19 traits in which A. afarensis resembles H. habilis more closely than any other species of Australopithecus which implies that this species of Australopithecus is the immediate ancestor of Homo. These include moderate robusticity and weak or absent anterior projection of the zygomatic arch, low origin of the masseters, distinct nasocanine and nasoalveolar contours of the anterior facial profile seen in norma lateralis, distinct subnasal and intranasal components of nasoalveolar clivus, moderate to wide separation of vomer and anterior nasal spine, sharp and continuous margins of the piriform aperture, absence of anterior pillars, relatively large upper canine, anterior origin of zygomatic usually above M1 or M1/P4, tapering and

narrow distal crown profile of the deciduous lower first
molar, strong to moderate mesial appression of the lower Ml
and M2 hypoconulids, moderate size of the postcanine teeth,
basal contour of the mandibular corpus frequently everted,
prominent to moderate hollowing above and behind the mental
foramen, ramus of mandible arising high on the corpus,
narrow extramolar sulcus, weakly developed superior
transverse torus of mandibular symphysis, and moderate
robusticity of the mandibular corpus. In all of these traits
A. africanus is most similar to A. robustus+boisei. In
these respects A. africanus appears to be too derived in the
direction of the robust australopithecines to be a very good
candidate as the immediate ancestor of H. habilis (Johanson
and White, 1979; Kimbel, 1984; Kimbel et al., 1984; White et
al., 1981).

How can one account for the conflicting inferences
derived from these lists of traits? If A. africanus is the
immediate ancestor of H. habilis, then the 19 traits that
link it to A. robustus+boisei and not to H. habilis or to A.
afarensis must be due to homoplasy. But to link A.
afarensis with H. habiliis to the exclusion of A. africanus
implies parallel or convergent evolution of 42 traits.

Perhaps homoplasy does explain the distribution of
traits. This is made more probable by the consideration
that the traits are highly intercorrelated: a single
genetic change could give a pleiotropic effect which would
alter a great number of traits simultaneously. The list of
19 or 42 traits does not imply that many genetic changes,
which would have to have occurred independently in two
species. In another study we reduce these and other traits
to 12 morphoclines which group traits whose character
states follow the same pattern of change from species to
species (Skelton et al., in preparation). Grouping traits
by known functional units is another method which avoids the
undesirable effects of atomizing biologically covarying
systems, but is very difficult to do. Another consideration
which makes homoplasy more likely is the fact that the
developmental pathways to the adult morphology in any
group of organisms are probably limited in number. If
selection favors individuals with greater ability to grind
tough food, for example, there are only a limited number of
ways a primate masticatory system can adapt. Hence
Hadropithecus resembles Theropithecus (Jolly, 1970).

Grouping the traits into biologically meaningful functional units can only be done imperfectly. There are certain obvious clusters of characteristics which vary among early hominids. The upper canine-premolar complex incorporates over a dozen traits listed above and thereby a disproportionate number of characters are added to the list which makes A. afarensis distinctly primitive and inappropriate for the station of immediate ancestor to Homo. Likewise, a disproportionate number of traits in this study are related to the size of the cheek-teeth and supporting structures. These traits show the derived nature of A. africanus in the direction of the robust australopithecines. Collapsing of the lists of traits does not, however, substantially affect the answer to the question at hand. There is a fundamental conflict between two alternatives and the resolution rests with the amount of homoplasy which can be accommodated.

Could the 42 traits (or much smaller number of functional complexes) shared by H. habilis and the non-afarensis australopithecines be due to homoplasy? Kimbel et al. (1984) and White et al. (1981) explore the possibility that many of these traits may well have evolved independently in two separate lineages. They note that flexion of the cranial base and reduction in facial prognathism may have occurred independently and would account for such traits as the vertical tympanic plates and the strong articular eminence of the mandibular fossa. The loss of the asterionic notch and the reduction of the pneumatization of the base of the temporal squama in both H. habilis and the non-afarensis australopithecines might be due to parallel evolution. In this case the similarity in structure evolved for different reasons: in A. africanus these changes occurred as part of the increasing emphasis on the anterior fibers of M. temporalis whereas in H. habilis the changes occurred as part of the overall reduction of the masticatory system. Perhaps all of the resemblances between H. habilis and non-afarensis australopithecines are the result of similar parallel evolution, but it is more difficult to understand how, for example, the many dental similarities came to be.

Could the 19 traits shared by H. habilis and A. afarensis and not by other australopithecines be the result of homoplasy? McHenry (1984), Skelton et al. (1984), and Wolpoff (1983) explore aspects of this possibility. Most of

the characteristics that unite A. afarensis and H. habilis
are related to the degree to which the cheek teeth have
specialized for heavy grinding: in all non-afarensis
australopithecines there is marked postcanine megadontia,
heavy buttressing of the face, jaw, and surrounding
structures, a migration forward of the chewing muscles,
and a hypertrophy of these muscles. A. afarensis and H.
habilis have less strongly developed masticatory systems.
If A. africanus is the immediate ancestor of H. habilis,
then the resemblance between A. afarensis and H. habilis is
due to convergent evolution: masticatory hypertrophy
evolved from moderate in A. afarensis to heavy in A.
africanus back down to moderate in H. habilis. Although
this seems like an unparsimonious scenario, it involves only
one functional complex and it does not require parallel
evolution of the 42 traits (or fewer functional complexes,
but probably related to many) shared by the non-afarensis
hominids. In this view the origin of the Homo lineage is
seen as the point of intersection of two trends: the trend
of increasingly heavy use of the cheek teeth from A.
afarensis to A. africanus to A. robustus+boisei which is
greater than expected from the accompanying body size
increase (McHenry, 1984) and trend to smaller teeth seen
through time in the Homo lineage (Wolpoff, 1971).

Not all of the traits which appear to exclude A.
africanus from ancestry of Homo are obviously related to the
hypertrophied posterior masticatory apparatus, however.
Upper canine size, for example, may be independent of cheek-
tooth size, but Wolpoff (1978) argues that relative canine
size is characterized by negative allometry (although see
Wood, 1979, for a contrary view). Actually, neither the
mesiodistal nor the buccolingual diameter of the upper
canine is statistically different in the small sample of A.
afarensis, A. africanus or H. habilis teeth. Considering
the biased sample of upper canine teeth of A. africanus and
the possibility that they were lighter in body weight than
either other hominid, it is doubtful that upper canine size
really is a valid character uniting H. habilis and A.
afarensis.

An alternative is that the immediate ancestor of H.
habilis is as yet unknown. This imaginary hominid would
have those characteristics shared by H. habilis and A.
robustus+boisei listed above and thereby would look most
like A. africanus of the known species of hominids.

However, it would not have those characteristics related to masticatory hypertrophy which A. africanus shares with the robust australopithecines and not with H. habilis.

Johanson,DC, White, TD (1979) A systematic assessment of early African hominids. Science 202: 321.

Jolly, CJ (1970) The seedeaters: A new model of hominid differentiation based on a baboon analogy. Man, 5: 5.

Kimbel, WH (1984) Variation in the pattern of cranial venous sinuses and hominid phylogeny. Am J Phys Anthropol 63: 243.

Kimbel, WH, White, TD, Johanson, DC (1984) Cranial morphology of Australopithecus afarensis: a comparative study based on a composite reconstruction of the adult skull. Am J Phys Anthropol 64: 337.

McHenry, HM (1984) Relative cheek-tooth size in Australopithecus. Am J Phys Anthropol 64: 297.

Rak, Y (1983)"The Australopithecine Face".New York: Academic

Skelton, RR, McHenry, HM, Drawhorn, GM (1984) Phylogenetic analysis of Plio-Pleistocene hominids. Am J Phys Anthropol 63: 219.

Tobias, PV (1980a) "Australopithecus afarensis" and A. africanus: a critique and an alternative hypothesis. Paleontol. Africana 23: 1.

Tobias, PV (1980b) A study and synthesis of the African hominids of the late Tertiary and Early Quaternary Periods. In Konigsson, LK (ed)"Current Arguments on Early Man".Oxford: Pergamon. p. 86.

Ward, SC, Kimbel, WH (1983) Subnasal alveolar morphology and the systematic position of Sivapithecus. Am J Phys Anthropol 61: 157.

White, TD, Johanson, DC, Kimbel, WH (1981) Australopithecus africanus: its phyletic position reconsidered. S Afr J Sci 77: 445.

Wolpoff, MH (1971) Metric trends in hominid dental evolution. Studies in Anthropology, Case Western Reserve University 2:1-244.

Wolpoff, MW (1978) Some aspects of canine size in the australopithecines. J Hum Evol 7: 115.

Wolpoff, MH (1983) Australopithecines: the unwanted ancestors. In K Reichs (ed):"Hominid Origins". Washington, D.C.: University Press, p. 109.

Wood, BA (1979) Models for assessing relative canine size in fossil hominids. J Hum Evol 8: 493.

Hominid Evolution: Past, Present and Future, pages 227–232
© *1985 Alan R. Liss, Inc.*

A REVIEW OF THE DEFINITION, DISTRIBUTION AND RELATIONSHIPS
OF AUSTRALOPITHECUS AFRICANUS

Bernard Wood

Department of Anatomy and Biology as Applied
to Medicine,
The Middlesex Hospital Medical School, London.

INTRODUCTION

The hominid taxon which forms the subject of this paper
was proposed some sixty years ago, but the material which
makes up most of the currently accepted hypodigm of
Australopithecus africanus was discovered more recently and
at sites other than Taung. These remains have been the
subject of two detailed interpretative studies in this
decade (Tobias 1980; White et al 1981). The present
analysis has much in common with the previous studies, but
it differs in several important ways. It begins by briefly
reviewing the characteristics of the cranium and mandible of
A. africanus, and then seeks to define, and establish
diagnostic criteria for that taxon. These features are then
used to examine the evidence for A. africanus at East
African sites, in particular Koobi Fora and Olduvai.
Finally, cladistic methods are used to explore the
relationships between A. africanus and other early hominid
taxa. Whereas most other attempts at cladistic analysis of
this material have used a relatively restricted range of
characters, the present analysis is based on characters
chosen according to a clearly stated protocol.

HYPODIGM AND DEFINITION

The reference hypodigm of A. africanus is taken to be
the hominid remains recovered from Taung (Dart 1925),
Sterkfontein 4 and Makapansgat 3 and 4. A full list of
the characteristics of the A. africanus cranium cannot be

given here, but Wood (1985) can be consulted for a detailed
review of the characteristics of the vault, face and base.
However, a list of features which are consistently present
in the members of the hypodigm is a description of a taxon,
and not a definition of it. Any list of characters of A.
africanus may, and probably does, include some features that
are (1) common to both apes and hominids, (2) common to
all hominids, (3) common to two, or more, hominid taxa or
(4) confined to a single taxon. Features in categories (1)
and (2) are described as symplesiomorphic, or
plesiomorphies, those in category (3) synapomorphic, or
apomorphies, and features in category (4), autapomorphic.
Clearly, autapomorphic features are automatically included
in a definition, but do apomorphic and plesiomorphic
characters have any role to play? The status plesiomorphic
merely implies that the features were present in the common
ancestor; they need not necessarily be present in the
primitive form in all the taxa within the clade. Similar
arguments apply to apomorphic characters. Thus, while
individual apomorphic and plesiomorphic characters cannot
define a taxon, a particular combination of such features
may be unique, and thus diagnostic of the group.

When the characteristics of A. africanus are classified
in the way outlined above it is apparent that the A.
africanus has few, if any, autapomorphies. Therefore, any
definition must rely on combinations of variables belonging
to the two other categories. For example, although a broad
mid-face, a forward-facing and protracted malar region and
the presence of zygomatic tubercles are features which A.
africanus shares with Australopithecus robustus and boisei,
these latter taxa have a derived pattern of cranial base
anatomy compared to the more pongid-like cranial base of A.
africanus. Thus a pongid-like cranial base (an inferred
plesiomorphy) taken together with the cited facial features
(a series of inferred apomorphies) is a combination of
characters which is seen only in A. africanus. A fuller
list of such characters is given in Wood (1985), and an
example of the application of a 'combination' definition to
Homo erectus is given in Wood (1984).

A. AFRICANUS IN EAST AFRICA

Although the affinities between A. africanus and some
of the hominid remains from the Omo has been stressed,

controversy has concentrated on the problem of whether
remains from Olduvai, Koobi Fora, Laetoli and Hadar should
be assigned to A. africanus. The discussions about the
Laetoli and Hadar material are crucial and interesting ones,
but will not be dealt with in detail in this contribution.
Suffice it to say that, in the cladistic analysis which forms
the second part of this paper (and which was based on 39
characters), none of the characters was synapomorphic for A.
africanus and A. afarensis. Thus characters were either
found in one of the two taxa, and not the other, or, if they
were present in both, they were found also in at least one
other taxon.

Two relatively complete specimens from Koobi Fora, the
skull KNM-ER 1805 and the cranium, KNM-ER 1813, have at one
time or another been likened to the 'gracile'
australopithecines, either as Homo africanus (Olson 1978),
or as A. africanus (R.E. Leakey and Walker 1970). Details
of the cranial base (Dean and Wood 1982), mandible (Wood and
Chamberlain 1985a) and dentition (Wood and Abbott 1983;
Wood, Abbott and Graham 1983) of KNM-ER 1805 effectively
exclude it from Australopithecus africanus. Discrete
morphological differences between KNM-ER 1813 and A.
africanus are listed in detail in Wood (1985), and while
these have to be set against the considerable phenetic
similarities in the overall size and shape of the cranial
vault, the balance of evidence is strongly in favour of KNM-
ER 1813 being specifically, if not generically, distinct
from A. africanus. Similar considerations of the vault,
face and base effectively exclude crania like OH 24 from the
A. africanus hypodigm.

CLADISTIC ANALYSIS

The relationship between A. africanus and six other
hominid taxa has been studied using cladistic techniques
(Wood and Chamberlain 1985b). The analysis was confined to
the cranium and mandible, and the 39 characters that were
used were distributed across five anatomical regions; vault
and endocranium, face, palate, upper jaw and maxillary
dentition, cranial base, mandible and mandibular dentition.
The character analysis was based on quantitative data. The
'best fit' cladogram for each region was taken to be the
arrangement which resulted in the most intact
synapomorphies. The 'quality' of the regional cladograms

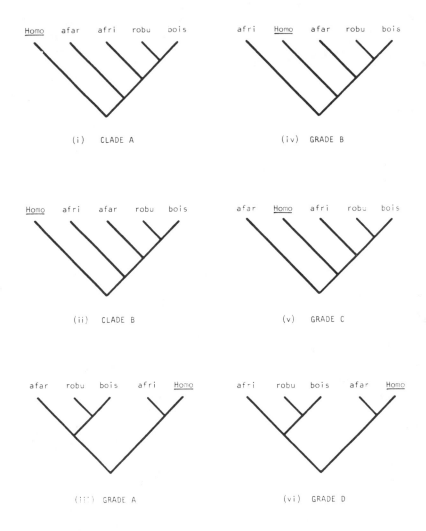

Fig. 1. i) CLADE A, CI=0.72 — supported in this study by
Mandible (CI=0.77), Palate (CI=0.65) and Face (CI=0.76).
ii) CLADE B, CI=0.73 — supported in this study by Face
(CI=0.76). iii) GRADE A, CI=0.73 — supported in this study
by Vault (CI=0.84). iv) GRADE B, CI=0.71 — supported in
this study by Base (CI=0.73). v) GRADE C, CI=0.70 — not
supported by any region in this study. vi) GRADE D, CI=0.71
— not supported by any region in this study.

was compared using the Consistency Index (CI) (Kluge and
Farris 1967). A CI score of unity is given to a cladogram
that has no broken synapomorphies i.e. each character state
appears the minimum number of times; progressively lower CI
scores indicate progressively less economical cladograms.
All characters are given equal value in this exercise.

The summary cladograms which are indicated by, or which
are compatible with, the distribution of the characters in
the five regional cladograms are shown in Fig.1.

Two additional cladistic schemes are also included in
the figure; one (Grade 'C') is consistent with a recently
published phylogenetic analysis. The other is a scheme that
has been raised in discussion, but not formally proposed.

Several general conclusions can be drawn from this
analysis and their order reflects the degree of confidence
attached to them:-

i.   H. habilis is strongly associated with a Homo clade
      that is defined by at least eight character states.
ii.   The degree of association between A. robustus and A.
      boisei is a good deal stronger than that between either
      of them and any other taxon.
iii. A. africanus shares character states with both the Homo
      and the 'robust' australopithecine clades. A.
      africanus can be associated with either of the clades
      with little effect on the soundness of the cladograms.
iv.   A. afarensis forms the sister group of the 'robust'
      australopithecine clade in several of the summary
      cladograms. Disruption of this arrangement results in
      lower CI scores.
v.   Of the two taxa that have been proposed as sister taxa
      of the remaining hominids (A. africanus and A.
      afarensis) the CI scores suggest that A. africanus is
      the more likely candidate.

ACKNOWLEDGEMENTS

Andrew Chamberlain has collaborated in the cladistic
analysis. I am grateful to the Director and Trustees of the
National Museums of Kenya, the Government of Tanzania, the
Director and Trustees of the Transvaal Museum and the Head

of the Anatomy Department, University of the Witwatersrand, for allowing me to study fossils in their care.

This research was supported by the NERC.

Dart RA (1925). Australopithecus africanus, the man-ape of South Africa. Nature 115:195.
Dean MC, Wood BA (1982). Basicranial anatomy of Plio-Pleistocene hominids from East and South Africa. Am J phys Anthropol 59:157.
Kluge AG, Farris JS (1969). Quantitative phyletics and the evolution of anurans. Syst Zool 18:1.
Leakey REF, Walker A (1980). On the status of Australopithecus afarensis. Science 207:1103.
Olson TR (1978). Hominid phylogenetics and the existence of Homo in Member 1 of the Swartkrans Formation, South Africa. J Hum Evol 7:159.
Tobias PV (1980). "Australopithecus afarensis" and A. africanus: critique and an alternative hypothesis. Palaeont afr 23:1.
White TD, Johanson DC, Kimbel WH (1981). Australopithecus africanus: its phyletic position reconsidered. S Afr J Sci 77:445.
Wood BA (1984). The origin of Homo erectus. Cour Forsch Inst Senckenberg 69:99.
Wood BA (1985). Early Homo in Kenya, and its systematic relationships. In Delson E (ed): "Ancestors: the Hard Evidence," New York: Alan R. Liss.
Wood BA, Abbott SA (1983). Analysis of the dental morphology of Plio-Pleistocene hominids. I. Mandibular molars – crown area measurements and morphological traits. J Anat 136:197.
Wood BA, Abbott SA, Graham SA (1983). Analysis of the dental morphology of Plio-Pleistocene hominids. II. Mandibular molars – study of cusp areas, fissure pattern and cross-sectional shape of the crown. J Anat 137:287.
Wood BA, Chamberlain AT (1985a). A reappraisal of variation in hominid mandibular corpus dimensions. Am J phys Anthropol (in press).
Wood BA, Chamberlain AT (1985b). Australopithecus: grade or clade? In Wood BA, Martin L, Andrews P (eds): "Major Topics in Primate and Human Evolution," Cambridge: Cambridge University Press.

Hominid Evolution: Past, Present and Future, pages 233–237
© 1985 Alan R. Liss, Inc.

SEXUAL DIMORPHISM, ONTOGENY AND THE BEGINNING OF
DIFFERENTIATION OF THE ROBUST AUSTRALOPITHECINE CLADE

Yoel Rak, Ph.D.

Department of Anatomy and Anthropology
Sackler Faculty of Medicine
Tel Aviv University, Ramat Aviv, Israel

Numerous morphologies have been said to characterize the masticatory system and face of Australopithecus africanus; the resemblance of these morphologies to those of A. robustus and their functional interpretation have lent support to the view that A. africanus is a member of the robust clade (Rak 1983). Those specimens in its hypodigm that exhibit fewer structures interpreted as derived are regarded here as females. As such, it is possible to view them, along with the young individuals, as a more generalized, morphologically intermediate group between A. africanus and A. afarensis. This perspective can provide us with some insights into the manner in which the differentiation of the robust lineage may have begun. For the purpose of this discussion, four morphological features have been chosen: the anterior pillars, situated laterally adjacent to the nasal opening; the flat naso-alveolar clivus, stretching between the lower part of the two pillars; the maxillary furrows -- defined grooves running along the lateral aspect of the anterior pillars; and, finally, the straight, steep zygomatico-alveolar crest. The anterior pillars and the clivus combine to form a distinct topographical unit, the naso-alveolar triangular frame, which is set off from the more peripheral part of the face by the maxillary furrows.

Each of these four traits has been interpreted as part of a specialized masticatory system. The anterior pillars constitute a structural response to the increase in the occlusal load on the front of the palate, which results from the change in function of the premolars undergoing molarization. The biomechanical need for the two pillars to support the palate at its most anterior end is what causes the flattening of the naso-alveolar clivus. The maxillary furrow is actually a common canine fossa undergoing modification because of the infra-orbital region's change in inclination, which advances the masseter muscle anchored on its inferior margins. The straightening and steepening of the zygomatico-alveolar crest stem from its role as a beam connecting the heavily loaded occlusal surface and the origin of the masseter muscle and in this fashion maximize the vertical occlusal load through the long axis of this (lateral) pillar. All of these morphologies are found in A. robustus, but there they represent the more advanced stage in the robust sequence. However, the same morphologies do distinguish between A. africanus and A. afarensis. The latter lacks the anterior pillars; the lateral nasal margins are sharp and thin; the clivus juts out and is convex in both the sagittal and transverse cross sections; the common canine fossa is present; and the zygomatico-alveolar crest is curved. These facial characters testify that the face of A. afarensis is the most generalized of all the australopithecine species, both because it lacks these four and many other explainable robust morphologies and because it shares its facial topography with many other primates, among them early and modern Homo. The masticatory system of A. africanus, of which the four described features are part, contributes significantly to placing that species at the base of the robust lineage. No other taxon has been proposed to fill the gap between A. africanus and the more generalized A. afarensis, the common ancestor of the two hominid lineages.

Not all the specimens of the A. africanus hypodigm, however, fully adhere to the above characterization of the robust clade. Several lack some of these four characters, and others exhibit them to a much lesser degree. It is suggested, on the basis of what is commonly known about recent primates, that the specimens lacking the more derived morphologies are probably females or young A. africanus. This group (including STS 52, STS 53 and TM 1512) lacks pillars and maxillary furrows, and their clivus is not flat. The distinct naso-alveolar triangular frame is thus necessarily absent. As such, these specimens can perhaps reveal the more primitive morphological stage of the robust morphocline, the more generalized morphology between the bulk of A. africanus and of A. afarensis. The more generalized facial appearance of the presumptive females, therefore, is morphologically closer to that of the generalized A. afarensis and, in this respect also, of early Homo than the face of the more derived part of the A. africanus hypodigm is. The morphological resemblance of the females and young of specialized species to the more generalized closely related species is well known and has long been documented (see, for example, Cramer 1977; Schultz 1960, 1962). The differences that separate the face of this group from that of A. afarensis are of the greatest interest, for it is they which afford us a glimpse of the manner in which differentiation of the robust clade began. The steep, straight zygomatico-alveolar crest clearly seen in the presumptive female specimens STS 52 and STS 53 and also the juvenile from Taung suggests that the morphological manifestation of the differentiation of the robust clade commenced with a change in the shape and orientation of the crest, which in turn indicates the beginning of the increase in the forces acting upon the masticatory system and the architectural need for the modification of the so-called lateral pillar. A curved zygomatico-alveolar crest in the generalized face seems to support Endo's reconstruction of the fundamental architecture of the generalized facial skeleton (1966), in which the lateral pillar is considered of almost no importance. Hence, the changes in the crest are most significant, as they indicate a shift in the basic architecture of the face. They apparently preceded other kinds of change: an alteration in premolar size and shape led to the development of the anterior pillars, and a shift in the orientation of the infraorbital region resulted in the transformation of the canine fossa into the maxillary furrow.

Much of the generalized morphology is seen, as expected, in the face of the young individual from Taung. The naso-alveolar clivus is convex in the sagittal and transverse planes. The canine fossa is of the primitive type, that is, not modified into a maxillary furrow. The presence of anterior pillars, however, is most curious, especially in light of the fact that they are absent in the face of the adult females. It is certainly unusual among the primates that a juvenile face exhibit a morphological specialization while the adult females are lacking it; ordinarily, specialized traits are either missing in both, or present in a modest form in the adult females, while still absent in the juvenile males. Therefore, assignment of this juvenile to the male gender, which I would advocate, does not solve the problem satisfactorily. A possible explanation that merits consideration is that the species under discussion is characterized by extreme sexual dimorphism and by such distinct specialization of the male that cranial differences between male and female are already detectable at a young age. However, this is inconceivable; such a phenomenon does not occur even in primates with a greater degree of sexual dimorphism than that found in the present hypodigm of A. africanus. An alternative interpretation suggests that the presence of pillars in such a young individual demands that that particular specimen be placed in a more advanced position in the robust sequence. On the basis of chronological considerations, Tobias (1973a,b) has proposed that the Taung specimen may represent a "late surviving A. robustus". Morphology, however, does not appear to support this interpretation (Clarke 1977; Grine 1981; Rak 1983). A decision to move this specimen farther along in the morphological sequence by no means implies advocating its removal from the species A. africanus.

The preliminary observations presented here suggest that much information about the early stages of robust differentiation is indeed hidden in the regrettably small and somewhat neglected fossil group of putative females. A closer look at the female and juvenile A. africanus specimens is surely in order.

Clarke RJ (1977). The cranium of the Swartkrans hominid, SK 847, and its relevance to human origins. Unpublished Ph.D. dissertation. University of the Witwatersrand, Johannesburg.

Cramer DL (1977). Craniofacial morphology of Pan paniscus. Contrib Primatol 10, Basel: S. Karger.

Endo B (1966). Experimental studies on the mechanical significance of the form of the human facial skeleton. J Fac Sci Univ Tokyo, Section 5, 3:1.

Grine FE (1981). Trophic differences between "gracile" and "robust" australopithecines: a scanning electron microscope analysis of occlusal events. S Afr J Sci 77:203.

Rak Y (1983). "The Australopithecine Face." New York: Academic Press.

Schultz AH (1960). Age changes and variability in the skulls and teeth of the Central American monkeys Alouatta, Cebus, and Ateles. Proc Zool Soc Lond 133:337.

Schultz AH (1962). Metric age changes and sex differences in primate skulls. Z Morphol Anthropol 52:239.

Tobias PV (1973a). Implications of the new age estimates of the early South African hominids. Nature 246:79.

Tobias PV (1973b). New developments in hominid paleontology in South and East Africa. Ann Rev Anthrop 2:311.

Hominid Evolution: Past, Present and Future, pages 239-245
© *1985 Alan R. Liss, Inc.*

TAUNG FACIAL REMODELING:
A GROWTH AND DEVELOPMENT STUDY

Timothy G. Bromage

Department of Anatomy & Embryology, University
College London, WC1E 6BT, London, England
Department of Anthropology, University of Toronto,
Toronto, M5S 1A1, Ontario, Canada

The focus and indeed the challenge of this study are
conveniently expressed by Moore and Lavelle (1974):

> The changes in the remodelling processes that led to the
> characteristic form of Homo sapiens must...have occurred
> within the Hominidae at some point after their
> divergence from the Pongidae. This possibility could be
> tested only if studies of growth sites and remodelling
> processes became possible in fossil hominid forms. This
> is, of course, unlikely...(p. 213).

Thus it was clear that 1) a non-destructive means of
acquiring remodeling information from intact specimens would
have to be developed and 2) against what seemed to be all
odds, this information would have to be acquired from the
craniofacial remains of Plio-Pleistocene fossil hominids.
This challenge has been met and is complemented by an equally
important advance, namely the determination of Taung's age at
death, so providing a means of measuring the rate of
maturation and growth of the Taung child.

Bone growth remodeling is a fundamental mechanism
responsible for skeletal morphogenesis which involves
coordinated surface patterns of bone resorption and
deposition. Indeed, in addition to displacement, bone
remodeling is the key mechanism by which species specific
craniofacial growth patterns are achieved. For instance, the
differences between modern human remodeling patterns and
those of Macaca reflect unique differences in their patterns
of facial growth (Enlow 1966).

The bone cells responsible for the two remodeling
activities have characteristic microscopic formative effects
on the bone surfaces they directly overlie (Boyde & Jones
1972) and hence the development of a non-destructive strategy
for the study of bone remodeling has naturally centered on
the utility of the scanning electron microscope (SEM)
designed for the imaging of surfaces. So as not to subject
Taung to SEM preparation procedures and the electron beam,
the facial skeleton was replicated with a high resolution
silicone-based dental impression material. Positive epoxy
resin replicas were then prepared for and observed by
conventional SEM (see Bromage (1985a) for a detailed
description of the SEM/replica technique).

The second challenge to this study recognizes that bone surface changes at the microscopic level will naturally have occurred and these must be considered in order to permit their characterization and their distinction from remodeling features. This subfield is often referred to as microscopic taphonomy or, as I like to call it, taphonomy with a small "t". Particular attention has been paid to the abrasion of forming surfaces (Bromage 1984) because of their notable fragility. Fortunately many forming bone surfaces are characterized by a gross microscopic feature called intervascular ridging (IVR) bone which is relatively resistant to abrasion and which lends itself to an integrated SEM and light binocular microscope interpretation. Large Howship's lacunae, characteristic of resorbing bone surfaces, are also identifiable although their interpretation has yet to be fully appreciated in light of specific abrasion etiologies.

Figure 1 shows frontal and lateral views of the resultant remodeling interpretation (resorption is shown by light stippling and deposition by dark stippling). Space does not permit an illustration showing the patchy nature of the actual remodeling map, but the field characteristics of remodeling patterns justify expansion to the remodeling status shown, with two exceptions that will be discussed below. Sutural growth is also a form of remodeling and responds in compensatory fashion, as do many surface bone remodeling regions. As such, sutural growth can be predicted on strict morphogenetic grounds but will not be dealt with here. A more detailed analysis of Taung's sutural growth can be found in Bromage (1985b).

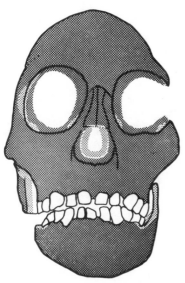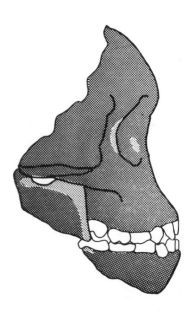

The orbital region. The inner orbital aspect is predominantly depository. These surfaces have an anterior component to their orientation and hence deposits of bone on the inner surface combined with contralateral resorption cause the orbit to drift anteriorly. At the same time, expansion of the orbit can be accommodated by sutural growth and, whereas deposition on the superior wall is keeping pace with downward vectors of growth due to the expanding overlying frontal lobes of the brain, deposits on the inferior aspect relocate the orbital floor upward, compensating for downward growth influences from the nasal capsule and palate. This maintains the orbit in the same relative position during the growth period. Deposits on the superior and inferior rims permit these free edges of the orbit to drift anteriorly in concert with orbital growth.

A resorptive field in the vicinity of the supero-medial part of the lacrimal permits this corner of the orbit to relocate infero-medially, keeping pace with downward nasal capsule growth. That portion in the vicinity of the inferior part of the lacrimal is depository and relocates supero-laterally to accommodate the rapidly expanding ethmoidal sinuses below it. Resorption at the antero-inferior lateral corner of the orbital floor relocates this area outward and is connected to a hypothesized resorptive field over the anteriorly-facing aspect of the lateral rim and extending somewhat along the inner aspect of the lateral wall. This area could not be mapped due to excessive abrasion, but this field is presumed to exist due to the notched configuration of the rim as seen in lateral view (Enlow 1966) and because the posteriorly-facing cortex (its contralateral cortex) adjacent to anterior temporalis, in the temporal fossa, is depository. This indicates a posteriorly directed mode of growth of the lateral orbital rim.

The nasal and frontal region. All periosteal anteriorly-facing surfaces are depository. This contributes to the maintenance of a facial depth and profile congruent with the anteriorly drifting orbits. Resorption on the inner aspects of the nasal capsule signifies an outward mode of growth: expansion of the nasal capsule laterally and inferiorly. Thus, at the same time, the capsule is growing in width and drifting downward, thereby increasing its separation vertically from the orbits. The expanding orbital and nasal functional matrices not only influence their own immediate bony surroundings, but require this separation because of medial encroachment of the orbits, related to the expanding temporal lobes of the brain, on to the root of the nose. Thus the roof of the nasal capsule must also be depository to some degree with contralateral resorption on the meningeal aspect (inferior to the frontal lobes) so that this region can participate in the downward mode of growth.

The premaxillary, maxillary and zygomatic region. Virtually all external periosteal surfaces are depository. Deposits on the laterally-facing aspects along with mid-palatal sutural growth and deposits on the anteriorly-facing aspects, combined with sutural growth in the coronal plane, contribute to lateral and anterior components of growth in this region. The still patent premaxillary sutures, in part, support this growth sequence: providing space for sizable incisors that are certainly developing in this area and promoting the apparent physiological spacing presently between the deciduous incisors. Resorption on the nasal floor signifies a depository palatal surface (the contralateral cortex) and a downward mode of growth of the nasomaxillary complex. Thus deposits on the inferiorly directed surfaces of this region, ie. the zygomatico-alveolar crest, zygomatic and alveolar arches, also participate, with sutures in the transverse plane, in facial height increase.

The mandible. A resorptive field identified in the retromolar space is characteristic of the remodeling of ramus into corpus, thereby lengthening the corpus to accommodate the developing permanent teeth. This being the case, it can be determined that the resorptive field may have extended up the anterior aspect of the ramus and across the lateral aspect of the coronoid process to the neck of the mandible. Note that the coronoid process is oriented such that its medial surface faces slightly posteriorly and superiorly (on the original). This would be the contralateral depository surface. This surface also faces medially and so at one time this remodeling feature accommodates drift relocation posteriorly, lengthening vertically and medial relocation into what will become part of the lengthening corpus. Resorption across the mandibular notch permits this region to relocate inferiorly, thus enlarging the notch while, in addition, resorption extending onto the condylar neck permits the growing wide end of the condyle to successively relocate into the narrower neck, while the whole mandible is displaced inferiorly and anteriorly, commensurate with its postero-superior articulation with the cranial base. Deposits on the posterior aspect of the ramus confirm its posterior drift and so lend support to the presence of the remodeling fields described above.

The labial surface of the corpus is entirely depository, accentuating its anterior and lateral modes of growth. On the lingual aspect, patches of deposition imply that the mandible as a whole was increasing in size, in pace with the developing permanent teeth. A resorptive field over part of the infero-posterior aspect, resulting in relocation of the posterior regions of the mandible into wider positions, cannot be ruled out, however (this area being missing).

Remodeling fields may be considered to have 4 primary attributes: their size, shape, placement and their rate of activity. Preliminary studies of chimpanzee growth and remodeling suggest that in the first 3 attributes Taung and the chimpanzee are the same. But we may ask, why then isn't Taung an ape? It may be that the differences are due to alterations in the rates of certain remodeling activities. Before we can convincingly demonstrate this, however, we need an accurate age at death for Taung in order to assess the maturation rate for this individual.

Superficially the state of the developing dentition of Taung is typical of both modern man and the three great apes, in that the first permanent molars are close to occlusion and no deciduous teeth have been shed. However, on closer examination, the near-coincident eruption of the permanent incisors and first permanent molars in modern man and 'robust' australopithecines (Paranthropus) contrasts with the disparate eruption times of these teeth in the great apes and 'gracile' australopithecines (Dean & Wood 1981; Dean 1985a,b). Specimens of Australopithecus, LH-2 and Sts 24+24a, illustrate eruption of their first permanent molars before any significant root development of the permanent incisors and certainly well in advance of incisor eruption, which suggests an ape-like pattern of dental development. Taung also shares this pattern, which is corroborated by the modicum of root on the recently erupted first permanent molars (Sperber, this symposium), and may also be predicted to have very little or no permanent incisor root formed. So both in their eruption pattern and in development, the teeth of Australopithecus resemble the great apes more closely than modern Homo.

Recently, Bromage and Dean (1985) have estimated the absolute age at death of Sts 24+24a. Estimates of the age at death of this specimen are important because its dental formation status is just very slightly behind that of Taung and its dental eruption status is just very slightly ahead of Taung (Skinner 1975). Therefore, for purposes of determining the age at death for Taung, Sts 24+24a and Taung have been considered to be the same age. Employing the unworn right upper permanent incisor of Sts 24+24a, the age at death was calculated from 1) the observed number of perikymata (which represent surface manifestations of Striae of Retzius and between which an average of 7-day increments of growth occur), 2) estimates of hidden growth increments and 3) the time of onset of calcification and was estimated to be 3 years 4.1 months, or 3.3 years. Practical and technical limitations will not permit the direct evaluation of the surface features of Taung's permanent incisors, and so Taung has also been considered to represent an individual 3.3 years old at the age at death by comparison.

From what was said above, concerning the ape-like dental eruption and maturation pattern of Australopithecus, we may also corroborate the 3.3 year result by simply comparing Taung to dental eruption standards for the great apes. Nissen and Reisen's (1964) Pan data on the mean eruption times of the female and male first permanent molars, when combined, also give an age of 3.3 years. Thus in the light of these data Taung is proposed to be 3.3 years with a range of 2.7 to 3.7 (the range is derived from eruption times for the first permanent molars in Pan (Nissen & Reisen 1964)).

Thus in addition to similarities in facial remodeling it appears that Taung, and Australopithecus generally, had maturation periods similar to those of the extant chimpanzee. Differences in Taung's facial morphology, therefore, must be due to uniquely derived modifications in the rates of remodeling activities at particular sites, relative to Pan. One of these is the diminished rate of deposition on the naso-alveolar clivus of Taung, especially nearer the nasal aperture, resulting in a less prognathic profile relative to a chimpanzee of the same dental age. This has the additional effect of making the canine eminence and its pillar-like extension (anterior pillar in the language of Rak (this symposium)) a prominent feature. It should be noted also that the anterior pillar in Taung is the result of increased rates of mesial drift of the canine (and its associated pillar) over that of anteriorward drift of the incisor region. In Pan, mesial canine drift results in the squared configuration of the anterior dental arcade of the maxilla, which is also more or less reminiscent of the condition in adult australopithecines, but because of relatively greater anterior drift of the incisor region, the pillars are less distinct. Diminished rates of lateral growth and remodeling components of the premaxillary aspect of the dental arch in Taung, due to a relative reduction in size of the permanent incisors, are also indicated, so bringing the anterior pillars slightly medially and more confluent with the lateral walls of the nasal aperture. This growth and remodeling sequence tends to confirm the claim by Rak (this symposium) that Taung is a male, as the anterior pillars have progressed slightly forward despite Taung's young dental age.

In Taung, no supra-orbital torus has developed as in a chimpanzee of the same age. Thus the relative rates of depositional activity over the supra-orbital margin and the frontal squama have operated to preclude the development of the ophryonic groove, or depression, separating these anatomical parts of the frontal bone. It is likely that this occurs as a result of a more forward drift of the frontal squama, conforming to the more vertical disposition of the frontal lobes of the brain over the orbits.

In addition to these inferences derived from comparison to the chimpanzee, the taxonomic valency of facial remodeling can be demonstrated by comparing Taung to other immature hominid taxa. Preliminary remodeling interpretations of two immature 'robust' australopithecine mandibles, SK-61 and SK-64, indicate the bilateral expression of resorptive fields on the labial aspect of alveolar bone in the canine region. These surfaces are depository on Taung, which demonstrates not only its distinctiveness from the 'robust' australopithecine, but also the value that such interpretations have in assessments of immature individuals. One commonly held claim is that juveniles of closely related forms look similar, are not representative of their adult "type", and so are not very useful for taxonomic purposes. Theoretically, however, this claim cannot be justified. Before the craniofacial skeleton has the "ontogenetic time" to significantly respond to environmental factors, the differences which do exist between juveniles of closely related taxa must largely be attributed to genetic differences in their growth patterns. These differences, which are expressed early in ontogeny, should be able to be separated out as a group from those which may simply be responses to unique environmental situations (or niches) and occur much later on. Inasmuch as differences in facial remodeling reflect differences in craniofacial morphogenesis, their study, between taxa, should demonstrate the taxonomic value of young individuals.

In conclusion it may be said that the remodeling processes characterizing Taung specifically, and extrapolated to *Australopithecus africanus* generally, have diverged from the pongid condition as represented by the extant chimpanzee. It is clear that whereas the remodeling patterns are similar, alterations in the rates of remodeling activities have occurred relative to Pan. This finding, along with the recorded genetic similarity between modern Homo and Pan (Chiarelli, this symposium), suggests that the development of the distinctive human craniofacial region is, in part, probably the result of changes in regulatory genes influencing the rates of remodeling at particular growth sites. Also, particular anatomical regions of the australopithecine face have undergone change while others have not, suggesting that the global concept of neoteny cannot account for the origin of the ontogenetic sequence resulting in the modern human face.

Presently, it is instructive to note that Taung's growth and maturation were very similar to those of an ape and yet it was not an ape. This study projects Taung and its species as a harbinger of mosaic evolution: man-ape in aspects of its growth, but in other features such as bipedal posture and morphology of the teeth and dental arches, more like an ape-man.

I am indebted to P.V. Tobias for permission to study the original and for the opportunity to present these results on this occasion. I am grateful also to A. Boyde, M.C. Dean and S.J. Jones for their help during the course of this study. This work has been kindly supported by the L.S.B. Leakey Foundation and University of Toronto doctoral grants.

REFERENCES

Boyde A, Jones SJ (1972). Scanning electron microscope studies of the formation of mineralized tissues. In Slavkin HC, Bavetta LA (eds): "Developmental Aspects of Oral Biology." New York: Academic Press, pp 243-274.

Bromage TG (1984). Interpretation of scanning electron microscopic images of abraded forming bone surfaces. Am J Phys Anthropol 64:161.

Bromage TG (1985a). Systematic inquiry in tests of negative/positive replica combinations for SEM. J Microsc 137:209.

Bromage TG (1985b). "A Comparative Scanning Electron Microscope Study of Facial Remodeling in Homo sapiens, Pan sp. and Plio-Pleistocene Hominidae." Ph.D. Thesis, University of Toronto (in preparation).

Bromage TG, Dean MC (1985). Re-evaluation of the age at death of immature fossil hominids. Nature (submitted).

Dean MC (1985a). The eruption pattern of the permanent incisors and first permanent molars in Australopithecus (Paranthropus) robustus. Am J Phys Anthropol (in press).

Dean MC (1985b). Similarities in the cranial base and developing dentition of Homo and Paranthropus. In Wood BA, Martin LB, Andrews P (eds): "Major Topics in Primate and Human Evolution." Cambridge: Cambridge University Press (in press).

Dean MC, Wood BA (1981). Developing pongid dentition and its use for ageing individual crania in comparative cross-sectional growth studies. Folia primatol 36:111.

Enlow DH (1966). A comparative study of facial growth in Homo and Macaca. Am J Phys Anthropol 24:293.

Moore WJ, Lavelle CLB (1974). "Growth of the Facial Skeleton in the Hominoidea." New York: Academic Press.

Nissen HW, Reisen AH (1964). The eruption of the permanent dentition in the chimpanzee. Am J Phys Anthropol 22:285.

Skinner MF (1975). "Dental Maturation, Dental Attrition and Growth of the Skull in Fossil Hominidae." Ph.D. Dissertation, University of Cambridge.

Hominid Evolution: Past, Present and Future, pages 247–253
© *1985 Alan R. Liss, Inc.*

DENTAL MORPHOLOGY AND THE SYSTEMATIC AFFINITIES OF THE
TAUNG FOSSIL HOMINID

Frederick E. Grine

Depts. Anthropology and Anatomical Sciences
State University of New York
Stony Brook, NY 11794

Sixty years ago a nearly complete skull of a juvenile
individual from the Buxton Quarry at Taung was designated by
Professor Raymond A. Dart (1925) as the holotype of a novel
genus and species, *Australopithecus africanus*. His pro-
nouncement that this comparatively small-brained creature
from southern Africa represented a 'pre-human stock' ran
counter to prevailing notions of human origins, and it
evoked immediate and sometimes rather strong response
(Hrdlička 1925; Keith 1925, 1931; Osborn 1927; Abel 1931).

Although the fundamentally hominid features of the
craniofacial skeleton and dentition of the Taung specimen
were noted by Dart (1925, 1934), Broom (1925, 1929) and
others, it was only the subsequent discoveries of specimens
from the Transvaal caves that served to convince the
palaeoanthropological fraternity of the primitive hominid
status of *Australopithecus* (Broom 1946; Clark 1947, 1952;
Robinson 1956).

More than a half century after its unveiling, Dart's
little fossil is still the subject of controversy. Not
only has its geological age been the subject of debate, but
its traditionally accepted taxonomic association with the
'gracile' australopithecines of Sterkfontein and
Makapansgat has been questioned.

Partridge (1973) attempted to date the Taung hominid
site, as well as several of the Transvaal localities, by
estimates of nick-point migration and valley widening.
By this method he estimated the 'Taung cave' to have opened

at about 870,000 yrs BP; the estimates recorded for
Makapansgat (3.7 m.y.),Sterkfontein (3.3. m.y.) and
Swartkrans (2.6 m.y.)were substantially older. The appli-
cability of this method to the determination of cave
opening has been questioned by DeSwardt (1974), Butzer
(1974) and Bishop (1976), who noted that the assumptions
employed by Partridge are even less relevant to Taung than
to the Transvaal caves.

Geological investigations by Butzer (1974, 1980) have
suggested that the limestone tufa within which the Taung
hominid cave apparently formed can be linked with unit A of
the Vaal Younger Gravels. While this purported association
may set an upper terminus by way of regional stratigraphy,
Butzer has emphasized that this does not provide a 'date'
for the Taung hominid. More recent analyses of the tufas
exposed in the Buxton Quarry have suggested an 'absolute'
maximum age of approximately 1.0 m.y. for the australopithecine specimen (see contributions by Partridge and Vogel in
this volume). A degree of circumspection should be attached
to this 'date', however, insofar as the efficacy of the
technique utilized in its derivation has yet to be demonstrated conclusively.

Early interpretations of the correlative geochronological age of the Taung hominid, based upon analyses of the
'Taung fauna', suggested an approximation to the older
Transvaal sites to some workers, while for others this
fauna held a closer resemblance to elements from the
younger Transvaal caves (see Tobias (1978) for a review of
this literature). Although there are potential problems
concerning not only the possible heterogeneity of elements
comprising the 'Taung fauna', but also the relationship
between these elements and the hominid specimen (Grine
1984), recent analyses of the non-hominid fossils from
Taung have suggested an age between Sterkfontein Member 4
and Swartkrans Member 1 (Cooke 1983; Delson 1984; Vrba
this volume). More particularly, the abundant and well-
correlated cercopithecoids from Taung indicate an age
'just older' than 2.0 m.y.(Delson 1984).

The disparate results obtained by recent geological
and palaeontological investigations would appear to indicate that the question of the geochronological age of the
Taung hominid has yet to be resolved satisfactorily. It is
possible, however, that these apparently conflicting results

might both be correct if the animal remains entombed within the Taung hominid cave some 1.0 m.y. ago were derived from an anachronistic, relict (e.g. refugia) community.

Following Partridge's (1973) proposal of a young geochronological age for the Taung hominid, Tobias (1973, 1974) suggested that this specimen may have been associated with the 'robust australopithecine lineage', rather than with the australopithecines from Sterkfontein and Makapansgat. In this context, it is important to note that Tobias employs 'robust lineage' synonymously with *A. robustus* (as represented by the Kromdraai and Swartkrans fossils) or 'robust australopithecine' (Tobias 1973:83; 1974:43). This argument was amplified by him in a later paper (Tobias 1978), in which several morphological differences between the Taung and Sterkfontein fossils observed by Broom and Robinson, amongst others, were cited in support of the notion that the Taung skull represents a juvenile 'robust' australopithecine.

In view of this interesting possibility, and in light of the substantially larger samples of Plio-Pleistocene hominid specimens available since Robinson's (1956) meticulous study of australopithecine teeth, the author has undertaken a comparative re-examination of the Taung dentition. Details concerning the metrical and non-metrical features of these teeth have been recorded elsewhere (Grine 1981, 1982, 1984, 1985) and need not be elaborated upon here. Rather, only the salient features concerning the systematic affinities of the Taung fossil hominid will be summarized in the present paper.

The deciduous and permanent tooth crowns of the Taung specimen tend to be notably dissimilar to eastern African 'robust' australopithecine homologues for most of the metrical features recorded. The first and second mandibular deciduous molars of the Taung specimen are relatively broad buccolingually, and while these Taung values tend to differ from those recorded for Sterkfontein specimens and resemble Kromdraai and Swartkrans 'robust' australopithecine values, they are also comparable to those obtained for *A. afarensis* homologues. In other instances the Taung dental values display a closer resemblance to those of Sterkfontein and Makapansgat specimens than to those of the Transvaal 'robust' australopithecines. For example, the deciduous canines possessed by the Taung child are, like

those from Sterkfontein (and Laetoli and Hadar) relatively
larger than 'robust' australopithecine homologues, and this
difference may have been functionally significant (Grine
1981). In a number of instances, however, the metrical
values of the Taung crowns fall within the zones of overlap
between the ranges circumscribed by the southern African
'gracile' and 'robust' australopithecine samples.

Overall, whilst the metrical values of the Taung dental
crowns appear to show a closer resemblance to those
recorded for Sterkfontein and Makapansgat specimens than to
those evinced by Kromdraai and Swartkrans specimens,these
gross dimensions and indices do not serve to clearly define
the affinities of the Taung hominid. Odontometric data,
comprising continuously varying features, are perhaps of
more limited value than non-metrical characters in the
systematic assessment of fossil hominid specimens. Not
only are the available samples small, but a rather marked
degree of sexual dimorphism in size appears to have been
evinced by Plio-Pleistocene taxa in both eastern and
southern Africa. If such dimorphism applied also to the
deciduous teeth, this could have an appreciable effect upon
the interpretation of metrical data, where the samples
commonly comprise only a few specimens. In modern hominoid
populations, the metrical features of the deciduous teeth
show stronger degrees of sexual dimorphism than do the
non-metrical characters (Grine 1984).

Cladistic analyses of the non-metrical characters of
the deciduous teeth and first permanent molars have
revealed that the 'robust' australopithecines from
Kromdraai, Swartkrans and eastern Africa share a suite of
synapomorphies that serve to distinguish them not only from
the 'gracile' australopithecine specimens of Sterkfontein
and Makapansgat, but also from the Laetoli and Hadar
fossils (Grine 1982, 1984, 1985). The Taung dentition does
not evince any of the character states that serve to
define the 'robust' australopithecines as a group. Rather,
the morphological traits possessed by the Taung specimen
are similar to those shown by Sterkfontein and Makapansgat
specimens. That is not to say, however, that the Taung
and Sterkfontein teeth are identical in every respect.
Indeed, differences in detailed dental traits between these
specimens were employed by Broom (1950) and Robinson (1954)
to separate them taxonomically.

While Broom (1950) proposed to distinguish the Taung and Sterkfontein specimens at the species (and even generic) level, on the basis of the presence of a mesial stylid on the lower dc of the Sterkfontein specimen and its apparent absence from the Taung homologue, Robinson (1954) employed this feature, together with the presence of a well-developed tuberculum sextum on the lower molars of Taung, to suggest that the Sterkfontein and Makapansgat australopithecines should be separated from the Taung specimen at a subspecific level. Other differences noted by Robinson (1956) include the weak development of the Carabelli trait on the Taung maxillary molars, and the comparatively weak lingual surface relief on the lower deciduous canine of the Taung specimen. The Taung and Sterkfontein specimens differ also in the relative reduction of the hypocone shown by the upper dm2 and M1 of the Taung individual, and in the 'triangular' occlusal outline of the lower dm1 of the Taung specimen (Grine 1984).

All but two of the features in which the Taung and Sterkfontein (and Makapansgat) dentitions differ are variable within known australopithecine samples and are, therefore, of questionable validity (Grine 1982, 1984). The Taung and South African 'gracile' specimens would appear to differ most impressively in the presence or absence of a mesial stylid on the lower dc and in the relative size of the hypocone on the upper dm2 and M1. The presence of a mesial stylid on the canine appears to be unique to the Sterkfontein specimen, whilst the relatively small size of the hypocone seems to be unique to the Taung individual. In neither instance does the condition evinced by the Taung specimen indicate a special relationship between it and the South African 'robust' australopithecines.

Thus, the overwhelming body of dental evidence points to the similarity of the Taung, Sterkfontein and Makapansgat specimens, and to the fact that they share a number of characters that serve to distinguish them from the 'robust' australopithecine specimens of southern and eastern Africa.

As aptly observed by Dart (1929), in response to criticisms that the purported age of the Taung skull was inconsistent with its being considered as a human precursor: "Geological age does not affect the question of the proximity in relationship of the Taungs (specimen) to man .... the answer to that question lies in its anatomical characters."

The dental characters possessed by the Taung specimen do not conform to the hypothesis that it represents a juvenile 'robust' australopithecine.

I would like to thank Professor P.V. Tobias for inviting me to present this paper at the Taung Diamond Jubilee International Symposium, and for the gracious hospitality shown throughout the symposium.

REFERENCES

Abel W (1931). Kritische Untersuchungen über
    Australopithecue africanus Dart. Morph Jhbk 65:539.
Bishop WW (1976). Comparison of australopithecine-bearing
    deposits in eastern and southern Africa - a new look at
    a sixteen-year-old problem. Ann S Afr Mus 71:225.
Broom R (1925). Some notes on the Taung skull.
    Nature 115:569.
Broom R (1929). Note on the milk dentition of
    Australopithecus. Proc Zool Soc Lond 1928:85.
Broom R (1946). The occurrence and general structure of
    the South African ape-men. Mem Tvl Mus 2:1.
Broom R (1950). The genera and species of the South
    African fossil ape-men. Am J Phys Anthropol 8:1.
Butzer KW (1974). Paleoecology of South African
    australopithecines: Taung revisited. Curr Anthrop 15:
    367.
Butzer KW (1980). The Taung australopithecines:
    contextual evidence. Palaeont Afr 23:59.
Clark WE Le Gros (1947). Observations on the anatomy of
    the fossil Australopithecinae. J Anat 81:300.
Clark WE Le Gros (1952). Hominid characters of the
    australopithecine dentition. J R Anthrop Inst 80:37.
Cooke HBS (1983). Human evolution: the geological
    framework. Canad J Anthrop 3:143.
Dart RA (1925). *Australopithecus africanus*: the man-ape
    of South Africa. Nature 115:195.
Dart RA (1929). A note on the Taungs skull. S Afr J Sci
    26:648.
Dart RA (1934). The dentition of *Australopithecus
    africanus*. Folia Anat Jap 12:207.

Delson E (1984). Cercopithecid biochronology of the
    African Plio-Pleistocene: correlation among eastern and
    southern hominid-bearing localities.
    Cour Forsch Inst Senckenberg 69:199.
DeSwardt AMJ (1974). Geomorphological dating of cave
    openings in South Africa. Nature 250:683.
Grine FE (1981). Trophic differences between 'gracile' and
    'robust' australopithecines: a scanning electron micro-
    scope analysis of occlusal events. S Afr J Sci 77:203.
Grine FE (1982). A new juvenile hominid (Mammalia,
    Primates) from Member 3, Kromdraai Formation,Transvaal,
    South Africa. Ann Tvl Mus 33:165.
Grine FE (1984). The deciduous dentition of the Kalahari
    San, the South African Negro and the South African Plio-
    Pleistocene hominids. PhD thesis, University of the
    Witwatersrand.
Grine FE (1985). Australopithecine evolution: the
    deciduous dental evidence. In Delson E (ed)
    "Ancestors: the Hard Evidence", New York: Alan R. Liss,
    (in press).
Hrdlička A (1925). The Taungs ape. Am J Phys Anthrop 8:379.
Keith A (1925). The fossil anthropoid ape from Taungs.
    Nature 115:234.
Keith A (1931). "New Discoveries Relating to the Antiquity
    of Man." London: Williams and Norgate.
Osborn HF (1927). Recent discoveries relating to the
    origin and antiquity of man. Science 65:481.
Partridge TC (1973). Geomorphological dating of cave
    opening at Makapansgat, Sterkfontein, Swartkrans and
    Taung. Nature 246:75.
Robinson JT (1954). The genera and species of the
    Australopithecinae. Am J Phys Anthropol 12:181.
Robinson JT (1956). The dentition of the
    Australopithecinae. Mem Tvl Mus 9:1
Tobias PV (1973). Implications of the new age estimates of
    the early South African hominids. Nature 246:79.
Tobias PV (1974). The Taung skull revisited.
    Nat Hist 83:38.
Tobias PV (1978). The South African australopithecines in
    time and hominid phylogeny, with special reference to
    the dating and affinities of the Taung skull. In
    Jolly C. (ed): "Early Hominids of Africa",
    London: Duckworth, p. 45.

Hominid Evolution: Past, Present and Future, pages 255–263
© *1985 Alan R. Liss, Inc.*

ASIAN AUSTRALOPITHECINES ?

Jens Lorenz Franzen

Forschungsinstitut Senckenberg

Senckenberganlage 25, D 6000 Frankfurt am Main

Africa is known today as the only continent which has yielded australopithecines. Since their first discovery by Raymond Dart 60 years ago at Taung, these early hominids have been unearthed from the far south up to the north-east of Africa. Did they ever leave this continent? If not, were they unable to do so, or have they simply remained unrecognized in other parts of the world?

Robinson (1953, 1954) was the first to point out the possibility that Meganthropus palaeojavanicus from Java could be an Asian representative of the robust australopithecines. After careful comparisons he stated, "... every single important character ... is typically australopithecine in nature, resembling P. crassidens in particular" (1953: 36). In this context he mentioned especially the large overall size; the relatively small canine, judged from its empty socket in the Meganthropus type mandible, which is known today as Sangiran 6; the root structure, although the premolars are only 1-rooted in Meganthropus; the asymmetry of the occlusal surface of P_4; the greater length than width of the occlusal surface of M_1; the presence of a tuberculum sextum in M_1; the massiveness of the corpus mandibulae, and the median sagittal contour of the symphysis. Consequently in 1954, Robinson proposed to transfer the species palaeojavanicus, created by Weidenreich in 1945, to Broom's genus Paranthropus.

Already in the same year Koenigswald (1954a,b) criticised and rejected the idea that his Meganthropus be regarded as an australopithecine. For him Meganthropus

looked too primitive for that. In this context he pointed
particularly to an undivided ridge between the protoconid
and the metaconid in the premolars, and in the M_1; the
presence of a paraconid in an isolated second lower deciduous molar and a more angular outline of the P_3. Besides,
he remarked that the P_4 of Australopithecus displayed two
roots instead of one as in Meganthropus (1954b: 796).

Robinson (1955) tried to explain the inconsistencies
of his theory by variability, but he was immediately
rebutted by Koenigswald, who, at the Third Panafrican Congress on Prehistory, provided the following additional
arguments (1957a: 160): The canine of Meganthropus, which
is known from an isolated tooth only, "... is larger than
in the Australopithecinae with canines of a similar type",
and the "first premolar is larger than the second". But
Koenigswald admitted also that the "second premolar [of
Meganthropus] is of a more primitive type than in Sinanthropus, and in its asymmetrical posterior part similar to the
same tooth in the Australopithecinae and the higher Anthropoids." (loc.cit., 160).

After some years of armistice which saw the extraordinary new discoveries of Mary and Louis Leakey at
Olduvai Gorge and at other places in East Africa, Tobias
and Koenigswald again picked up the problem in 1964 in
their famous "comparison between the Olduvai hominines
and those of Java" based upon the original specimens.

They concluded (1964: 517), "... that Meganthropus of
Java shows certain advances over the australopithecine
grade of hominid organization." And they drew attention
particularly to the symphyseal region stating, "There is no
deep pit nor massive inferior transverse torus as in the
Natron jaw and the large A. robustus crassidens ... jaws",
and to"the relatively high position of the mental foramen".
Hence they tentatively included Meganthropus together with
Homo habilis in their 2nd grade of hominization, between
Australopithecus as the 1st, and Olduvai 13, Telanthropus
and Pithecanthropus modjokertensis as their 3rd grade of
hominid evolution. One should add that Olduvai 13 and
perhaps part of the hypodigm of Telanthropus are now regarded as Homo habilis, too, whereas Pithecanthropus modjokertensis represents a robust and somewhat primitive-
looking Homo erectus.

It was only later in his life, that Koenigswald accepted Meganthropus as "australopithecoid" (1973a, b), but it was not earlier than 1980 that he conceded finally that Meganthropus palaeojavanicus had to be regarded as an australopithecine.

The present state of affairs is characterized by two trends. One tendency is to call every human fossil from Asia which is older than Upper Pleistocene Homo erectus, and the other is to regard all hominids from Asia as younger than 1 m.y. (e.g. Pope 1983; Pope, Cronin 1984; Zhang Yinyun 1984). In this way we seem indeed to be getting a very clear-cut picture: Homo erectus would be the first hominid arriving in the Far East, and this not earlier than 1 m.y. ago. But does such a scheme really solve the problems? Are all facts covered and explained by such an hypothesis?

My interest was drawn to the question of Asian australopithecines when I tried to find out the taxonomic and phylogenetic position of so-called Pithecanthropus dubius (Franzen 1985). This species was created by Koenigs-wald in 1950 to accommodate the anterior fragment of a mandible discovered by Javanese collectors in 1939, known today as Sangiran 5. The specimen comprises part of the right ramus with M_2, a heavily worn M_1, the roots of P_4, and the relics of the sockets of P_3, C and the 2 incisors. Remarkable features are the robustness of the ramus, which is considerably more massive than that of mandible B of Homo erectus, but less than that of Meganthropus; the double-rooted premolars; a small canine to judge from the remnant of its socket; very large molars in comparison to those of Homo erectus, with flatly worn occlusal surfaces, and the median sagittal contour of the symphysis, resembling in its robustness and distinctive superior torus those of Meg-anthropus and Australopithecus robustus, although the inferior torus is much less pronounced (fig. 1).

All these traits are typical of the australopithecines, particularly of their robust members. To include such a specimen in Homo erectus would require either the expansion of the definition of that species to such an extent that it would comprise almost everything and nothing, or the expla-nation of the extraordinary characters of this specimen as aberrant variations.

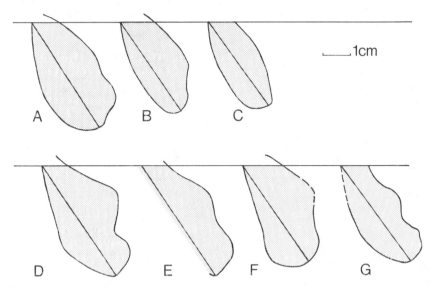

Fig. 1. Sagittal contour of the symphysis (partly recon-
structed)of A Meganthropus palaeojavanicus (S 6, holo-
type; Weidenreich 1945, fig.12); B "Pithecanthropus dubius"
(S 5, holotype; original); C Homo erectus (S 1B; original);
D A.robustus (SK23; Robinson 1954, fig.6); E A.robustus
(SK6; Robinson 1953, fig.6); F A. africanus (Sterkfontein;
Broom and Robinson 1950, fig. 13A); G A. afarensis (LH4;
White 1977, pl.4, fig.E).

    The second alternative was already eliminated by the
discovery of a second mandible, Sangiran 9, which was described in great detail by Sartono (1961). Preserved are the
right ramus with the last two molars, both premolars, the
canine and the sockets of one lateral and two central incisors. The specimen displays exactly the same characters as
mentioned above for Sangiran 5. Beyond that the mandible
shows a more V-shaped horizontal outline as is typical for
australopithecines, and a small canine in place.

    If indeed both mandibles cannot be included in the
genus Homo, because they lack apomorphic characters such as
one-rooted P_4; a more parabolic horizontal outline of their
corpus; a comparatively thin median sagittal section of the

symphysis, and the absence of a fossa genioglossi, where in fact do they belong ?

In my opinion there are two possibilities (fig. 2): one is to include the species dubius in Australopithecus, Meganthropus being derived independently from either the same species or an already recognized robust australopithecine. The other is to include dubius as a primitive member of the genus Meganthropus, from which Meganthropus palaeojavanicus would have been derived. For the moment there is no evidence ruling out either alternative. It seems easier to derive M. palaeojavanicus from dubius. All that would be needed is a small increase in size and a reduction of the roots of the premolars from two to one. But simplicity alone would not suffice as a test of such a hypothesis.

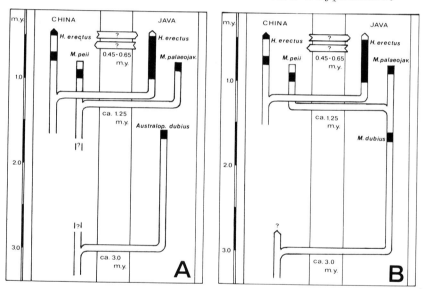

Fig. 2. Two alternative hypotheses: A including the species dubius in Australopithecus, B in Meganthropus.

Another possibility is that Meganthropus developed independently elsewhere and later migrated into Java together with Homo erectus. In any case, however, both alternatives accord with the facts of morphology, dating and palaeogeography better than the simplistic assumption that every early hominid from Asia belonged to <u>Homo erectus</u>:

1. Morphology.

The considerable differences in structure between
<u>Homo erectus</u> on the one hand and <u>M. palaeojavanicus</u> and the
species <u>dubius</u> on the other are explained as follows. It
is suggested that an early offshoot of <u>Australopithecus</u>
migrated into Asia and developed there in some ways in
parallel with the robust African australopithecines. This
concept would explain particularly the mixture of primitive and derived features in <u>Meganthropus palaeojavanicus</u>
as shown by Koenigswald (1954b, 1957a) on one hand and
Robinson (1953, 1954, 1955) on the other.

2. Dating.

Although the type specimen of <u>dubius</u> was collected
from the surface, the site was found by Koenigswald "to be
close to the lower marine intercalation" within the lower
part of the Black Clay (Koenigswald 1973a: 107). This would
correspond very well with a minimum age of 1.6 m.y. based
on foraminiferal analysis by Siesser & Orchiston (1978) of
the claystone adhering to the second mandible assigned to
that taxon. Pope & Cronin (1984: 379) are forced to talk of
"an inadequate and obviously reworked sample with highly
debatable stratigraphic ranges" in order to deny or at
least to cast a doubt on such a high age of the <u>dubius</u>
mandibles. But a recently announced cranium, called by
Sartono (1982) "<u>Homo palaeojavanicus</u>", is said to come
from the lowermost vertebrate-bearing horizon of the Sangiran section, which would again indicate a chronometric
date of more than 1.5 m.y. (Pope, Cronin 1984: 380, Fig.
2).

3. Palaeogeography.

A chronometric age of more than 1.5 m.y. would make
<u>dubius</u> in any case a member of the first immigration wave
that entered Java at about 3.0 m.y. ago, when the Sunda
Shelf was exposed for the first time by eustatic sea-level
lowering (Berggren, Van Couvering 1979). This in turn would
be in keeping with such primitive characters of this species
as the 2-rooted premolars and the presence of a fossa genioglossi (Franzen 1985).

What about the mainland? Are there any candidates for
australopithecine status from the Asian continent too? There
are! Already in 1957 Koenigswald described a couple of iso-

lated teeth hunted up by him in old Chinese drugstores as
belonging to a new genus and species, which he called
Hemanthropus peii, pointing to "the close resemblance of
the upper molars with those of Paranthropus robustus from
South Africa" (1957b: 158; 1957c). The drugstore material
was of unknown provenance as usual, but in 1975 Gao Jian
described the same type of teeth as a new but unnamed
species of Australopithecus from the "Dragobone cave" of
Jianshi district in Hupei province. There these teeth had
been found associated with Gigantopithecus and other
mammals. Thus the geological age could be inferred from
the accompanying fauna as late Early Pleistocene (1975:
88), corresponding to a chronometric range between 0.8
and 1.2 m.y. (Pope, Cronin 1984: 380, Fig. 2). Although
Zhang Yinyun (1984: 92) recognized that "the molars are
quite similar to those ... of Meganthropus of Indonesia
in size and shape", he included them together with Meganthropus palaeojavanicus, dubius, and mandible B in Homo
erectus, thus recognizing, as did Pope (1983) and Pope and
Cronin (1984), very large variability of that species in
Asia (fig. 3). I should prefer to consider the teeth from

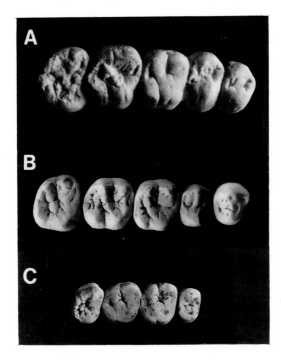

Fig. 3. M^3-P^3 of A.
robustus (Swartkrans;
cast); B "Hemanthro-
pus" peii (China;
compiled originals),
and C Homo erectus
(Sangiran/Java; com-
piled originals)
showing the resem-
blance of A and B
contrasting with the
dissimilarity of B
and C. -- Photo:
Senckenberg-Museum,
E. Pantak.

Jianshi, together with Koenigswald's drugstore material,
as australopithecines from the Asian continent and to call
them Meganthropus peii (Koenigswald 1957). In spite of a
few minor differences of proportions there seems to be no
difficulty in deriving M. peii together with M. palaeojavanicus from the same common ancestor.

In this way we arrive at the conclusion that the
recognition of australopithecines in Asia would not confuse
but could help to clarify the early evolution of hominids
on that continent. This concept would explain the scanty
remains from Java and China as relics of an Asian offshoot
of an early radiation of Australopithecus, which was
followed much later by an immigration of Homo erectus, and
which finally became extinct after a period of coexistence.

References:
Berggren WA, Van Couvering JA (1979). The Quaternary. In
   Robinson RA, Teichert C (eds): "Treatise on Invertebrate
   Paleontology", Lawrence: Geol Soc America, 505.
Broom R, Robinson JT (1950). Further Evidence of the Struc-
   ture of the Sterkfontein Ape-Man Plesianthropus. In
   Broom R, Robinson JT, Schepers, GWH: "Sterkfontein
   Ape-Man Plesianthropus", Transvaal Mus Mem 4: 7.
Franzen JL (1985). What is Pithecanthropus dubius Koenigs-
   wald 1950? In Delson E, Tattersall I (eds): "Ancestors",
   New York: A.Liss, in press.
Gao Jian (1975). Australopithecine teeth associated with
   Gigantopithecus. Vertebrata Palasiatica 13 (2): 87.
Koenigswald GHR von (1950). Fossil hominids from the Lower
   Pleistocene of Java. Proc Internat Geol Congr Gr Brit
   1948 9: 59.
-- -- -- (1954a). The Australopithecinae and Pithecanthropus. III. Proc Kon Nederl Akad Wet (B) 57 (1): 85.
-- -- -- (1954b). Pithecanthropus, Meganthropus and the
   Australopithecinae. Nature 173: 795.
-- -- -- (1957a). Meganthropus and the Australopithecinae.
   Proc Third Pan-African Congr Prehist, Livingstone 1955:
   158.
-- -- -- (1957b). Remarks on Gigantopithecus and other
   hominoid remains from Southern China. Proc Kon Nederl
   Akad Wet (B) 60: 153.
-- -- -- (1957c). Erratum: Hemanthropus n.g. not Hemi-
   anthropus. Proc Kon Nederl Akad Wet (B) 60: 416.

Koenigswald GHR von (1973a). The oldest hominid fossils
from Asia and their relation to human evolution.
Accad Naz Lincei 182: 97.
-- -- -- (1973b). Australopithecus, Meganthropus and Ramapithecus. J Hum Evol 2: 487.
-- -- -- (1980). Neue Einblicke in die Geschichte der Homi-
niden. Ann Naturhist Mus Wien 83: 181.
Pope G (1983). Evidence on the age of the Asian Hominidae.
Proc Nat Acad Sci USA 80: 4988.
Pope G, Cronin JE (1984). The Asian Hominidae. J Human
Evol 13: 377.
Robinson JT (1953). Meganthropus, australopithecines and
hominids. Am J Phys Anthropol (ns) 11 (1): 1.
-- -- -- (1954). The genera and species of the Australopithecinae. Am J Phys Anthropol (ns) 12: 181.
-- -- -- (1955). Further remarks on the relationship
between "Meganthropus" and australopithecines. Amer J
Phys Anthropol (ns) 13 (3): 429.
Sartono S (1961). Notes on a new find of a Pithecanthropus
mandible. Publikasi Teknik Seri Paleontologi 2: 1.
-- -- -- (1982). Characteristics and chronology of early
man in Java. Congr Internat Paléont Humaine, Nice Prétirage 2: 491.
Siesser WG, Orchiston DW (1978). Micropaleontological reassessment of the age of Pithecanthropus mandible C
from Sangiran, Indonesia. Mod Quatern Res SE-Asia
4: 25.
Tobias PV, Koenigswald GHR von (1964). A comparison between
the Olduvai hominines and those of Java and some implications for hominid phylogeny. Nature 204 (4958):
515.
Weidenreich F (1945). Giant Early Man from Java and South
China. Anthropol Pap Amer Mus Natur Hist 40 (1):
1.
White, TD (1977). New fossil hominids from Laetoli, Tanzania. Amer J Phys Anthropol 46 (2): 197.
Zhang Yinyun (1984). The "Australopithecus" of West Hubei
and some Early Pleistocene hominids of Indonesia. Acta
Anthropol Sinica 3 (2): 92.

Hominid Evolution: Past, Present and Future, pages 265–269
© 1985 Alan R. Liss, Inc.

TOWARD A SYNTHETIC ANALYSIS OF HOMINOID PHYLOGENY

Jeffrey H. Schwartz

Department of Anthropology
University of Pittsburgh
Pittsburgh, PA 15260

INTRODUCTION

The sixtieth anniversary of Raymond Dart's landmark
publication marks a critical period in the assessment of
"Man's place in Nature". The honoring of a paleontological,
and thus morphological, triumph comes at a time when the
very adequacy of these pursuits is being questioned. Largely
through molecular, biochemical and chromosomal studies,
Huxley's (1864) and later Schultz's (1936, 1968) construct
of a great ape group, to which humans are broadly related,
was replaced by the notion that the African apes alone are
more closely related to humans. Of late, the morphological
association of the African apes is being rejected in favor
of a molecular grouping of the chimpanzee with humans.
There is very little morphologically that supports the
grouping of humans with either the African apes or the
chimpanzee alone (Andrews, Cronin 1982). However, morphology
is suggestive of the unity of humans and the orang-utan
(Schwartz 1984a, b). Since morphological and molecular/
chromosomal interpretations of hominoid relationships are in
apparent discord, a review of the latter and their underlying assumptions is timely.

MOLECULES, CHROMOSOMES AND PHYLOGENETIC RECONSTRUCTION

Over the years, the notion of "genetic distance =
evolutionary distance" has been applied broadly among vertebrates. However, initial justification of this assumption
came from an apparent corroboration from a morphological
theory of relatedness. For example, Nuttall (1904) measured

the success of his "demonstration of certain blood-relation-
ships amongst animals" by its compatibility with a morphologically based grouping of humans with the African apes.
However, the preferred theory was Huxley's.

Sarich and Wilson (1966: 1564) concluded that immunological distance reflects evolutionary distance because
their initial comparisons of humans, chimpanzee, rhesus,
and capuchin were consistent with the notion that "these
species form an approximate evolutionary series, the chimpanzee being the closest relative of man and the capucin
(sic) monkey the most distant." They subsequently aligned
the African apes more closely than the orang-utan with
humans not only because the albumin of the African apes
reacted more strongly with human antiserum, but because they
saw their data as "in qualitative agreement with the anatomical evidence, on the basis of which the apes, Old World
monkeys, New World monkeys, prosimians, and nonprimates are
placed in taxa which form a series of decreasing genetic
relationships to man" (p. 1565). Thus, validation came
from a particular morphologically based theory of relatedness.

Most recently, Sibley and Ahlquist (1984) pushed the
notion of "genetic distance = evolutionary distance" to the
limit, concluding from their study of DNA hybrids that
humans and the chimpanzee are sisters. They (1983: 254)
argue that the most thermally stable hybrids are those that
share "'primitive' or 'plesiomorphous' sequences," but, since
only derived homologous features reflect closeness of relatedness, it would seem that thermal stability of DNA
hybrids would not faithfully reproduce phylogenetic relationships. Thermal stability of DNA hybrids, as well as degrees of
immunological reactivity or undifferentiated assessments of
overall sequence similarity, can suggest homology but cannot
distinguish between primitive and derived states of homology.
It is therefore impossible, without additional information,
to know what any given comparison represents.

Choice of particular molecular phylogenies has also
been tempered by a predisposition toward "acceptable" morphological phylogenies. For example, Baba et al (1982) outline procedure when the difference in sequence nucleotide
replacements (NR) among competing phylogenies is statistically
insignificant: "Molecular anthropologists are generally
skeptical, for example, of lowest NR trees that save a few

NR's but depict relationships among lineages for which there is very little or no *a priori* support from comparative anatomy or the fossil record" (p. 160). Since the large hominoids differ in their hemoglobin, at most, by a few NRs (Goodman et al 1983), various phylogenies are thus available. Goodman et al (1983) also concluded that myoglobin and carbonic anhydrase data support the grouping of humans with the African apes. However, sequences for the latter are lacking for all hominoids, and the most parsimonious myoglobin grouping is actually that of humans with the orangutan, which differ only at positions 23 and 110.

Electrophoresis has been used to support a human-African ape group (Bruce, Ayala 1979). Since this conclusion was based on an interpretation of molecules and morphology, the calculations of genetic identity (GI) and genetic distance (GD) are revealing. From their table 5, one derives a series, from humans, of pygmy chimpanzee (.312 GD, .732 GI), the orang-utan "subspecies" (.347 and .350 GD, .707 and .705 GI), gorilla (.373 and .689), and, finally, common chimpanzee (.386, .680). However, in table 6, where the species are lumped, the chimpanzee and orang-utan are equally "removed" from humans (.349 GD, .706 GI) with gorilla the outlier (.373, .689).

Mitochondrial (mt) DNA cleavage map data can be interpreted to yield a human-chimpanzee-gorilla trichotomy, a human-African ape nested set, and a human-orang-utan group (see Schwartz, 1984a). MtDNA sequence data (Brown et al 1982) has been used to support the second theory. However, one must first "[choose] closely related species for analysis" (p. 225), and, because mtDNA appears to evolve 5-10 times more rapidly than nuclear DNA, a temporal constraint is also demanded. Thus, in order to use mtDNA, one must have prior knowledge of the phylogenetic history of the group to be analyzed.

Immunologic analysis of collagen has supported an orangutan origin of the Piltdown jaw and canine (maximum binding [MB] = 1) (Lowenstein et al 1982). Using the MBs also published for humans, chimpanzees and rhesus, the data on the jaw yield a series, from the orang-utan, of pygmy chimpanzee (.76MB), human (.61), common chimpanzee (.56), and then rhesus (.35). The canine data yield a sequence from the orang-utan of human (.66), common chimpanzee (.62), pygmy chimpanzee (.42), and rhesus (.34). Since averaging is

accepted practice, the series from the orang-utan thus derived is human (.635), both chimpanzees (.59), and then
rhesus (.345).

Yunis and Prakash (1982) concluded from their study of
chromosomes that humans and the chimpanzee are sisters, to
which the gorilla and then the orang-utan are related. However, they (p. 1528) assumed first that "the orangutan is a
more primitive species" of hominoid. Thus, they demonstrated, not relatedness, but an interpretation of those
chromosomal changes which would have occurred given a particular phylogeny. A review of their data reveals that the
pairs human-orang-utan and chimpanzee-gorilla, respectively,
share the greatest number of homologous chromosomes
(Schwartz 1984b). Mai (1983) concluded that there are five
equally likely chromosomal phylogenies, of which three unite
humans with the orang-utan.

WHERE DO WE GO FROM HERE?

One of the seemingly least assailable morphological
groups among the hominoids is the knuckle walking African
apes. Yet, recent molecular and chromosomal studies--es-
pecially a recalibration of a mtDNA clock, which pictures
human and chimpanzee sister lineages diverging only 2.5
million years ago (Hasegawa et al ms.)--force a rejection of
such a group. The mtDNA study even demands an abandonment
of *Australopithecus*. However, the theory that the large
hominoids differentiate into human-orang-utan and chimpanzee-
gorilla sister groups is consistent with morphology and is
actually supported by some molecular and chromosomal evidence. The uniquenesses that support the unity of these two
groups can be interpreted as primitive retentions or paral-
lelisms, but only as a consequence of first accepting another
set of hominoid relationships. Whatever the final outcome,
however, it is obvious that there must be a melding of
theory applicable to both the molecular and organismal.
Sophisticated technology does not produce more accurate
phylogenies than conventional means. Phylogenetic interpretation is ultimately a reflection of the theoretical
predisposition of the investigator.

Andrews P, Cronin J (1982). The relationships of *Sivapithe-cus* and *Ramapithecus* and the evolution of the orang-utan. Nature 297:541.

Baba M, Weiss ML, Goodman M, Czelusniak J (1982). The case of tarsier hemoglobin. Syst Zool 31:156.

Brown WM, Prager EM, Wang A, Wilson AC (1982). Mitochondrial DNA sequences of primates: tempo and mode of evolution. J Mol Evol 18:225.

Bruce, EJ, Ayala FJ (1979). Phylogenetic relationships between man and the apes: electrophoretic evidence. Evolution 33:1040.

Goodman M, Braunitzer G, Stanzl A, Schrank B (1983). Evidence of human origins from haemoglobins of African apes. Nature 303:546.

Hasegawa M, Kishino H, Yano T (n.d.). Dating of the human-ape splitting by a molecular clock of mitochondrial DNA.

Huxley TH (1864). "Evidence as to Man's Place in Nature." London: Williams and Norgate.

Lowenstein JM, Molleson T, Washburn SL (1982). Piltdown jaw confirmed as orang. Nature 299:294.

Mai LL (1983). A model of chromosome evolution and its bearing on cladogenesis in the Hominoidea. In Ciochon RL, Corrucini RS (eds): "New Interpretations of Ape and Human Ancestry," New York: Plenum, pp 87-114.

Nuttall, GHF (1904). "Blood Immunity and Blood Relationship." Cambridge: at the University.

Sarich V, Wilson AC (1966). Quantitative immunochemistry and the evolution of primate albumins: micro-complement fixation. Science 154:1563.

Schultz AH (1936). Characters common to higher primates and characters specific to man. Quart Rev Biol 11:259,425.

Schultz AH (1968). The recent hominoid primates. In Washburn SL, Jay PC (eds): "Perspectives on Human Evolution, vol 1," New York: Holt, Rinehart, Winston, pp 122-195.

Schwartz JH (1984a). The evolutionary relationship of man and orang-utans. Nature 308:501.

Schwartz JH (1984b). Hominoid evolution: a review and a reassessment. Curr Anthrop 25:655.

Sibley CG, Ahlquist JE (1983). Phylogeny and classification of birds based on the data of DNA-DNA hybridization. Curr Ornith 1:245.

Sibley CG, Ahlquist JE (1984). The phylogeny of the hominoid primates, as indicated by DNA-DNA hybridization. J Mol Evol 20:2.

Yunis JJ, Prakash O (1982). The origin of man: a chromosomal pictorial legacy. Science 215:1525.

Hominid Evolution: Past, Present and Future, pages 271–278
© *1985 Alan R. Liss, Inc.*

HOMINIDS AND HOMINOIDS, LINEAGES AND RADIATIONS

Charles Oxnard, B.Sc., M.B., Ch.B., Ph.D., D.Sc.

University Professor and Professor of Anatomy
and Biology
Department of Anatomy and Cell Biology, School
of Medicine, University of Southern California,
Los Angeles, California, 90033, USA)

MORPHOMETRIC STUDIES OF PRIMATES

Though initially morphometrics were used in few
studies (Ashton, Healy and Lipton 1957) today they are
accepted as most useful for examining biological structure
(Reyment, Blackith and Campbell 1984). Our morphometric
investigations of primates are aimed at functional anatomy, at assessments of individual species, and at relationships within the entire Order (Oxnard 1983a, 1984).
They provide the first anatomical evidence from living
species supporting the close molecular link of humans and
African apes (Oxnard 1983b).

In contrast morphometric studies of fossils remain
controversial though progress has been made. Early
cranial and dental studies showed that australopithecines
had ape-like features (Zuckerman 1970). Later postcranial investigations showed that they also possessed orangutan-like features and features different from all living
forms (Oxnard 1975; Ashton 1981). The most recent work
presents three orders of conclusion about the australopithecines (Oxnard 1983a, 1984).

The first-order conclusion states that australopithecine postcrania are uniquely different from both apes and
humans. After initial scepticism even opposing investigators now confirm this (e.g. McHenry, Corruccini
1978). The second-order conclusion states that these
unique structures are most consonant with unique functions. Though bipedal they were also accomplished in the

trees. This second-order conclusion is now supported by
many functional studies (e.g. Stern, Susman 1983).
The third-order conclusion states that, if a different
morphology reflects a different locomotion, then it is
possible that creatures that displayed them arose independently of humans. This is not yet accepted. The alternative, that the curious anatomy of the australopithecines indicates a curious evolutionary mosaicism as it
evolved towards the human condition, is generally held.

One problem is that the morphometric studies rely upon
few fossils. It can scarcely be expected, suggests one
critic (McHenry 1976), that studies of a few individual
bones will change the minds of many anthropologists. This
criticism has validity but can be met by studies of teeth.
Can we examine tooth dimensions, not so much to learn
about teeth, but to learn about species? Such studies
distinguish sexual dimorphism. It is, therefore, necessary to summarize current knowledge of sexual dimorphism in
primates.

Males are generally regarded as typifying the species
(McCown 1984). In Elliot's (1913) review only 4 of 477
type specimens are female. In most museums female specimens are outnumbered by males two to one. Female form is
often expressed as a percentage of a "standard" form: the
male. Avoiding bias often falls into the opposite error:
regarding female form as the guide. Some studies use the
mean between the sexes to represent the species. As long
as female-male differences are viewed as lying on an axis
from small to large sexual dimorphism this seems not unreasonable. Mean position on the axis characterises a
given species; dispersion along the axis describes sexual
variation.

But recent studies of extant primate teeth (summarised
in Oxnard, 1985) show that dental sexual dimorphism is
complex. It involves simple scalar quantities: means and
dispersions for each sex, ratio quantities: relative numbers of each sex, and complex vector quantities: amount and
direction of multivariate and pattern differences. Recent
morphometric studies of primate body proportions (Oxnard
1983c) further imply: that sexual dimorphism is not unidimensional, that two major patterns exist in primates,
and that each hominoid has a unique sexual dimorphism.
These phenomena are important for studies of hominoid

fossils.

NEW MORPHOMETRIC STUDIES OF TEETH

Study of tooth dimensions in extant hominoids (Oxnard
1985) confirms that humans have small sexual dimorphism
and apes have large. It also shows that though African
apes have very large sexual dimorphism of variance, orangutans and humans do not. And it shows major differences
in multivariate patterns of sexual dimorphism in every
species. The fossils can be interpolated into these
results.

Measures of the lengths and breadths of a total of
2,295 teeth representing Homo sapiens (modern), Homo
sapiens neanderthalensis, Homo erectus and Homo habilis
have been analysed using both univariate and multivariate
statistical methods. The results show the basic similarity of sexual dimorphism in all these species. Bimodal
distributions exist at only a few tooth positions in each
group. Each group displays equal spreads for each mode
where bimodal distributions exist. The multivariate
amount and direction of differences between the modes
(putative males and females) are similar. The equal sizes
of the modes indicate that the number of specimens in
each mode (putative sex) is the same. For modern humans,
this is because the sex ratio is 1:1. If the fossil bi-
modality also represents sexual dimorphism, then their sex
ratios, too, are 1:1 (figs 1, 2).

Measures of the lengths and breadths of a total of 742
teeth representing four australopithecines have been
similarly analysed. The results show fundamental differences between all australopithecines and all Homo. Bimodal distributions with big distances between the modes
imply big sexual dimorphism in these forms. The spreads
of each mode differ, those for upper modes (putative
males) being greater than those for lower modes (putative
females). Multivariate amounts and directions of differences between putative sexes are different from both
modern humans and extant apes. The relative sizes of each
mode indicate a 2:1 ratio or more. Samples of extant apes
in which the sexes are artificially adjusted to a 2:1 fe-
male/male ratio are similar (figs 1, 2).

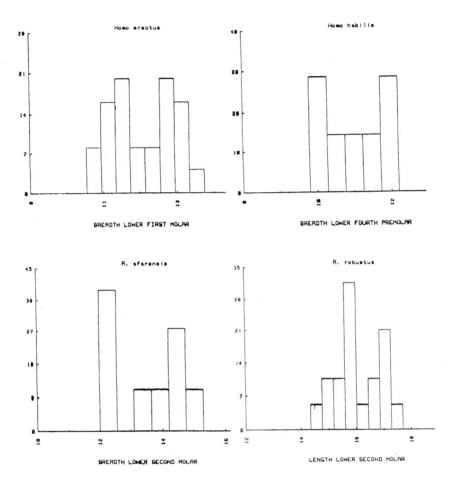

Figure 1 (above): Examples of univariate results with equal bimodal distributions in fossil <u>Homo</u> and unequal in Australopithecus. Similar results exist for many tooth positions and all fossils.

Figure 2 (following): Multivariate results comparing males (on the right) and females for each group. Top frame, pattern in <u>Homo</u> (H) contrasted with extant apes (C= <u>Pan</u>, G= <u>Gorilla</u>, and O= <u>Pongo</u>); middle frame: extant <u>Homo</u> (S) and some fossils (N= <u>neandertal</u>, E= <u>erectus</u>, H= <u>habilis</u>, R= <u>Ramapithecus</u>); bottom frame: australopithecines (R= <u>robustus</u>, B= <u>boisei</u>, A= <u>africanus</u>, F= <u>afarensis</u>).

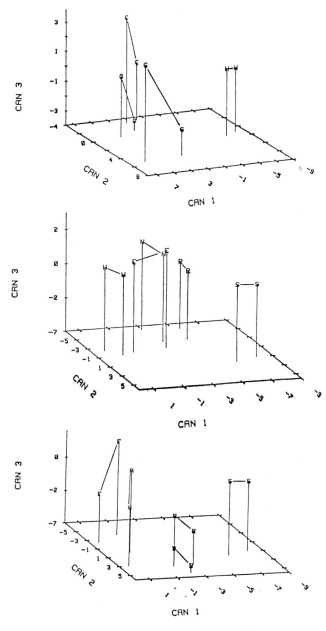

Fig 2

Measures of the lengths and breadths of a total of 954 teeth representing ramapithecines from China are available. Results (Wu, Oxnard 1983) discount the possibility that they are an ape-like species with small females and large males (Lipson, Pilbeam 1982). There are bimodal distributions with big distances between the modes for 20 dimensions in Sivapithecus, and bimodal distributions with small distances between the modes for 11 dimensions in Ramapithecus. The spreads of each mode are equal in both forms. Multivariately, Sivapithecus falls with orangutans, Ramapithecus with Homo erectus. And the number of specimens in each mode (putative sex) shows a ratio of 2:1 or more in Sivapithecus, but 1:1 in Ramapithecus. Sivapithecus resembles orangutans in which the ratio between the sexes has been artificially adjusted to 2:1. Ramapithecus resembles Homo (Figs 1, 2).

The data for Gigantopithecus from China include lengths and breadths from 736 teeth. Almost all dimensions show enormous distances between the modes of each distribution, but no differences in the spreads of each mode. The multivariate positions and patterns for Gigantopithecus are different from any other group so far examined. The numbers of specimens in each mode are the same, suggesting that Gigantopithecus had a 1:1 sex ratio.

DISCUSSION: PATTERN AND TIME

Fitting these findings into the view that includes australopithecines as on the human lineage, but excludes all early Chinese fossils as related to apes, requires three separate evolutionary parallels: the complex pattern of sexual dimorphisms and sex ratios that is common to Ramapithecus, Gigantopithecus, and Homo. Placing these findings into the scheme that recognises Ramapithecus as hominid and Sivapithecus as pongid, but still with australopithecines as prehuman ancestors, requires three evolutionary reversals: from ape-like sexual dimorphisms and sex ratios before Ramapithecus to human-like in Ramapithecus, back to ape-like in australopithecines, and back again to human-like in fossil and living Homo.

But interpolation of these findings into a system that sees australopithecines as sibling lines in a hominoid radiation avoids excessive parallels or reversals.

Human-like sexual dimorphisms and sex ratios in
Ramapithecus, Gigantopithecus, and fossil and living Homo
form a hominid radiation. Orangutan-like sexual dimorphisms and ratios in Sivapithecus are an off shoot of a
Pongo radiation. African ape-like sexual dimorphisms and
sex ratios are yet another. And the special combination
of sexual dimorphisms and sex ratios of australopithecines
are further extinct radiations.

This last system has an apparent problem to do with
time; "molecular clock" estimates link humans and African
apes at 4 m.y.BP (Gribbin , Cherfas 1983).   But recent
work (Goodman 1983; Cronin 1983; Sibley, Ahlquist
1984) give upper limits for that link at 7, 8 and 10
m.y.BP respectively. Discounting linear molecular evolution pushes it to 11 or 13 m.y.BP (Gingerich, 1984). But
another group has returned to a link at 1 m.y.BP
(Hasegawa, Kishino, Yano 1984).   These new figures
provide sufficient ambiguity to allow Ramapithecus to be
included in a radiation close to humans and not apes.

Finally these findings mean that structural sexual
dimorphism in hominoids should be re-evaluated. Different
types and patterns of sexual dimorphism exist rather than
just sexual dimorphism of small or large degree. The idea
that human sexual dimorphism is primarily a simple
size-related phenomenon is unlikely. The idea that small
sexual dimorphism means simply monogamy and big sexual
dimorphism polygyny cannot be the whole story. The
age-old idea that human sexual dimorphism was much larger
in relatively recent progenitors, resembling that in the
African apes of today, seems to be incorrect.

Ashton EH (1981). The australopithecinae; their
   biometrical study. Symp Zool Soc Lond 46:67.
Ashton EH, Healy MJR, Lipton S (1957). The descriptive
   use of disriminant functions in physical anthropology.
   Proc Roy Soc B 146:552.
Cronin JE (1983). Apes, humans and molecular clocks: a
   reappraisal. In Ciochon RL, Corruccini RS (eds): "New
   Interpretations of Ape and Human Ancestry." New York:
   Plenum, p 115.
Elliot DG (1913). A review of the primates. New York:
   Monogr Amer Mus Nat Hist 1 (3 vols).
Gingerich P (1984). Rates of evolution; effects of time
   and temporal scaling. Am J Phys Anthropol 63:163.

Goodman M (1983). "Macromolecular Sequences in Systematic and Evolutionary Biology". New York: Plenum.

Gribbin J, Cherfas J (1983). "The Monkey Puzzle". New York: Pantheon.

Hasegawa M, Yano T, Kishino H (1984). A new molecular clock of mitochondrial DNA and the evolution of the hominoids. Proc Jap Acad 60:95.

Lipson S, Pilbeam D (1982). Ramapithecus and hominoid evolution. J Hum Evol 11:545.

McCown ER (1984). Sex differences: the female as baseline for species descriptions. In Hall RL (ed): "Sexual Dimorphism in Homo sapiens: a Question of Size", New York: Praeger, p 37.

McHenry HM. (1976). A view of the hominid lineage. Science 189:988.

McHenry HM, Corruccini R S (1978). The femur in human evolution. Am J Phys Anthropol 49:473.

Oxnard CE (1975). "Uniqueness and Diversity in Human Evolution: Morphometric Studies of Australopithecines". Chicago: University Press.

Oxnard CE (1983a, 1984). "The Order of Man: a Biomathematical Anatomy of the Primates". Hong Kong: University Press: New Haven: Yale University Press.

Oxnard CE (1983b). Anatomical, biomolecular and morphometric views of the primates. Prog Anat 3:113.

Oxnard CE (1983c). Sexual dimorphisms in the overall proportions of primates. Amer J Primatol 4:1.

Oxnard CE (1985). "Fossils, teeth and sex: a new perspective on human evolution". Hong Kong: University Press.

Reyment RA, Blackith RE, Campbell N A (1984). "Multivariate Morphometrics". London: Academic Press.

Sibley CG, Ahlquist JE (1984). The phylogeny of the hominoids as indicated by DNA-DNA hybridization. J Mol Evol 20:2.

Stern JTJr, Susman RL (1983). The locomotor anatomy of Australopithecus afarensis. Am J Phys Anthropol 60:279.

Wu R, Oxnard CE(1983). Ramapithecus and Sivapithecus from China: some implications for higher primate evolution. Am J Primatol 5:303.

Zuckerman S (1970). "Beyond the Ivory Tower". New York: Taplinger.

Human Hominids That Say and Do

Hominid Evolution: Past, Present and Future, pages 281–286
© *1985 Alan R. Liss, Inc.*

EVOLUTION OF THE HOMINID UPPER RESPIRATORY TRACT:
THE FOSSIL EVIDENCE

Jeffrey T. Laitman, Ph.D.

Department of Anatomy
Mount Sinai School of Medicine
New York, New York 10029 USA

The human upper respiratory system comprises the
structures of the larynx, pharynx, nasal and oral cavities.
The region is the crossroads of our respiratory and
alimentary tracts, as well as the site for the production of
articulate speech. Proper coordination of the constituent
parts of this region is a prerequisite for normal human
existence.
     Despite the central role the upper respiratory system
has in human physiology, little is known about its role in
human evolution. The upper respiratory region of our
ancestors has been difficult to investigate primarily
because of the lack of preserved soft tissue structures. As
a result, the configuration of this region in fossil
hominids has remained a mystery.
     Although direct evidence such as a fossilized larynx
does not remain, our studies have shown that the basicranium
of fossil hominids can serve as a guide to reconstruct their
upper respiratory tracts. This paper will briefly review
some of our recent work which has addressed the question of
what this region may have been like in our ancestors.

THE MAMMALIAN UPPER RESPIRATORY TRACT

     Both classic postmortem examination of the upper
respiratory tract in mammals (e.g.Negus 1965), as well as
our cineradiographic studies of the region (Laitman et
al.,1977; Laitman,Crelin 1980a), have shown that the position
of the larynx is of prime importance in determining the way
an individual can breathe, swallow or vocalize. In most
mammals the larynx is positioned extremely high in the neck,

usually corresponding to the level of the first to third or fourth cervical vertebrae. This high position permits the epiglottis to pass up behind the soft palate to lock the larynx into the nasopharynx, providing a direct air channel from the nose through the nasopharynx, larynx and trachea to the lungs.

Due to this high larynx, liquids, or in certain species semi-solid material, can flow around the interlocked larynx and nasopharynx, via channels known as the piriform sinuses, en route to the esophagus. This condition permits the patency of the airway while liquid is transmitted around the larynx during deglutition. In essence, two separate pathways are created: a respiratory tract from the nose to the lungs, and a digestive tract from the oral cavity to the esophagus.

While this basic mammalian pattern - found with variations from dolphins to apes - enables an individual to breathe and swallow simultaneously, it severely limits the array of sounds an animal can produce. The high position of the larynx means that only a small supralaryngeal portion of the pharynx exists. In turn, only a very limited area is available to modify the initial sounds generated at the vocal folds (vocal cords). Most mammals thus depend primarily on altering the shape of the oral cavity and lips to modify laryngeal sounds. While some animals can approximate some human speech sounds, they are anatomically incapable of producing the range of sounds necessary for complete, articulate speech.

THE HUMAN UPPER RESPIRATORY TRACT

The upper respiratory tract of the human newborn is remarkably similar to the condition found in most other primates. The larynx is located high in the neck, permitting separate breathing and swallowing pathways to exist. Our studies have demonstrated that newborns and young infants indeed breathe, swallow and vocalize in much the same fashion as monkeys or apes (Laitman, Crelin 1980b).

The infant larynx remains high in the neck until approximately one and one-half to two years of age. Sometime around the second year a dramatic change begins to occur. The larynx begins to descend in the neck, an alteration which will change considerably the way the child breathes, swallows and vocalizes. The precise time this shift occurs or the physiologic mechanisms which underlie it are still poorly understood. The end result, however, is a condition in which the larynx has migrated to a much lower position in the neck than is found in any other mammal.

In adult humans, the larynx extends from the level of the 4th almost to the 7th cervical vertebra. The low position of the larynx means that it can no longer lock into the nasopharynx and separate the respiratory and digestive pathways. After laryngeal descent these pathways cross above the larynx. This crossing can, and all too often does, have serious drawbacks. For example, a bolus of food can easily lodge in the entrance to the larynx and, if not expelled rapidly, lead to suffocation.

The descent of the larynx does produce a feature of enormous positive value: a greatly expanded supralaryngeal portion of the pharynx. The ability of the pharynx to modify laryngeal sounds is thus much greater than that possible for newborns or any nonhuman mammal. In essence, it is this expanded pharynx which gives us the unique ability to produce a full range of speech sounds.

THE BASICRANIUM AND UPPER RESPIRATORY TRACT

The position of the larynx obviously has a critical effect in determining the way an individual can breathe, swallow or vocalize. The problem posed for us was how its position in fossil hominids could be determined. As noted, soft structures are not preserved in fossil specimens and certain bony landmarks, such as the hyoid bone or styloid process, which can give clues to the position of upper respiratory structures, are also often missing.

Fortunately, one portion of the upper respiratory region which often does remain is its "roof", the base of the skull. Our investigation of this region in extant primates has shown that a relationship exists between the degree of flexion of the basicranium and the location of structures such as the larynx and pharynx. The amount of flexion of the skull was determined through a series of nine linear measurements taken between five craniometric points on the exocranial surface of the skull base. These measurements generated a basicranial line that indicated the degree of flexion present (see Laitman et al., 1978; Laitman, Heimbuch 1984 for details.)

Our analyses indicated that two basic patterns exist concerning the relationship of the skull base to the larynx and associated structures. In the first, the basicranium is relatively "flat" or nonflexed and corresponds to a larynx positioned very high in the neck. This pattern was found to exist consistently in all the mammals we studied except for

older humans. In the second pattern the basicranium is
found to be markedly arched, or flexed, and corresponds to a
larynx positioned quite low in the neck. This pattern is
found only in humans after the early years of life.

RECONSTRUCTING THE REGION IN FOSSIL HOMINIDS

The relationships observed in extant mammals between
the angulation of the basicranium and the position of the
larynx proved to be a valuable tool in reconstructing what
this area may have been like in fossil forms. If, for
example, the skull of a fossil hominid was found to have
little or no basicranial flexion, similar to the condition
found in living chimpanzees or monkeys, we would reconstruct
the position of the larynx high in the neck. Conversely,
fossil hominids exhibiting marked basicranial flexion,
similar to the condition found in modern adult humans, would
be reconstructed with an upper respiratory tract much like
our own. Once the position of the larynx in these early
hominids is determined, it is possible to infer their
breathing, swallowing and vocalizing patterns.

Based upon these relationships in extant primates, we
have examined the basicrania of fossil hominids ranging from
Plio-Pleistocene australopithecines through late Pleistocene
forms and have offered reconstructions of what their upper
respiratory tracts may have been like (Laitman et al., 1979;
Laitman, Heimbuch 1982; Laitman 1983). Our analysis of the
early Plio-Pleistocene hominids shows that they exhibit
basicrania which closely approximate the general nonflexed
condition of the extant apes, while being considerably
different from the highly flexed skulls of modern humans.
The basicranial similarities between the australopithecines
and extant apes suggest that their upper respiratory tracts
were also similar in appearance. As with the living
nonhuman primates, the early hominids probably exhibited a
larynx positioned high in the neck. This, in turn, would
have allowed them to breathe and swallow liquids
simultaneously. The high laryngeal position would also
greatly restrict the supralaryngeal portion of the pharynx
available to modify sounds. As a result, these early
hominids probably had a very restricted vocal repertoire
compared with that of modern adult humans. For example, the
high larynx would have made it impossible for them to
produce some of the universal vowel sounds found in human
speech patterns (see also Lieberman 1984).

The upper respiratory tract of the australopithecines

Figure 1. Reconstruction of the upper respiratory tract of
an australopithecine during normal respiration. Note the
high position of the larynx which provides a direct airway
from the nose to the lungs.

thus may have been essentially like those of the living
apes. The key question which follows is when did change
begin to occur en route to the pattern shown today by modern
humans? Our preliminary data on the crania of some Homo
erectus specimens such as KNM-ER 3733 are providing valuable
clues. It is among specimens of Homo erectus that we begin
to find the first instances of basicranial flexion, away
from the nonflexed pattern of the extant apes and
australopithecines, and towards that of modern Homo sapiens.
These basicranial changes suggest that a restructuring of
upper respiratory anatomy away from the ape-like pattern of
the australopithecines and towards that of modern humans may
have begun with this group.
    The crucial restructuring of the hominid upper
respiratory tract may thus have begun some 1 1/2 million
years ago with Homo erectus. The incipient degree of
basicranial flexion in this group indicates that the larynx
may have begun its descent into the neck, thus changing the

basic mammalian upper respiratory pattern retained by our australopithecine ancestors. The lowering of the larynx would also increase the pharyngeal area available to modify laryngeal sounds. While change may have begun at this time, the first instances of full basicranial flexion similar to that of modern humans does not appear until the arrival of Homo sapiens some 300,000 to 400,000 years ago. It may have been at this time that hominids with upper respiratory tracts similar to ours first appeared.

ACKNOWLEDGEMENTS

Supported by NSF grant BNS 8025476 and by grants from the Foundation for Research into the Origins of Man.

REFERENCES CITED

Laitman JT (1983). The evolution of the hominid upper respiratory system and implications for the origins of speech. In de Grolier E (ed): "Glossogenetics: The Origin and Evolution of Language", Paris: Harwood Academic Publishers, p 63.

Laitman JT, Crelin ES (1980a). Tantalum markers as an aid in identifying the upper respiratory structures of experimental animals. Lab Anim Sci 30:245.

Laitman JT, Crelin ES (1980b). Developmental change in the upper respiratory system of human infants. Perinatol/Neonatol 4:15.

Laitman JT, Crelin ES, Conlogue GJ (1977). The function of the epiglottis in monkey and man. Yale J Biol Med 50:43.

Laitman JT, Heimbuch RC (1982). The basicranium of Plio-Pleistocene hominids as an indicator of their upper respiratory systems. Am J Phys Anthropol 59:323.

Laitman JT, Heimbuch RC (1984). A measure of basicranial flexion in the pygmy chimpanzee, Pan paniscus. In Susman RL (ed):"The Pygmy Chimpanzee: Evolutionary Biology and Behaviour", New York: Plenum Press, p 49.

Laitman JT, Heimbuch RC, Crelin ES (1978). Developmental change in a basicranial line and its relationship to the upper respiratory system in living primates. Am J Anat 152:467.

Laitman JT, Heimbuch RC, Crelin ES (1979). The basicranium of fossil hominids as an indicator of their upper respiratory systems. Am J Phys Anthropol 51:15.

Lieberman P (1984). "The Biology and Evolution of Language", Cambridge: Harvard University Press.

Negus VE (1965). "The Biology of Respiration", London: E and S Livingstone.

Hominid Evolution: Past, Present and Future, pages 287–298
© *1985 Alan R. Liss, Inc.*

EARLY ACHEULEAN WITH *HOMO HABILIS* AT STERKFONTEIN

R.J. Clarke, Ph.D.

Senior Research Officer, Dept. of Anatomy
Palaeoanthropology Research Group
University of the Witwatersrand
Johannesburg   2001   South Africa

Since 1956 when C.K. Brain first discovered apparently
ancient artefacts in overburden at Sterkfontein (Robinson
1957), the site has been divided into two breccia areas. The
Type Site at the eastern end (*Figure 1*) has yielded large
numbers of *Australopithecus africanus* fossils but no stone
tools (Broom *et al* 1950). The second area west of the Type
Site, designated the Extension Site by Robinson (1957), has
yielded numerous ancient artefacts and a few hominid remains,
notably the Stw 53 cranium of *Homo habilis* (Hughes, Tobias
1976). Excavations of the Extension Site overburden conducted
by R.J. Mason and C.K. Brain in 1957 yielded Middle and Late
Stone Age artefacts (Mason 1957). Excavations of the
underlying breccia by J.T. Robinson later in 1957 yielded an
apparently archaic industry consisting of crude cores and
flakes. Mason (1957) considered the artefacts to have some
similarities to the Oldowan, but also noted that the more
complex flaking techniques showed marked similarities to the
African "Chelles-Acheul". He further observed that although
the sample was small, the presence of a handaxe-like tool
suggested that further excavations might reveal more advanced
tool types. When Robinson continued the excavation of the
red-brown breccia--later classified as Middle Breccia
(Robinson 1962), now known as Member 5 (Partridge 1978)--he
recovered among other crude artefacts a proto-handaxe. This
handaxe, plus the technology and typology of the other
artefacts excavated by Robinson, suggested to Mason (1962)
that the industry should be assigned to what he then termed
the Earlier Chelles-Acheul, *i.e.*, Early Acheulean. Mason
could not agree with those who classed the Sterkfontein
artefacts as Oldowan and he has steadfastly adhered to that

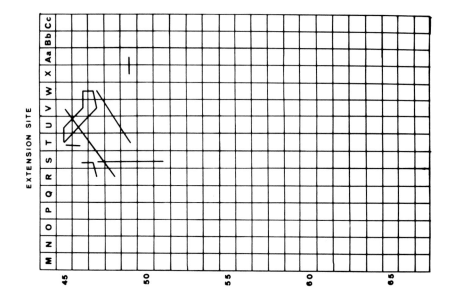

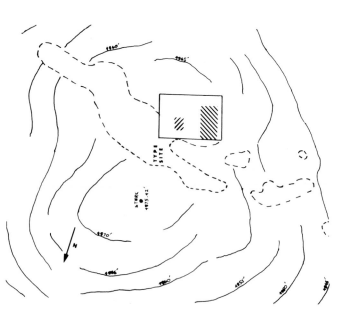

Figure 1. Sterkfontein. Large square is part
of Hughes's grid used in this study, enlarged at
right (see text for explanation). Small hatched
area is Mason and Brain's excavation. Large
hatched area is Robinson's grid.
Scale: large rectangle is 45 x 72 feet.

opinion (Mason 1976).

Mary Leakey (1970, 1971), after studying the early collection of artefacts from Sterkfontein, concluded that they compared well with the Developed Oldowan B from Olduvai Bed II. She stated that sites at Olduvai were classified as Acheulean if bifaces comprised 40 percent or more of the tools (Leakey 1971: 2) and observed that at Sterkfontein bifaces occurred in almost the same proportions as in the Developed Oldowan (p. 273). An attribute analysis by Stiles (1979) on the Sterkfontein artefacts showed that Mason's handaxe, plus four handaxes from the overburden and one from a dump (recovered by A.R. Hughes), clustered exclusively with the Early Acheulean handaxes of Olduvai EF-HR. Volman (1984) considered that, on the data available at the time the Sterkfontein artefacts may be comparable to Oldowan/Developed Oldowan A assemblages from East Africa, and not Acheulean/ Developed Oldowan B. All the true bifaces were found in unconsolidated deposits or where derivation from Member 5 was uncertain, and he thought it possible that Member 5 may contain one Early Stone Age entity and the later deposits another, or that more than one archaeological entity was represented in Member 5. Clarke (1985) considered that the presence in Member 5 of the *Homo habilis* cranium Stw 53, so similar to OH 24 that is contemporary with Oldowan artefacts at about 1.8 m.y., plus the absence of *Paranthropus* known to be existing at Swartkrans *ca*. 1.5 m.y. ago, suggested that the Sterkfontein industry was older than 1.5 m.y. and therefore could represent a facies of the Oldowan. Although I agree with Mason (1976) that Mary Leakey's separation of Acheulean from Developed Oldowan on handaxe percentages is not satisfactory, I did question whether there were any handaxes at all from the *in situ* deposits. Mason's proto-handaxe from *in situ* breccia could also be considered an elongated chopper. The other bifaces studied by Stiles had all come from overburden and dumps. I therefore decided to examine and plot vertical distributions of all artefacts recovered 'from *in situ* deposits by Mr. A.R. Hughes.

CURRENT RESEARCH

The more than one thousand artefacts examined have been recovered by Mr. Alun Hughes during his excavations through overburden, decalcified breccia and solid breccia. The material from the overburden and at the top of decalcified

breccia pockets consists mainly of Middle Stone Age artefacts,
but beneath this in the solid breccia and lower decalcified
breccia there are artefacts of a more primitive kind. With
the exception of material near the surface of the excavation,
there is no reason to doubt that the artefacts recovered from
deeper regions of the unconsolidated, decalcified breccia are
contemporary with the surrounding consolidated breccia. The
calcified and decalcified breccias are one unit, as is
demonstrated by the fact that fragments of the Stw 53 cranium
were recovered from a decalcified pocket of Member 5 breccia,
while the rest of the cranium was recovered from the
adjacent solid breccia (Tobias 1978: 248).

In addition to several hundred manuports and quartz
chips and chunks which have not been included in the plots,
the following artefacts made out of quartz, quartzite, chert
and diabase have been examined for this study: 682 flakes,
166 cores, 15 core fragments, 21 retouched flakes, 19
choppers, 3 polyhedrons, 3 sub-polyhedrons, 2 cleavers, 2
possible cleavers and 2 handaxes. A small number of these
pieces diagnostic of the Middle Stone Age are concentrated
near the top of the excavation. Of the remaining artefacts,
most of the choppers and cores would not be out of place in
the Oldowan, but similar artefacts occur also at Olduvai
Bed II site EF-HR with the Early Acheulean (Leakey 1971:
124-37). Among the choppers and cores from deep levels in
Member 5 are two definite and two possible cleavers made on
side-struck flakes, one handaxe on a flake, and one chert
bifacial handaxe (*Figure 2*). As I made moulds and coloured
plastic casts of the Olduvai artifacts for Mary Leakey in
1967-68, I have first-hand knowledge of them. I can thus
state that technologically and typologically the Sterkfontein
handaxes and cleavers are like those of the Olduvai ER-HR
Early Acheulean. The Sterkfontein handaxe on a flake is
virtually identical to the EF-HR handaxe on a flake
illustrated by Leakey (1971: 130).

The right of *Figure 1* shows the area of the Hughes
excavation grid used for this study. Each square measures 3
ft. by 3 ft. Towards the eastern end, large quantities of
A. africanus remains were found but without artefacts,
whereas at the western end large numbers of artefacts and 13
hominids which are attributable to *Homo* have been recovered.
Each thick line across the grid represents the distribution
in horizontal distance of fragments of a single individual
of *A. africanus*. The greatest horizontal distance separating

Figure 2. Member 5 Early Acheulean artefacts. Top, left to right: 2 cleavers on flakes with bulbs trimmed away and a core. Bottom, left to right: unifacial handaxe on a flake, bifacial handaxe, chopper. Bifacial handaxe is made on chert and all other pieces on quartzite. Scale in centimeters.

parts of one individual is 15 ft. and the greatest vertical
separation is 6 ft. 11 inches. The closed lines show the
distribution of 25 fragments of one individual over a depth
of 2 ft. 1 inch.

In the East-West vertical distribution plot along the T
axis (*Figure 3*), there is a well-defined separation between
the eastern area containing *A. africanus* but no stone
tools and the western area containing numerous artefacts.
The dotted line marks the approximate separation between
Members 4 and 5. Another section drawn along the R axis
showed the same configuration. A composite vertical
distribution plot along the East-West axes M to V (*Figure 4*)
shows some of the more diagnostic artefacts, excluding the
Middle Stone Age. These are some of the cores and choppers
and all of the cleavers and handaxes, plotted in relationship
to all the hominid remains beginning with North-South axis 44.
Again, the separation between *Australopithecus* and the
artefacts is clearly seen. It is significant that all four
hominid humeral distal ends found in the *Australopithecus*
area have their articular surfaces chewed away by a carnivore,
probably a leopard. Furthermore, the *Australopithecus*
accumulation is interspersed with large numbers of fossilised
sticks. The conformation of this accumulation is thus
suggestive of a talus cone and it seems probable that a
leopard and/or sabre-toothed cat was dropping most of the
Australopithecus remains into the cave, probably from trees
around a hole in the roof similar to those seen at
Sterkfontein today. Such a scenario was suggested by Brain
(1970) for the *Paranthropus* accumulation at Swartkrans.

The current excavation in the *Australopithecus* talus
cone has reached 28 ft. below datum and hominid remains are
still being retrieved in large quantity. The clear vertical
separation of the talus cone from the tool area and *Homo*
fossils and the absence of any *Paranthropus* fossils or
artifacts from the talus cone show that it can be considered
as one distinct unit. It is apparent, however, that the talus
cone accumulated over a very long period of time and it is
pertinent to wonder whether all the hominids within it are
of the species *A. africanus*. From the lowest levels reached
so far, there are very large molars and premolars typical of
Australopithecus and there are molars and premolars of very
small dimensions comparable to *Homo*. Although this material
has yet to be studied in detail and one must consider sexual
and individual variation, I believe it does raise the

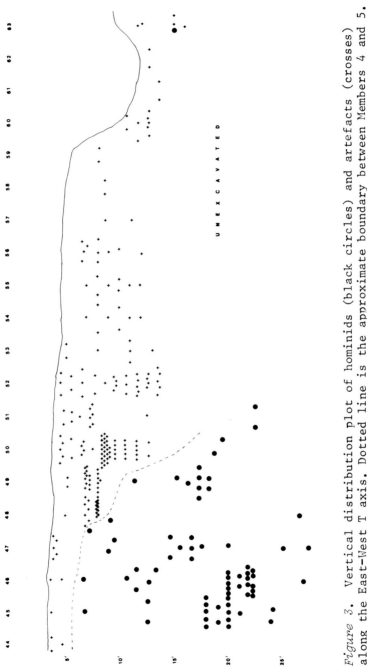

Figure 3. Vertical distribution plot of hominids (black circles) and artefacts (crosses) along the East-West T axis. Dotted line is the approximate boundary between Members 4 and 5.

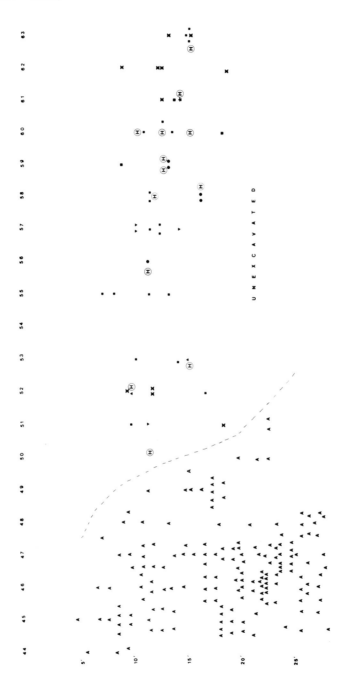

Figure 4. Composite vertical distribution plot, East–West along axes M to V, to show all *Australopithecus* fossils (A), all *Homo* fossils (H), and some diagnostic tools. Upright triangle = handaxe, inverted triangle = cleaver, X = chopper, square = core, black circle = flake, star = bifacial core.

question of whether a small-toothed, less-specialised form of
Australopithecus was contemporary with the large-toothed
A. africanus.

THE MEMBER 5 HOMINIDS

The first hominids associated with artefacts were six
specimens recovered in 1957-58 by Robinson (1958). These
comprised a juvenile maxilla fragment and isolated teeth and
fragments. The following table lists the hominids and
directly associated artefacts excavated from Member 5 by
A.R. Hughes.

Stw Number	Specimen	Grid Square	Depth	Associated Artefacts
53	cranium	V-60	12'8"	---
75-78	P3,C,I2,I1	V-60	9'7"-10'7"	---
80-83	crushed mandible and teeth	Q-60	14'8"-15'8"	core (below)
84	L. mandibular corpus	M-61	13'8"-14'8"	bifacial core
85	R. mandibular corpus	R-56	10'9"-11'9"	flake, core
86	molar chip	T-63	11'9"-12'7"	2 cores, sub-polyhedron
87	R. P4̄	R-58	15'8"-16'8"	2 flakes
88	R. talus	R-59	12'3"-13'3"	flake
89	L. 2nd metatarsal	S-59	12'0"-13'4"	flake
94	L. M3̄	V-58	11'2"-12'0"	---
95	R. C and P3	V-52	9'4"-9'11"	2 flakes (below)
169	L. I2	V-50	10'8"-11'8"	---
311	R. femur head	R-53	14'5"-15'5"	chopper, core

Stw 53, Stw 80, and Stw 84 can be assigned to *Homo habilis*,
whilst there is no reason to exclude the other more
fragmentary or postcranial remains from that species. In 7
cases out of the 13, there is a direct association of
artefacts with hominids, with both coming from the same
square and depth.

Clarke (1985) considered that the presence in Member 5
of a *Homo habilis* cranium similar to the 1.8 m.y. OH 24 and,
the absence of *Paranthropus* which is so common at nearby
Swartkrans, would indicate that Member 5 is older than
Swartkrans, *i.e.*, older than 1.5 m.y. The now proven

association of Early Acheulean artefacts with *Homo habilis*
fossils in Member 5 is of great significance. Firstly,
although the dating of the Early Acheulean from East Africa
is uncertain, it does seem to have appeared after 1.5 m.y.
based on the estimated age of 1.4 m.y. for EF-HR (Isaac 1982:
174). The possibility also exists, however, that the
Acheulean could be much older than has so far been realised.
It would certainly seem that Sterkfontein Member 5 is older
than 1.5 m.y. and may date to about 1.6 m.y. based on the
presence of *H. habilis* and the absence of *H. erectus* and
Paranthropus. Secondly, Sterkfontein provides the first
indication that *H. habilis* was making Early Acheulean
artefacts. At Olduvai, *Homo habilis* is associated only with
Oldowan artefacts (Leakey 1971: 3, 272). The earliest *Homo
erectus* is known from the same deposits that yielded
Paranthropus at both Swartkrans and East Lake Turkana at *ca.*
1.5 m.y. (Clarke 1985). The most recent *H. habilis* is OH 13
from Olduvai MNK II and that could be about 1.6 m.y. based on
its position just above Tuff IIa dated to about 1.7 m.y.
(Hay 1971: 14). At no site is there any indication of *Homo
habilis* existing at the same time as or more recently than the
earliest *Homo erectus* and all indications are that *H. habilis*
probably developed into *H. erectus* some time before 1.5 m.y.
Thus it seems highly improbable that the Sterkfontein Early
Acheulean interspersed with *Homo habilis* fossils could have
been made by *Homo erectus* living at the same time. The
possibility that the handaxes and cleavers were made at a
later date by *H. erectus* and fell onto a talus slope
containing *H. habilis* and Oldowan artefacts is also
improbable. The handaxes and cleavers are randomly dispersed
among the other artefacts and no *Paranthropus* fossils (known
to be abundant at nearby Swartkrans at the time of *H. erectus*)
are found in Member 5.

That the Stw 53 cranium is so similar to the OH 24 *Homo
habilis* of 1.8 m.y. ago does not mean that it is also of that
age. A left mandibular corpus (Stw 84) directly associated
with a bifacial core in Member 5 has a low, thick corpus very
similar to that of OH 13, the most recent *H. habilis*. The
OH 13 mandible fits very well the cranium of the *H. habilis*
KNM ER 1813 from East Lake Turkana, and the maxillary
dentition and cranial vault sections of OH 13 are very
similar to 1813. Thus it is probable that Stw 53 and OH 24
are males of *H. habilis* and OH 13, KNM ER 1813 and Stw 84
are females.

DEPOSITION OF MEMBER 5 BRECCIA

The large number of flakes in mint condition suggest that they did not move far from the point of origin. Entrance to caverns in the Sterkfontein area today is usually by means of vertical slots with deep drops into the passage beneath. These slots are mostly surrounded by trees and bushes. It is thus probable that early *Homo* used such shaded areas for tool making and tool using activities and the fresh waste flakes, cores, and occasional handaxes, etc. were washed by rain into the slots, together with soil and bone. Excavations in Member 5 have not reached the same depth as in the Member 4 talus and deeper levels may yet yield Oldowan artefacts. At present, however, there is a temporal disconformity between the Member 4 talus cone with *A. africanus* and no tools and the Member 5 breccia with *H. habilis* and early Acheulean artefacts.

CONCLUSION

Sterkfontein Member 4 comprises a talus cone containing *Australopithecus* remains probably dropped by carnivores. Member 5 abuts vertically against Member 4 and contains *Homo habilis* specimens associated with Early Acheulean artefacts, possibly 1.6 m.y. old.

ACKNOWLEDGEMENTS. My special thanks are due to my wife, Kathleen Kuman, for assistance with the sorting, cataloguing and analysis of artefacts, for her time and care in collating data, for preparing the vertical distribution plots and the diagrams, and for comments on and typing of this article. I also extend my gratitude to Miss Janet Clarke and Dr. Yvette Deloison for assistance with sorting and cataloguing. I thank Professor P.V. Tobias and Mr. Alun Hughes for providing the opportunity for me to study the material from their excavations. *Figure 1* (left) is adapted from a plan drawn by Mr. Hughes.

Brain CK (1970). New finds at the Swartkrans australopithecine site. Nature 225:1112.
Brain CK (1981). "The Hunters or the Hunted." Chicago: University of Chicago Press.
Broom R, Robinson JT, Schepers GWH (1950). "Sterkfontein

Ape-man, Plesianthropus." Pretoria: Transvaal Museum
    Memoir No. 4.
Clarke RJ (1985). *Australopithecus* and early *Homo* in southern
    Africa. In Delson E (ed): "Ancestors: the Hard Evidence."
    New York: Alan R. Liss.
Hay RL (1971). Geologic background of Beds I and II. In
    Leakey 1971, p. 9.
Hughes AR, Tobias, PV (1977). A fossil skull probably of the
    genus *Homo* from Sterkfontein, Transvaal. Nature 265:310.
Isaac G L1 (1982). The earliest archaeological traces. In
    Clark JD (ed): "The Cambridge History of Africa, Vol 1."
    Cambridge University Press.
Leakey MD (1970). Stone artefacts from Swartkrans. Nature
    225:1222.
Leakey MD (1971). "Olduvai Gorge, Vol 3." Cambridge Univ Press.
Mason RJ (1957). Occurrence of stone artefacts with
    Australopithecus at Sterkfontein. Nature 180:523.
Mason RJ (1962). The Sterkfontein stone artefacts and their
    maker. So Afr Arch Bull 17:109.
Mason RJ (1976). The earliest artefact assemblages in South
    Africa. Nice: IXth Cong Union Intl des Sciences Prehist et
    Protohist, p 140.
Partridge TC (1978) Re-appraisal of lithostratigraphy of
    Sterkfontein hominid site. Nature 275:282.
Robinson JT (1957). Occurrence of stone artefacts with
    Australopithecus at Sterkfontein. Nature 180:521.
Robinson JT (1958). The Sterkfontein tool-maker. The Leech
    28:94.
Robinson JT (1962). Australopithecines and artefacts at
    Sterkfontein. So Afr Arch Bull 17:87.
Stiles D (1979). Recent archaeological findings at the
    Sterkfontein site. Nature 277:381.
Tobias PV (1978). The earliest Transvaal members of the
    genus *Homo* with another look at some problems of hominid
    taxonomy and systematics. Z Morph Anthrop 69:225.
Volman TP (1984). Early prehistory of southern Africa. In
    Klein RD (ed): "Southern African Prehistory and
    Paleoenvironments." Rotterdam: AA Balkema, p 169.

Hominid Evolution: Past, Present and Future, pages 299–301
© *1985 Alan R. Liss, Inc.*

STERKFONTEIN MEMBER 5 STONE ARTEFACT IDENTIFICATION AND THE
KLIPPLAATDRIF GRAVELS

Revil J. Mason, PhD (Cape Town)

Director, Archaeological Research Unit

University of the Witwatersrand, 1 Jan Smuts Ave,
Johannesburg

C.K. Brain's 1956 discovery of stone artefacts in the
Sterkfontein breccia rubble was a critically important
discovery.   For the first time the world of the australopithecines was placed in relation to the world of Stone Age
Man in terms of material evidence in a stratified deposit.
Brain's discovery was recorded by J.T. Robinson.   I conducted the 1957 excavation of the Sterkfontein site and
noted marked similarities between the Sterkfontein breccia
artefacts and the Acheulean assemblages from the Klipplaat-
drift site 121 km south of Sterkfontein (Robinson, Mason
1957).   Brain's discovery of the Sterkfontein Member 5
artefacts posed a problem of identification.   In the Transvaal Early Stone Age, artefact identification depends on the
only well dated Early Stone Age site yet excavated here which
is the Cave of Hearths Early Stone Age Bed 1-3 accumulation.
Cave of Hearths Early Stone Age Acheulean Bed 1-3 accumulation
has 2215 stone artefacts, apart from thousands of waste fragments, associated with a late Pleistocene fauna which H.B.S.
Cooke and R.F. Ewer found to be similar to faunas from other
Acheulean sites between Hopefield and Olduvai, but lacking the
sivatheres, elephants and suids at the latter sites.   The
Cave of Hearths Early Stone Age Acheulean assemblage provides
a model for assessing similarity or lack of similarity to
other local assemblages including Sterkfontein Member 5 and
Klipplaatdrif.

The Olduvai Gorge assemblages have been used by Leakey
(1970), Bower (1977) and Stiles (1979) to assess and identify
the Sterkfontein Member 5 artefacts.   These writers made two
critical errors.   They ignored the local, dated, Early Stone
Age assemblages such as Cave of Hearths Beds 1-3 assemblages.

Their second error was to totally ignore or make only partial
or oblique reference to publications on Sterkfontein Member 5
artefacts by local workers such as myself (Mason in Robinson
and Mason 1957; Mason 1961, 1962, 1976).

The Cave of Hearths Early Stone Age Bed 1-3 assemblage is
directly comparable with the Sterkfontein Member 5 assemblages,
because it is located less than 10 days' walking time from
Sterkfontein, whereas the Olduvai assemblages are located six
months' to a year's walking time from Sterkfontein and were
therefore located in a substantially different environment.

The Sterkfontein Member 5 assemblage available to me in
1962 was limited to 98 shaped stone tools and 188 hand-sized
natural rocks foreign to the cave dolomite.   The Sterkfontein
Member 5 assemblage represents the same eight artefact classes
present in both the Cave of Hearths and Klipplaatdrif
assemblages.   The Klipplaatdrif assemblage is located about
two days' walking time due south of Sterkfontein (Fig 1).   The
Klipplaatdrif assemblage consists of 882 artefacts representing eight distinct classes of shaped stone artefacts and one
class of random shaped stone tools.   The Cave of Hearths,
Sterkfontein and Klipplaatdrif artefact assemblages consist
of the same artefact classes.   The only difference between
the Cave of Hearths and the Sterkfontein-Klipplaatdrif
assemblages is the increased use of a large number of secondary
trimming flakes for the shaping of the Cave of Hearths handaxe, cleaver and discoid classes.   The close similarity
between the Cave of Hearths, Sterkfontein and Klipplaatdrif
assemblages suggests membership of the same technological
tradition, identified as the Acheulean tradition.   Because
the Sterkfontein Member 5 assemblage is associated with a
fauna including archaic animals which are absent from the Cave
of Hearths Bed 1-3 faunal assemblage, I located the Sterkfontein Member 5 assemblage in the Early Acheulean stage of the
Acheulean tradition.   The Klipplaatdrif site enables us to
see a broader context for the Sterkfontein Member 5 tool-
makers, who undoubtedly moved between Klipplaatdrif and
Sterkfontein and very probably further afield (Mason 1962).

These observations were recorded in papers published in
1957, 1961 and 1962 and, in substantial detail, in 1976
(Mason 1961, 1962, 1976). This is worth stressing again
since two subsequent studies of the Sterkfontein implements,
those of Bower (1977) and of Stiles (1979), have unaccount-
ably and regrettably failed to acknowledge these much earlier

identifications and extensive discussion of the Sterkfontein Member 5 artefacts.

REFERENCES

Bower JRF (1977). Attributes of Oldowan and Lower Acheulean tools: 'tradition' and design in the Early Lower Paleolithic. S Afr Archaeol Bull 32:113.

Leakey MD (1970). Stone artefacts from Swartkrans. Nature 225:1124.

Mason RJ (1961). The Acheulean culture in South Africa. S Afr Archaeol Bull 26:107.

Mason RJ (1962). "Prehistory of the Transvaal". Johannesburg: Witwatersrand University Press

Mason RJ (1976). The earliest artefact assemblages in South Africa. IXe Congres Union Internationale des Sciences Préhistoriques et Protohistoriques Colloque V. 140.

Robinson JR, Mason RJ (1957). Part 2 - Occurrence of stone artefacts with Australopithecus at Sterkfontein. Nature 180:523.

Stiles DN, Partridge TC (1979). Results of recent archaeological and palaeoenvironmental studies at the Sterkfontein Extension Site. S Afr J Sci 75:346.

Hominid Evolution: Past, Present and Future, pages 303–309
© *1985 Alan R. Liss, Inc.*

THE KLIPPLAATDRIF GRAVELS : MORPHOLOGY, AGE AND DEPOSITIONAL
ENVIRONMENT WITH SPECIAL REFERENCE TO COMPARATIVE EVIDENCE
FROM THE STERKFONTEIN EXTENSION SITE

T.C. Partridge, Ph.D.

University of the Witwatersrand,
Johannesburg.

THE KLIPPLAATDRIF GRAVELS

    A profile recorded by the author in trench CD 9-10 of
RJ Mason's 1960 excavation through the Klipplaatdrif gravels
(Mason 1962) is shown in Fig 1. The excavation is in the
6m terrace of the Klip River near Vereeniging, some 1.5 km
above its confluence with the Vaal, and was taken down to
bedrock, which consists of shales of the Volksrust Formation (Karoo Supergroup). The alluvial gravels of the Klip
River are overlain locally by some 55cm of colluvium deposited after the abandonment of the terrace following
channel incision. The gravels contain an upper, clast
supported horizon some 15cm thick, which represents a late
phase of reworking of the underlying deposits. The main
body of the terrace consists of some 70cm of matrix supported gravels within which the bulk of the Acheulean industry is preserved. The matrix is a red, well structured and
ferruginized loam containing about 18% clay and 27% silt;
chemical analyses show it to be eutrophic, indicating that
the ferruginization is due to the addition of iron oxides
rather than to relative enrichment through leaching.
Chert and quartzite are dominant lithologies in the gravel
and cobble fractions, which, although mostly subrounded,
contain a significant proportion of subangular elements.
Finer gravel lenses, with a higher clay content in the
fine matrix, are present locally within the coarser deposits. Stratification and imbrication are absent throughout.

PROFILE RECORDED IN R.J. MASON'S 1960 EXCAVATION (TRENCH CD9-10)

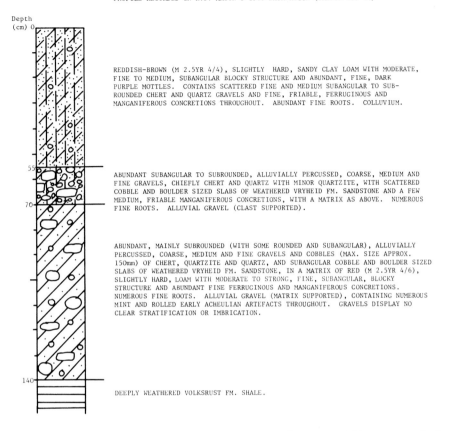

REDDISH-BROWN (M 2.5YR 4/4), SLIGHTLY HARD, SANDY CLAY LOAM WITH MODERATE, FINE TO MEDIUM, SUBANGULAR BLOCKY STRUCTURE AND ABUNDANT, FINE, DARK PURPLE MOTTLES. CONTAINS SCATTERED FINE AND MEDIUM SUBANGULAR TO SUB-ROUNDED CHERT AND QUARTZ GRAVELS AND FINE, FRIABLE, FERRUGINOUS AND MANGANIFEROUS CONCRETIONS THROUGHOUT. ABUNDANT FINE ROOTS. COLLUVIUM.

ABUNDANT SUBANGULAR TO SUBROUNDED, ALLUVIALLY PERCUSSED, COARSE, MEDIUM AND FINE GRAVELS, CHIEFLY CHERT AND QUARTZ WITH MINOR QUARTZITE, WITH SCATTERED COBBLE AND BOULDER SIZED SLABS OF WEATHERED VRYHEID FM. SANDSTONE AND A FEW MEDIUM, FRIABLE MANGANIFEROUS CONCRETIONS, WITH A MATRIX AS ABOVE. NUMEROUS FINE ROOTS. ALLUVIAL GRAVEL (CLAST SUPPORTED).

ABUNDANT, MAINLY SUBROUNDED (WITH SOME ROUNDED AND SUBANGULAR), ALLUVIALLY PERCUSSED, COARSE, MEDIUM AND FINE GRAVELS AND COBBLES (MAX. SIZE APPROX. 150mm) OF CHERT, QUARTZITE AND QUARTZ, AND SUBANGULAR COBBLE AND BOULDER SIZED SLABS OF WEATHERED VRYHEID FM. SANDSTONE, IN A MATRIX OF RED (M 2.5YR 4/6), SLIGHTLY HARD, LOAM WITH MODERATE TO STRONG, FINE, SUBANGULAR, BLOCKY STRUCTURE AND ABUNDANT FINE FERRUGINOUS AND MANGANIFEROUS CONCRETIONS. NUMEROUS FINE ROOTS. ALLUVIAL GRAVEL (MATRIX SUPPORTED), CONTAINING NUMEROUS MINT AND ROLLED EARLY ACHEULIAN ARTEFACTS THROUGHOUT. GRAVELS DISPLAY NO CLEAR STRATIFICATION OR IMBRICATION.

DEEPLY WEATHERED VOLKSRUST FM. SHALE.

Fig. 1. Klipplaatdrif: unpublished typical profile, recorded by the author, through deposits of 6m terrace.

Remnants of the 6m terrace in the vicinity of Klip-plaatdrif are shown in Figure 2. These are generally of limited extent, but a higher 16m terrace is better repre-sented, particularly along the west bank of the Klip River between Duncanville and Redan. Excavations, undertaken by Mason in this older terrace have revealed it to be sterile of artefacts at depth, although the surface layers contain stone tools of varying ages (Mason 1962). The mor-phological characteristics of these gravels are similar to those of the 6m terrace.

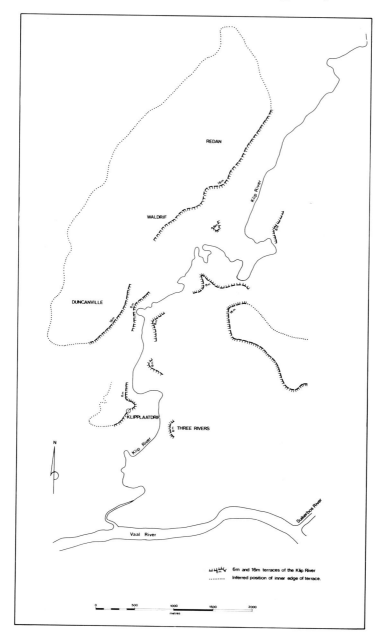

Fig. 2. Map showing distribution of terrace remnants in lower Klip River valley near Vereeniging.

The gravels of the 6m terrace were clearly deposited as bars in the high energy environment of a mature river channel. Lenses with markedly differing average clast sizes indicate a fluctuating flow regime. In these respects the Klipplaatdrif gravels are comparable with the primary alluvial gravels of the Windsorton area in the lower Vaal River basin (Partridge, Brink 1967). A notable difference is the preponderance of large lava clasts in the lower Vaal material in response to local bedrock lithology, but even more striking is the nature of the fine matrix. While that of the lower Vaal terraces is preponderantly sandy and frequently calcified, the matrix at Klipplaatdrif is notably clayey, rubefied and ferruginized, with well developed ferruginous and manganiferous concretions as well as more general cementation of particles.

The relatively high clay content and poorly sorted nature of the fine matrix argues for the alluvial transportation, over fairly short distances, of a poorly sorted mixture of colluvial debris, earlier alluvial deposits and chert residuum, a probability confirmed by the large proportion of subangular clasts in the gravel and cobble fractions. All of the constituent materials are present within a distance of 4 kilometers upstream of the site. Such limited transportation could readily explain the abrasion which characterizes approximately one third of the Acheulean artefacts found within the gravels; the more frequent fresh artefacts were probably discarded in situ by generations of hominids which camped on the exposed surface of the gravel bar and exploited it as a convenient source of raw material for tool making, between periods of inundation and renewed deposition. Flooding and minor reworking would effectively destroy any direct evidence of such occupations in the form of camp sites or living floors.

The ferruginization of the deposit is obviously a post-depositional phenomenon. The very marked reddening associated with advanced mobilization and redeposition of ferric oxides is indicative of a more humid local environment that is prevalent in the area today.

EVIDENCE FROM STERKFONTEIN

In the absence of any possibilities of radiometric dating or of fossil material for calibration, it is

difficult to assign an age to the Klipplaatdrif gravels except in the broadest terms. Mason (1962) has claimed that the typology of the stone industry compares closely with that found in situ in Member 5 of the Sterkfontein Formation, which was excavated in the Extension Site area at Sterkfontein, first by Mason and more recently by AR Hughes. Member 5 was deposited on the eroded westward sloping surface of a debris cone which forms Member 4 of the Sterkfontein Formation; Member 4 has yielded numerous hominid remains which belong exclusively to A. africanus; no stone artefacts have been recovered from this stratigraphic unit (Partridge 1978; Stiles, Partridge 1979). A number of hominid specimens, assigned by Tobias (1978) to Homo habilis or H. aff. habilis, have been found in Member 5 in association with stone artefacts and numerous manuports of foreign stone. Tobias (pers comm) has raised the possibility that specimens recently excavated by Hughes from above the Member 4/5 contact in the Extension Site area may include remains belonging to A. robustus.

The surface of Member 5 contains numerous deep solution pits and channels (makondos), which are filled either with the distinctive chocolate-brown, calcified deposits of the much younger Member 6, or with soil and rubble residual from the weathering and decalcification of Member 5. Artefacts and faunal remains which were not recovered from controlled excavations within the in situ breccia can be assigned to Member 5 or Member 6 with confidence, if there are fragments of the calcified deposit adhering to them.

Tobias (1978) considered it likely that H. habilis was responsible for the stone industry recovered from Member 5 at Sterkfontein. This authorship is certainly favoured by the direct association of both within the same stratigraphic unit. However, at Swartkrans, approximately one kilometer to the south-west of Sterkfontein, CK Brain has recovered both stone flake tools and bone artefacts from the earliest unit of the Swartkrans Formation (here called Member 1); this stratigraphic unit has yielded the remains both of A. robustus and of a hominid which RJ Clarke (pers comm) regards as early Homo erectus. The remains of more advanced H. erectus have been found in the succeeding unit, also in association with A. robustus; overlying deposits have yielded Acheulean artefacts. Since stone artefacts have nowhere been found in association with Australopithecus at sites where a more advanced hominid was not also present, context suggests that, at Swartkrans, H. erectus was the maker of these implements.

How do these various stratigraphic units relate in time?
Analyses of the fossil Bovidae from both Sterkfontein Member 5
and Swartkrans Member 1 have been carried out by ES Vrba, who
finds consistent evidence that the forms from Swartkrans
belong to the period 1.8-1.5 m.y. (Vrba 1982). The majority of
the bovids from Sterkfontein Member 5 are considered to be of
similar age, but perhaps marginally later (Vrba ibid), a conclusion supported by the comparison of the taxonomic affinities
of the associated stone industry with well dated counterparts
from East Africa undertaken by Stiles (Stiles, Partridge 1979).
Stiles concluded that the Sterkfontein artefact assemblage
equates best with Early Acheulean assemblages found above the
Lemuta Member in Bed II of Olduvai Gorge, which have been dated
to around 1.5 m.y.

Available faunal and artefactual evidence thus suggests
that Swartkrans Member 1 and Sterkfontein Member 5 are of
roughly similar age, with the Swartkrans unit apparently ranging a little further back in time than its Sterkfontein equivalent. This inferred temporal relationship is, however, at
variance with evidence provided by hominid remains from dated
localities in East Africa: H. habilis from Omo, East Turkana
and Olduvai Gorge ranges in age from 2.3-1.6 m.y., while
H. erectus makes its first appearance in the latter area
between 1.6 and 1.5 m.y. We are thus confronted by a paradox in assessing the evidence from Sterkfontein and Swartkrans:
the artefacts in Swartkrans Member 1 are associated with a
more advanced hominid than is represented   in the apparently
slightly younger implementiferous deposits at Sterkfontein.
Better dating resolution than has been forthcoming on the
basis of present comparative evidence from these sites may
eventually explain these apparent anomalies. In the meanwhile
the following hypotheses are put forward for consideration:

1)   Member 5 of the Sterkfontein Formation may span a longer
     period of time than has been considered likely hitherto
     - perhaps many hundreds of thousands of years.

2)   H. erectus may have been responsible for the Sterkfon-
     tein Acheulean industry, although no remains of this
     hominid have been recovered from this site.

There is now considerable faunal, palynological and sedimentological evidence to suggest that, during at least part of
the time spanned by Swartkrans Member 1 and Sterkfontein Member
5 (and, by analogy, probably also by the Klipplaatdrif deposit),

conditions were significantly drier than those which prevailed during the preceding stage when Australopithecus
africanus occupied the interior plateau of southern Africa
(Vrba 1980; Horowitz 1975; Stiles, Partridge 1979; Partridge
1982). This change to more arid conditions was associated
with an increase in grassland at the expense of bush savannah. The morphology of the Klipplaatdrif deposit is certainly
compatible with the flow regime of a semi-arid river, but the
advanced pedogenesis which followed implies a return to wetter conditions; several authors have suggested, on the basis
of independent evidence, that such a reversion to a more
mesic climate did, in fact, occur during the latter part of
the Lower Pleistocene (e.g. Partridge 1982).

That the Acheulean population was a resourceful and
stable one, well adapted to the vicissitudes of major environmental change, is demonstrated by the extraordinary persistence of this industrial complex through eastern and southern
Africa for nearly 1.5 million years.

Horowitz A (1975). Preliminary palaeoenvironmental impli-
    cations of pollen analysis of Middle Breccia from Sterk-
    fontein. Nature 258:417.
Mason RJ (1962). "Prehistory of the Transvaal". Johannesburg:
    Witwatersrand University Press.
Partridge TC, Brink ABA (1967). Gravels and terraces of the
    lower Vaal River basin. S Afr Geog J 49:21.
Partridge TC (1978). Re-appraisal of lithostratigraphy of
    Sterkfontein hominid site. Nature 275:282.
Partridge TC (1982). The chronological positions of the
    fossil hominids of southern Africa. Proc 1st Int Cong
    Human Palaeont (Nice), Paris: CNRS 2:617.
Stiles DN, Partridge TC (1979). Results of recent archaeo-
    logical and palaeoenvironmental studies at the Sterkfontein
    Extension Site. S. Afr J Sci 75:346.
Tobias PV (1978). The earliest Transvaal members of the
    genus Homo with another look at some problems of hominid
    taxonomy and systematics. Z Morph Anthrop 69:225.
Vrba ES (1980). The significance of bovid remains as indi-
    cators of environment and predation patterns. In Behrens-
    meyer AK, Hill AP (eds): "Fossils in the Making", Chicago:
    Chicago University Press: p247.
Vrba ES (1982). Biostratigraphy and chronology, based partic-
    ularly on Bovidae, of southern hominid-associated assem-
    blages. Proc 1st Int Cong Hum Palaeont, Paris: CNRS 2:707.

Humanization Con Brio: Accelerando, Largamente, Vivace

Hominid Evolution: Past, Present and Future, pages 313–318
© *1985 Alan R. Liss, Inc.*

RECENT RESEARCH ON HEIDELBERG JAW
Homo erectus heidelbergensis

Reinhart Kraatz

Geologisch-Paläontologisches Institut der
Universität Heidelberg (abbrev. GPIH)
D-6900 Heidelberg 1 Federal Rep.of Germany

In 1907 the jawbone of the "Heidelberg Man" was discovered in the former "Grafenrain sandpit" (Schoetensack 1908). This is at the northern fringe of the village of Mauer, which is situated in Southwestern Germany 10 km southeast of the city of Heidelberg in the hinterland of the east rim of Upper Rhine Graben.

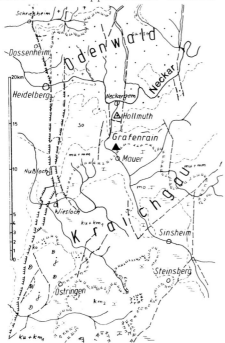

Fig. 1. Geological map showing location of former "Grafenrain sandpit" north of the village of Mauer in a region called Kraichgau. North of Kraichgau: Odenwald Mts.; west of this mountain topography,Heidelberg, and Kraichgau as well: Upper Rhine Graben. su, sm, so = L. Triassic ("Buntsandstein"); mu, mm, mo = M.Triassic ("Muschelkalk"); ku, kml, km2 = U. Triassic ("Keuper")

Schoetensack (late Prof. of Paleontol. and Paleoanthropol.,
GPIH) had predicted for more than 20 years before 1907
that a fossil human bone would be found there (Kraatz,
Querner 1967). He put this mandible in the genus Homo. By
doing so, Schoetensack was the first paleoanthropologist
worldwide to rank such an old fossil among real hominids,
that walked on two legs; thus he made himself many enemies
in his lifetime.

For almost 20 years the existing material has been studied
anew with a variety of methods and viewpoints:

1) sedimentation of the former meander of the Neckar river
   and evolutionary history of the Neckar: Rücklin,Schweizer
   1971; Meier-Hilbert 1973; Schweizer (with Kraatz) 1982;
2) the Mauer Fauna: Schütt 1969a, 1969b, 1970;
   Wv Koenigswald 1973, 1982;
3) dating of the locality by its fauna and matrix: Oakley
   1958, 1964; Fleischer et al. 1965; Oakley et al. 1971;
   Wv Koenigswald 1982;

Fig. 2. Lowest part of the
fluvial sandbeds (the so-
called "Mauerer Sand") of the
former meander of the Neckar
river. Large peel (2.70 m H./
1.50 m W.) in the main fossil
bearing sandbeds produced in
July 1982 by R Kraatz and
Karen Riedelsberger (Kraatz
1983).
Bottom, current waters, change
of sedimentation and erosion,
with channels;
middle, stagnant water;
top, current ripple bedding.
Photo by K. Schacherl, GPIH.

4)   possible artifacts of the Heidelberg Man: Pelosse 1966;
Pelosse, Kraatz 1976;
5)   and – last but not least – the human mandible:
GHRv Koenigswald 1968; Day 1977; Runge 1977; Puech, Prone,
Kraatz 1980; Puech 1980,1982; Czarnetzki 1983; Roth 1983;
6)   and its classification and nomenclature: Howell 1960;
Heberer 1965, 1969; BG Campbell 1964, 1978; Querner 1968.

    I will give a brief overview of some select recent
studies from the above mentioned references:

    1 – Sedimentation – The former Grafenrain sandpit is
situated in a late Pliocene / Pleistocene meander of the
Neckar river. All the fossils collected there between 1887
and 1962 are housed in the GPIH. I recently finished the
manuscript of a card catalogue for the complete fossil
record. There are some 5000 bones, teeth, horns and antlers
of a tropical fauna, dated to the early middle Pleistocene
("Cromer II"), including one human fossil and some twenty
bone fragments and pebbles used or altered by hominids.

    4 – Possible artifacts of the Heidelberg Man – In the
GPIH Mauer Fauna collection there are a few bones as shown
in fig.3. Voelcker (1933) considered it to be a compound
tool like a spear. During the Taung 60 Internat'l Symposium
(Johannesburg and Taung 1985) I saw samples of
osteodontokeratic tools of the Dart collection (Bernh.Price
Institute for Paleontol.Res., Wits Univ.). They have exactly
the same shape as this Mauer bone fragment. In any case it
is a proof of human activity alongside the Neckar river.

Fig.3.Osteodontokeratic tool of the Heidelberg Man. L 18 cm.
GPIH Mauer Fauna M.1767. Photo by Karl Schacherl, GPIH.

    5 – The human mandible – PF Puech (1980, 1982; Puech,
Prone, Kraatz 1980) studied dental wear of the teeth at high
magnification. He produced small sized replicas, then took
pictures of them under the microscope in order to find out
the ratio between horizontal and vertical lineation. The
concept is: horizontal lineation results from a vegetarian

diet whereas vertical lineation is the result of chewing meat. The signs of abrasion indicate that Heidelberg Man was a vegetarian rather than a carnivore.

Fig. 4. Microphotographs of the buccal surface of the M 2 right. Peels and microphotos by PF Puech, Nimes/France. Photographic reproduction by K -Schacherl, GPIH.

In 1977 a number of original European Homo erectus mandibles were housed in the GPIH (by courtesy of H de Lumley, Paris) for a study session of comparison.

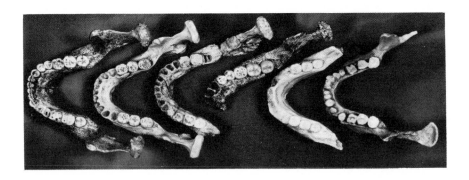

Fig. 5. The line of European Homo erectus mandibles, from left to right: Heidelberg; Montmaurin (Dordogne,France), Arago 2 (Tautavel,France), Arago 13 (Tautavel,France), Atapuerca (Burgos,Spain), Banolas (Barcelona,Spain).
Photo by Karl Schacherl, GPIH.

References:
Campbell BG (1964). Quantitative taxonomy and human evolution. In Washburn SL (ed.): "Classification and human evolution". London: Methuen and Co Ltd., p 50-74.
Campbell BG (1978). Some problems in hominid classification and nomenclature. In Jolly (ed): "Early hominids of Africa", London: Duckworth, pp 578-579.
Czarnetzki A (1983). Belege zur Entwicklungsgeschichte des Menschen in Südwestdeutschland. In Müller-Beck H (ed.): "Urgeschichte in Baden-Württemberg", Stuttgart: Konrad Theiss, pp 217-240.
Day MH (1977). "Guide to Fossil Man". London: Cassel.
Fleischer RL, Price PB, Walker RN (1965). Applications of Fission Tracks and Fission-Track Dating to Anthropology. Gen'l Electr Res Lab, Rep 65-RL-3878 M, Schenectady, N Y.
Heberer G. (1965). Die Abstammung des Menschen. In Gessner (ed.) "Handbuch der Biologie", Vol IX, 2nd part, Konstanz - Stuttgart: Akad. Verlagsges. Athenaion, pp 245-328.
Heberer G (1969). "Der Ursprung des Menschen. Unser gegenwärtiger Wissensstand". Stuttgart: Fischer.
Howell FC (1960). European and northwest African Middle Pleistocene hominids. Curr Anthrop 1: 195.
Koenigswald GHR von (1968)."Die Geschichte des Menschen"., 2nd ed Berlin-Heidelberg-New York: Springer.
Koenigswald W von (1973). Veränderungen in der Kleinsäugerfauna von Mitteleuropa zwischen Cromer und Eem (Pleistozän). Eiszeitalter und Gegenwart 23/24: 159.
Koenigswald W von (1982). Zur Gliederung des Quartärs. In Niethammer J, Krapp F(eds)"Handbuch der Säugetiere Europas"Vol 2/1 Rodentia II. Wiesbaden, pp 15-17.
Kraatz R (1983). Das urgeschichtliche Museum in Mauer an der Elsenz. Ruperto Carola 67/68: 199.
Kraatz R (1985). A review of recent research on Heidelberg Man - Homo erectus heidelbergensis. In Delson E (ed) "Ancestors: The Hard Evidence". New York: Liss.
Kraatz R, Querner H (1967). Die Entdeckung des Homo heidelbergensis durch Otto Schoetensack vor 60 Jahren. Ruperto Carola 42:178.
Meier-Hilbert G (1974). Sedimentologische Untersuchungen fluviatiler Ablagerungen in der Mauerer Neckarschleife. Heidelberger geogr Arb 40:201.
Oakley KP (1958). Application of fluorine, uranium and nitrogen analysis to the relative dating of the Rhünda Skull. Neues Jahrb Geol Pal, Mh, 1958: 130.
Oakley KP (1964)."Frameworks For Dating Fossil Man". 3rd ed, London: Weidenfeld and Nicholson.

Oakley KP, Campbell BG, Molleson TI (1971). Catalogue of Fossil Hominids. Part 2 Europe. London: Brit. Mus.

Pelosse JL (1966). Traces possible d'action humaines sur des os fossiles du gisement d'Homo heidelbergensis. Bull Assoc franc Et Quat 4: 247.

Pelosse JL, Kraatz R (1976). Une pièce osseuse faconnée du gisement de Grafenrain (Homo heidelbergensis). Quaternaria 19: 93.

Puech PF (1980). Homo heidelbergensis lointain ancêtre était droitier et vivait dans les bois. Montpellier: Midi Libre.

Puech PF (1982). L'usure dentaire de l'Homme de Tautavel. Coll Intern CNRS Nice, Vol 1: 249.

Puech PF, Prone A, Kraatz R (1980). Microscopie de l'usure dentaire chez l'Homme fossile: bol alimentaire et environnement. CR Acad Sci Paris, Vol 290, ser D: 1413.

Querner H (1968). "Stammesgeschichte des Menschen". Stuttgart: Kohlhammer, p 48.

Roth H (1983). Comparaison statistiques de la forme des arcades alvéolaire et dentaire des mandibules des hominides fossiles. Mus Nat d'Hist Natur, Musée de l'Homme, Mémoire 17, 2 Vol, Paris.

Rücklin H, Schweizer V (1971). The geology and geomorphology of Heidelberg and its surroundings. In G Müller (ed) "Sedimentology of Parts of Central Europe", Guidebook 7th Internat'l Sed Congr 1971, Frankfurt: Kramer, 337-344.

Runge B (1977). Morphologische Untersuchungen der Zahnkronen der Funde von Mauer, Oberkassel, Bad Nauheim und Trebur. In Schäfer U (ed)"Beiträge zur Odontol paläolith, meso- und neolith. Menschenfunde", Giessen: Anthropol Inst.

Schoetensack O (1908). Der Unterkiefer des Homo heidelbergensis aus den Sanden von Mauer bei Heidelberg. Ein Beitrag zur Paläontologie des Menschen. Leipzig: Engelmann.

Schütt G (1969a). Panthera pardus sickenbergi n.subsp. aus den Mauerer Sanden.Neues Jb Geol Paläont, Mh, 1969:299.

Schütt G (1969b). Untersuchungen am Gebiß von Panthera leo fossilis (v.Reichenau 1906) und Panthera leo spelaea (Goldfuss 1810). Neues Jb Geol Paläont, Abh, 134: 192.

Schütt G (1970). Nachweis der Säbelzahnkatze Homotherium in den altpleistozänen Mosbacher Sanden (Wiesbaden/Hessen). Neues Jb Geol Paläont, Mh, 1970: 187.

Schweizer V, with Kraatz R (1982). Kraichgau und südlicher Odenwald.In Gwinner MP (ed)"Sammlung Geologischer Führer", Vol 72, Berlin-Stuttgart: Gebr. Borntraeger.

Voelcker, I (1933). Knochenartefakte aus dem Altdiluvium des Neckars. Verh Naturhist-Med V Heidelberg, NF,17: 317.

Hominid Evolution: Past, Present and Future, pages 319–328
© *1985 Alan R. Liss, Inc.*

FOSSIL MAN FROM CUEVA MAYOR, IBEAS, SPAIN: NEW FINDINGS
AND TAXONOMIC DISCUSSION

E. Aguirre, Prof. Dr., A. Rosas, Lic.

Museo Nacional de Ciencias Naturales
CSIC
28006 Madrid,   Spain

There are multiple sites of anthropological interest
in the Sierra de Atapuerca, near Burgos,   Spain. The catalogue published by Martin Merino et al (1981) includes
27 cavities present within less than 1 km . Most of the
cavities are, in one way or another, related to ancient
and more important karst systems.   The current research
project "Atapuerca 2" initiates studies on two close but
different localities.   A set of filling breccias outcrop
on the railway trench, in the municipality of Atapuerca;
the outcrop of the fossil man-bearing deposits occurs in
an inner duct of the Cueva Mayor of Ibeas. In the trench,
"Trinchera", the exposed fillings of two cavities are being
sampled and excavated. Only the uppermost levels of both,
TD and TG sites, have been excavated systematically. In
the Cueva Mayor-Cueva del Silo system, an ancient mud deposit containing skeletal remains of over one hundred bears,
subsequently looted and removed, is exposed at the bottom
of a deep,   remote side cavity: we call it the SH site.
This secondary deposit was discovered before 1890 and has
been ravaged by amateur cave-explorers who liked collecting
fossil bear canines from it.   Almost 3 tons of the SH deposit were removed by the tusk hunters without their realizing
they were discarding hominid remains, along with thousands of
minor bones, fragments and teeth.   Trinidad Torres and·
his assistants were the first to recognize the fossil man
of Ibeas in 1976, while evacuating the overburden left
by the tusk hunters (Aguirre et al 1976).   Last summer,
1984,   we finished clearing away all the material looted
and abandoned by the amateurs, and started excavating systematically in that deep crevice.

THE "TRINCHERA" SEQUENCE

The TD section is 18.5m high from the present railway trench
bottom to the hill surface cut down by the trench work.
The filling deposits can be preliminarily divided into four
main units.   Unit TD-I (ca. 10m) consists of a complex sequence further subdivided into six levels, with evidence
of regime changes. Unit TD-II consists of banks of cross-
bedded, coarse, cemented sands, and includes bones of large
mammals.   Unit TD-III is a breccia with medium size lime
chunks and sandy matrix. Within the upper Unit TD-IV we
distinguish a thin member, or level TD-IVa, of fine-grained
sand unconformably overlying Unit TD-III; TD-IVb is a rich
bone- and tool-bearing, poorly stratified, unsorted breccia
with red clayey matrix; the conformably overlying TD-IVc
is a sterile, homometric, smallsized conglomerate with silty
matrix, known to us as "Paella conglomerate", including
dispersed large blocks of lime. This unit ends with deep
(up to 4m), pedogenetically altered horizons, including
vermiculite and caliches, on which a heavy erosion gives
place to the overlying Late Pleistocene colluvia and Recent
soils.

The vertebrate fauna represented in Unit TD-I is roughly
constant from level 3 to level 6; it includes Mimomys savini
Pliomys episcopalis, Pitymys gregaloides, Hystrix sp.,
Crocuta crocuta intermedia, Ursus praearctos, Panthera gombaszoegensis , Bison schoetensacki, Cervus ex gr. Elaphus,
Megacerini gen., and Equus (identifications communicated
by E. Gil, C. Sese·, E. Soto, E. Torres, J.F. Villalta).
This assemblage closely resembles that of La Pineta dated
0.73 m.y.BP (Coltorti et al 1982), although possibly including older stages of microtine evolution.   The association
of M. savini, Pliomys episcopalis and the earliest Pitymys
of the gregaloides group replaces the Allophaiomys fauna
between Monte Peglia and Le Vallonet, that is, nearly one
million years ago. On the other hand, the last dated appearance of Mimomys savini is found a few meters above the
Matuyama/Brunhes reversal in Hungary, and may therefore
be thought to occur ca. 0.7 m.y.BP. Consequently, we consider
Unit TD-I as including the earliest part of the Middle Pleistocene and probably the end of the Early Pleistocene as
well. Acheulean artefacts have been recovered from the
section at the upper levels of TD-I.   This part of the
section certainly deserves research on magnetic polarity;
this is scheduled for the present year.   Actually, Units

II and III are less rich in fossils; their faunal representation is not yet studied. Within the upper part of the
section, Unit TD-IV, the level TD-IV2 yields a fauna which
correlates with that of La Fage and Orgnac III in France.
Most of Unit TD-IV, at least, should correspond to late
Middle Pleistocene.

The TG section is lower (13.5m) and far less comprehensive than TD, but complementary as it contains more
information on several aspects. Above a residual breccia
attached to the outcropping bedrock on one side, the lowermost exposed levels at TG are related to ancient karst cons-
tructions and contain some tools as well as faunal remains
yet to be studied. The upper unit of TG includes, from
base to top: 1. a red breccia, poorly bedded with hetero-
metric limestone chunks in clayey matrix, rich in both
stone artefacts and faunal remains, being quite similar
in both lithogenetic character and faunal contents to TD-
IV2; 2. some 3 m of well stratified, gritty, homometric
"Paella" conglomerate of small angular limestone pieces
cyclothemically alternating with pink silts; in its lower
third, as many as nine occupation soils were found. Faunal
and floral remains indicate conditions similar to present,
that is continental, with warm summers following long frosty
seasons, near the end of the Middle Pleistocene; this subunit
probably corresponds to the level IV3 at TD, where it is
almost sterile, and to the upper laminated silts at a small
cavity (TZ) which opens on to the site. Finally, the upper
TG unit ends -- like the upper silts of TZ -- with carbonated
crusts and a stalagmitic floor, which has been sampled for
dating purposes.

There is, consequently, a rich sequence of fossil-
and tool-bearing cave deposits in the Atapuerca Trench,
bracketed between very late Early Pleistocene and the
beginning of the Late Pleistocene, that offers a unique
calibration of the whole Middle Pleistocene.

THE SH SITE

On the same southwestern slope of the Sierra de Atapuerca, no more than half a kilometer from the sites TD
and TG, and within the municipality of Ibeas, the 3 km.
long Cueva Mayor and Cueva del Silo karst system
gives us one of the richest and simultaneously the craziest

and most striking of fossil sites known.   Sima de
los Huesos (SH) is a vertical duct of more than 12m, which
opens at a small side gallery located 400m inside the
opening of the Cueva Mayor. The vertical duct goes further
down through a narrow, sloping funnel without issue, excavated
into an ancient bone breccia with numerous bear remains.

SH is a derived site. The old infilling of SH is a
mud flow containing from less than 5% to more than 90%
skeletal remains. The species represented, according to
present knowledge, are: Ursus deningeri (minimum number
of individuals 99), Cuon sp. (MNI, 1), Vulpes vulpes ssp.
(MNI, 6), Panthera gombaszoegensis (MNI, 1), Panthera leo
fossilis (MNI, 2), Lagomorpha Leporidae gen., and   fossil
man, Homo sapiens ssp., with a minimum number of 10
individuals represented. The single bone of an ungulate
among more than 7,000 fossils of bear is a phalanx of a
rhinocerotid. The present disposition of the SH deposits
suggests the hypothesis that we have before us remains of
a set of skeletons of hibernating bears, and of other animals,
mainly carnivores, which died inside a cave, later to be
dispersed by a stream and redeposited with mud deeper in a
hollow in the karst system. The matter deserves taphonomic-
al research.

Most of the human skeletal remains, as well as the
fossil Carnivora, were found relocated from the secondary
deposit by the bear tusk hunters. But the 1984 discovery
of a mandibular fragment (AT75), a tooth (AT74) and an epac-
tal bone (AT76) within undisturbed and untouched sediments
considered sterile proves that the fossil men of Ibeas were
dead before the skeletons or U. deningeri were dispersed.

An attempt at radiometric dating has been conducted
at the Laboratoire de faible radiactivité, Gif-sur-Yvette,
France, by Yokoyama and Van Nguyen (personal communication)
by 234U/238U, U/Th and U/Pa ratios detected by γ rays. A
date of ca. 350,000 y.BP and a minimum limit of 200,000
y.BP without a definite maximum, is the preliminary result
on the mandibular fragment AT75 recovered in situ. This
result is quite similar to that obtained on Tautavel fossil
man. On the other hand, the known life span of Ursus
deningeri runs from 730,000 y.BP at La Pineta to ca 200,000
y.BP or somewhere in the late Middle Pleistocene, when U.
deningeri evolved into U. spelaeus. This last species is
recognized in the "Paella" Conglomerate of the late upper
unit of TG in Atapuerca Trench, with a late Middle Pleistocene

faunal assemblage. Thus the hypothesis that the Ibeas men were contemporaneous with the U. deningeri population can be rejected only by force of prejudice.

There are grounds for considering any correlation with the lower unit of TD quite unlikely. On the other hand, correlation with upper levels of TG and their equivalents of TD is precluded by the presence of U. spelaeus (T. Torres in press). Therefore, the present evidence, although poor, is consistent enough to convince us that Ibeas Man lived in the Middle Pleistocene, excluding both its earliest and latest times.

A carbonate level at the top of the bone-bearing mud flow of SH is also being analyzed at Gif-sur-Yvette: it is hoped this will yield a latest date for the removal of Ibeas man bones from their original death (or burial?) place. Certainly every approach for making this estimation more precise will be attempted.

INVENTORY OF FOSSIL MAN REMAINS FROM SH

After the 1984 season, nearly ninety human remains from the SH site in the Cueva Mayor of Ibeas have been identified. These include one right and one left petrous; seven portions of right, seven of left parietal, five fragments of occipital bones; one complete mandibular body and fragments of four other lower jaws; two incomplete right humeri; several fragments of femoral shafts; two shafts and a distal portion of tibiae; five phalanges plus one phalangeal fragment. Adding the 13 teeth preserved on the mandibular remains to the 38 isolated teeth, the present hypodigm of Ibeas Man includes $4I^1$, $3I^2$, $2C$, $2P^1$, $2P^2$, $3M^1$, $3M^2$, $1M^3$ of the upper dentition, and $1dc$, $2I_2$, $3\bar{c}$, $2P_1$, $6P_2$, $6M_1$, $5M_2$, $5M_3$ of the lower dentition.

From the occlusal crown wear, the mesial/distal wear facets, and the differential size of these teeth, not fewer than 10 human individuals are represented in the collection of 51 teeth.

The study of these fossils is currently being carried out by P.J.Perez (tibia, pathology), J.L. Arsuaga (femur, phalanges, skull), I. Martinez Mendizabal (skull), J.M. Bermudez de Castro (dentition, humeri), A. Rosas (mandibles).

Let us restrict our present discussion to just a few of the traits that should be examined in view of their current diagnostic value.

POSITION OF IBEAS MAN AMONG EUROPEAN FOSSIL HOMINIDS

Summarizing the present results of the studies on the Ibeas man, although still in a very preliminary stage, we can say that a mixture of traits is observed, considered alternately primitive or modern, according to the current literature. We have tried to revise evidence in order to determine the polarity of several traits in fossil human mandibles to establish, whenever possible, the earliest appearance of a morphotype, and the relative frequency of morphotypes for different features. We have also attempted to identify evolutionary trends and/or particular deviations in the fossil samples arranged chronologically and chorologically. The same has been done for various dental traits.

Cranial Traits

The largest piece of a cranial vault of the Ibeas man is the nearly complete left parietal composed of fragments AT (B)17 and AT31a, 31b. This bone is very thick: the measured thickness varies from 13mm to 7.5 mm. Very thick parietal bones are known in the skulls from Salé, Swanscombe, Choukoutien. This could be considered as an apomorphic trait present in the Middle Pleistocene populations of Asia, Europe and Africa. The external transverse profile of the AT17+31 skull is smoothly curved, slightly flattened on its median part and shows a nonprominent parietal boss, similar to that of Neandertal man.

An overall glance at the imprints of the meningeal blood vessels of AT17+31 shows a disposition of the major branches similar to that observed on several Middle Pleistocene skulls from Java and the Mahgreb. A net of anastomosed, thin, third order furrows in the parietal of Ibeas is less developed than in Swanscombe and Biache-St.-Vaast skulls, more developed than in Fontéchevade, and similar to the Rabat Kebibat skull. This can, therefore, be considered a variable trait among Middle Pleistocene men, with still unknown evolutionary significance.

MANDIBLES

Many traits of the Ibeas mandibles resemble those of Arago, Mauer and Montmaurin and should be considered as distinctive of European anteneandertalians. Sequential variations of width, depth and thickness along the body, as well as a series of different partial longitudinal measurements on AT1, show close resemblance to those of Tautavel and Mauer.

The European mid-Pleistocene mandibles may be characterised as wide, comparatively short, robust and low, with thick, widely divergent bodies, and usually insignificant variations in depth along the body, with the exception of Mauer. Consistently low and even lower bodies are known also in the Krapina jaws and the Neandertalian hypodigm in France. On the other hand, in Krapina and among the Neandertalians, such mandibles are less thick and robust than those of Mauer, Montmaurin, Arago and Atapuerca. Among the latter there are no jaws which are mesially very deep and which taper distally, as in Krapina and the Neandertalians. The mandibles of Mauer, Montmaurin, Arago and Atapuerca closely resemble African Middle and Early Pleistocene mandibles OH22, OH23, ER730, ER992, SK15 and Cave of Hearths, in shape, body proportions and depth, being slightly less robust (Aguirre, Lumley 1977).

Among other greatly variable traits, the mandibles from Arago, Atapuerca, Mauer and Montmaurin show distinctly conspicuous external rims and intertoral furrow (except Arago XIII); moderate protuberantiae laterales, in a distal position, but never so extremely distal as on several Neandertalians; a highly variable internal torus mandibularis; variably defined, variably inclined but commonly well developed planum alveolare; commonly absent or inconspicuous sulcus mandibularis and trigonum mentale, except Arago II, recalling the appearance of incipient chin features on ER730. The European anteneandertalian mandibles, including those from Ibeas, show extreme degrees of backward and downward inclination of the symphyseal axis; Ibeas AT1 shows the maximum value, while Neandertal jaws vary widely in this trait. The inner, parabolic hiatus of the Ibeas mandibular body has an outline quite similar to those of Mauer, Atapuerca and other European anteneandertalians, a characteristic of this group, distinctive from australopithecines on one hand and Neandertalians and modern man on the other. Adult molars in a straight line are also distinctive of European anteneandertalians. Prominent tubercula marginalia anteriora and feeble or inconspicuous tubercula marginalia posteriora may also be counted diagnostic traits of this group, shared by AT1. Among the traits shown by AT1, AT75 and AT83, we may list also a thick lower margin, a well marked, deep masseteric fossa, and comparatively broad, low ramus with a shallow sigmoid notch.

Several unusual traits are exhibited by the AT1 lower jaw. The basal edge of the body of AT1 is quite flat and wide, as in Arago XIII. The digastric fossae of AT1 are 14 mm. apart. Known cases of widely separated digastric impressions are "Lucy" of Afar, 11mm.; Choukoutien H1, 9mm.; Ternifine I, 9 mm.; AL277.1, 7 mm. The interdigastric space of AT1 is flat, and there is no spina mentalis in it. The genial spines form a complicated, asymmetric pattern of branching rims. Such a pattern is seen only in ER730, although some of the trefoil structure of the Neandertalian genial apparatus can be seen in the complex design of AT1. Anterior flattening of the mandibular body distinguishes AT1 from other anteneandertalian mandibles. Several Neandertal mandibles are flattened anteriorly, as is ER 730. We may conclude that AT1, and other mandibular remains from Ibeas, belong morphologically to the anteneandertalian group of Mauer, Montmaurin, Arago II and XIII.

POSITION OF ANTENEANDERTALIANS AMONG OTHER FOSSIL HOMINIDS

It is not surprising that some traits in anteneandertalians partially resemble corresponding morphotypes in Neandertalians while others dissociate them from the latter, and associate them instead with old African morphotypes. Apart from mandibular robusticity, backward projection of gnathion, and other mentioned traits that make comparison among ER730, AT1 and Arago II (as others do between ER992 and Arago XIII) unavoidable, there is a strong resemblance between European anteneandertalians and African Middle Pleistocene hominids in the position of digastric fossae, number of mental foramina and relative size of molars (Aguirre et al. 1980).

The digastric fossae occupy part of the inner or lingual surface of the mandible from basal edge upwards, in modern man and Neandertalians. Although more basally located in A. (P) robustus, Australopithecus from Afar, and several jaws from the Pleistocene of Java, the digastric fossae also are located on the posterior part of the basal edge, turned slightly backwards and upwards. Only in Ibeas, Tautavel, Mauer, OH22

and ER730 are the digastric fossae plainly situated on the
basal edge, facing down (or almost, as per Rightmire, 1980).

Mental foramina are single or multiple in the Afar mandibles; three foramina are most common in Javanese Homo erectus.
The three holes morphotype is unknown to me among Middle Pleistocene mandibles of subsaharan Africa, where one or two is the
mode, as in Atapuerca. In Arago, one foramen is seen; in
Mauer two and three; in Palikao, 1, 2 and 3 occur on three
mandibles. Perhaps the samples are still too poor, but the
presently known distribution of morphotypes for this character coincides with other affinities between the European anteneandertalians and some older Central East African mandibles.

SOME DENTAL TRAITS

Cheek teeth of AT1 and AT75 are extremely small. Not only is the small size of M_3 and the presence of the + pattern
in some lower molars of Ibeas notable, but the reduction of
the M_2 is outstanding. In our opinion, the reduction of M_2
cannot simply be labelled a "modern" character, because M2s
shorter than M1s are observed in Ternifine III, OH22 and two
jaws from Choukoutien. This morphotype, therefore, appears
early in the Middle Pleistocene, soon reaching a high frequency.
Robustly rooted upper central incisors seem common in the Ibeas
hypodigm, independent of the robustness and size of the cheek
teeth. Considerable dimorphism of tooth size is seen in the
material from Ibeas. There are several very large teeth, including some premolars but, generally, the teeth of the European anteneandertalians are smaller than those of African Middle Pleistocene hominids referred to "H. erectus".

The anterior edge of the ramus covers the distal lower
corner of M3 crown in AT1, AT75. In this trait, the Atapuerca
jaws resemble those of Tautavel, and perhaps approach more
closely the Neandertal arrangement. In Atapuerca, the situation can be attributed to tooth size reduction as well as, or
more than, to the advance of the cheek bone.

CONCLUSIONS

The evidence of the Ibeas fossils contributes to a better
definition of European anteneandertalians and to an estimation
of their variability relative to certain traits. It shows

further that within this group ancestral characters of Neandertalians can be found, while other neandertalian traits
could have been inherited from Asian immigrants. Both common
traits and morphotypical differences among locally diverse
Middle Pleistocene populations can be detected. The separation
between H. sapiens and "H. erectus" vanishes. The authors pro-
pose that all populations from the Far East, Africa and Europe,
currently referred to as "H. erectus", should be considered
H. sapiens, and that "erectus", "rhodesiensis" and such names
could at most be used for varieties or races with a geographical and/or chronological meaning. The anteneandertalians
from Mauer, Tautavel and Ibeas could, under that criterion, be
named H.s. heidelbergensis.

This work was made possible by Grant N° 1849/82, from
CAICYT of the Spanish Government, and by substantial help from
Dirección del Patrimonio of Junta de Castilla-León. We deeply
appreciate the generosity of many people, mainly from Burgos
and Ibeas, whose work, supplies and hospitality are still un-
paid due to administrative inefficiency: their patience is
worthy of profound gratitude. We warmly thank the organisers
of the Taung Diamond Jubilee for the invitation to present this
paper.

Aguirre E, Basabe JM, Torres T (1976). Los fósiles humanos de
    Atapuerca (Burgos): nota preliminar. Zephyrus (Salamanca)
    26,27:489.
Aguirre E, de Lumley MA (1977). Fossil men from Atapuerca,
    Spain: their bearing on human evolution in the Middle Pleis-
    tocene. J Hum Evol 6:681.
Aguirre E, de Lumley MA, Basabe JM, Botella M (1980). Affinit-
    ies between the mandibles from Atapuerca and l'Arago, and
    some East-African fossil hominids. In Leakey RE, Ogot BA
    (eds): Proc 8th Panafrican Cong Prehist Quat Stud, 5-10 Sept
    1977. Nairobi: TILLMIAP, p 171.
Coltorti M, Cremaschi M, Delitala MC, Esu D, Fornaseri M, Mc-
    Pherron A, Nicoletti M, Otteloo R van, Peretto C, Sala B,
    Schmidt V, Sevink J (1982). Reversed magnetic polarity at
    an early Lower Palaeolithic site in Central Italy. Nature
    300:173.
Martin Merino MA, Domingo Mena S, Antón Palacios T (1981). Es-
    tudio de las cavidades de la zona BU-IV-A (Sierra de Atapuer-
    ca). Kaite (Burgos, Bibl Espeleol Burgalesa) 2:41.
Rightmire GP (1980). Middle Pleistocene hominids from Olduvai
    Gorge, Northern Tanzania. Am J Phys Anthropol 53:225.

Hominid Evolution: Past, Present and Future, pages 329–334
© *1985 Alan R. Liss, Inc.*

HUMAN REMAINS IN ISRAEL IN THEIR ARCHAEOLOGICAL CONTEXT

Avraham Ronen

Department of Archaeology
The University of Haifa, Israel

THE LOWER PALAEOLITHIC

Small fragments of a cranium and a few isolated teeth found at the site of Ubeidya possibly constitute the oldest human remains in Israel (Tobias 1966). However, the attribution of these fragments to the ca. 1.0 m.y. archaeological layers is uncertain. Aside from these remains, the long sequence of Lower Palaeolithic in Israel is practically devoid of human remains. Only its latest phase, the Yabrudian, yielded human skeletal material -- the Zuttiya skull (Turville-Petre 1927). Found in 1925, it was the first human remains ever to be unearthed in the Near East. Recent investigations at Zuttiya and Th/U dating place the Zuttiya frontal at ca. 130,000 years ago (Gisis, Bar-Yosef 1974). The massive, heavy eyebrow-ridged fragment is currently classified as archaic *H. sapiens* (Arensburg pers comm). Isolated human bones -- a femur and a molar -- were also mentioned from Layer E of Tabun (Garrod, Bate 1937: 67), a Yabrudian phase like Zuttiya. The complex stratigraphy of Tabun, however, renders these finds of dubious provenance.

THE MIDDLE PALAEOLITHIC

Israel has yielded numerous Middle Palaeolithic human remains, deliberately buried. These much discussed finds come from the Mount Carmel caves of Tabun and Skhul (Garrod, Bate 1937; McCown, Keith 1939), from Qafza (Neuville 1951; Vandermeersch 1981), Amud (Suzuki, Takai 1970), and Kebara (Smith, Arensburg 1977; Bar Yosef pers comm).

Altogether, close to 40 individuals have to date been re-covered.

The striking mixture of archaic and modern characters in the Middle Palaeolithic human remains was at first inter-preted as two distinct populations (Keith, McCown, 1937). Shortly after, the same students concluded in favor of a single, extremely variable population (McCown, Keith 1939). This notion became widely held for two decades, with but rare voices to the contrary (e.g. Stewart 1951). The single popu-lation view gave rise to two interpretations: a crucial point in evolution (McCown , Keith 1939; Keith 1948; Le Gros Clark 1955) or hybridization (Ashley-Montagu 1940; Dobzhansky 1944; Thoma 1957-58).

The single population concept was first questioned in the late 50's, mainly on chronological grounds (Howell 1958; Higgs 1961). Brothwell (1961) went as far as suggesting that the Skhul group belonged to *H. sapiens sapiens*. With the firm attribution of the Qafzeh and Skhul human remains to *H. sapiens sapiens* (Vandermeersch 1969), the neandertaloids of Tabun, Amud and Kebara became a group apart, and the old notion of two human groups thus returned. The old questions have also returned: how did two groups maintain their sepa-ration while sharing such a small geographical region as Mount Carmel and the Galilee?

The easiest solution would be a clear temporal separa-tion. However, the temporal relations between the groups are a matter of conflicting evidences. On the one hand, lithic technology studies indicate that the 'primitive' Tabun I skull is older than both the Skhul and Qafzeh cemeteries (Jelinek 1982:91). On the other hand, the assemblages of micro-fauna indicate that the Qafzeh burials are as old as, and perhaps older than, the Tabun human remains (Tchernov 1981: 77-78). The Tabun II mandible, of a "modern" type (Roth 1983), is worth mentioning here. Found in proximity to the Tabun I neandertaloid, this is the only case at present of the two human groups occurring at the same site. This fact does not clarify the temporal relationship between them, due to the unclear stratigraphical position of the Tabun finds.

The lithic industry associated with the two Middle Palaeolithic human groups seems to be similar, pending more minute studies. In any event, stone knapping was without doubt equally mastered by both. But a difference of behavioral nature exists between the two groups, which has passed unnoticed: more precisely, in their burial customs. *H. sapiens sapiens* was found in two sizeable cemeteries, Skhul and Qafzeh, with about 15 individuals in each case. The neandertaloids, on the other hand, are always found as single individuals or, at most, two or three per site. In addition, no grave objects were ever noted with any of the neandertaloids, whereas funerary objects were unquestionably associated with modern-type man: a boar's skull with Skhul V and deer horns with Qafzeh XI. Ochre, sea shell and ostrich egg shell, scattered around the Qafzeh burials, were probably connected with them (Vandermeersch 1981). As funerary rituals reflect the beliefs and customs of a culture group, a cultural barrier may well have separated the two Middle Palaeolithic human groups in Israel.

UPPER PALAEOLITHIC

In comparison with the relative abundance of Middle Palaeolithic human remains in Israel, their absence in the Upper Palaeolithic period (40,000 - 18,000 years ago) is all the more astonishing, and could perhaps be related to a different treatment of the dead. The only Upper Palaeolithic skeleton actually known is the Nahal En-Gev burial, dated to the very end of the period, a "Proto-Mediterranean" (Arensburg 1977).

KEBARAN

The Kebaran phase (18,000 - 14,000 BP) has to date yielded extremely scarce human remains: a single individual burial at Neve-David (Kaufman, Ronen, in press), a single burial at En Gev (Arensburg and Bar Yosef 1976) and isolated teeth at Hefziba (Ronen *et al* 1975). By the scarcity of its human remains the Kebaran continues the Upper Palaeolithic trend, though by the quantity and variety of its grinding stones the Kebaran already constitutes a prelude to domestication. The newly invented vegetal food processors very quickly found their way to become funerary objects, as attested by three fragments of mortars -- each of a different tool -- found in direct association with the Neve-David burial. This association may be taken to indicate that death was perceived by the Kebarans as a continuation of life. Their physical traits are "Proto-Mediterranean" (Arensburg 1977).

NATUFIAN

Apparently for the first time in human history, Natufian settlements were sedentary. Also for the first time, cemeteries now number tens of individuals, even close to 100. Objects of art -- sculpted and engraved stone and bone -- accompanied a few burials; necklaces and other ornaments were also found in burials, while other burials were devoid of any goods. Special "stone pipes" accompanied some of the burials and seem to have served as tombstones. These were heavy artifacts, ca. 1 m. high, roughly dressed into a cylindrical form and drilled across the long axis throughout. Being thus essentially stylised "killed" mortars, they constitute an elaboration of the Kebaran burial custom as seen in Neve David. On the basis of the repartition of grave goods, the Natufian society would seem largely hierarchical. About 300 individuals have been uncovered to date, of a proto-Mediterranean type (Arensburg *et al* 1975): small to medium stature, with a broad face and high vault.

SUMMARY

1. From the physical anthropologist's point of view, the population of Israel could have evolved largely locally from the Zuttiya man through the Middle Palaeolithic *H. sapiens sapiens* to the Kebaran and Natufian peoples (Arensburg pers comm).

2. Between a large number of Middle Palaeolithic individuals and a very large Natufian population in Israel, there is a lack of the Upper Palaeolithic human remains, as yet unexplained.

Arensburg B (1977). New upper Palaeolithic human remains from Israel. Eretz-Israel 13:208.
Arensburg B, Goldstein M, Nathan N (1975). The epipalaeolithic (Natufian) population in Israel. Arquivos de Anatomia e Antropologia 1:205.
Ashley-Montagu MF (1940). Review of the Stone Age of Mount Carmel Vol 2 Am Anthrop 42:518.
Brothwell DR (1961). The people of Mount Carmel. Proc Preh Soc 27:155.
Dobzhansky T (1944). On species and races of living and fossil man. Am J Phys Anthropol 2:251.

Garrod DAE, Bate DMA (1937). "The Stone Age of Mount Carmel",
Vol 1, Oxford: University Press.

Gisis I, Bar Yosef O (1974). New excavations in Zuttiyeh cave,
Wadi Amud, Israel. Paleorient 2:175.

Higgs ES (1961). Some Pleistocene faunas of the Mediterranean
coastal areas. Proc Preh Soc 27:144.

Howell FC (1958). Upper Pleistocene men of the southwest
Asian Mousterian. In Von Koenigswald GHR (ed): "Neanderthal
Centenary", Böhlau-Verlag-Köln-Grz, p 185.

Jelinek AJ (1982). The Middle Palaeolithic in the Southern
Levant, with comments on the appearance of modern *Homo
sapiens*. In Ronen A (ed): "The Transition from Lower to
Middle Palaeolithic and the Origin of Modern Man", Oxford:
BAR S151, p 54.

Kaufman D, Ronen A (in press). A Kebaran burial from Neve
David (Haifa, Israel).

Keith A (1948). "A New Theory of Human Evolution", London: Watts.

Keith A, McCown TD (1937). Mount Carmel man. His bearing
on the ancestry of modern races. In MacCurdy GG (ed):
"Early Man", Philadelphia: Lippincott, p 41.

Le Gros Clark WE (1955). "The Fossil Evidence for Human Evolution. An Introduction to Paleoanthropology," Chicago.

McCown TD, Keith A (1939). "The Stone Age of Mount Carmel",
Vol 2, "The Fossil Human Remains from the Levalloiso-
Mousterian", Oxford: University Press.

Neuville R (1951). Le Paléolithique et le Mésolithique du
désert de Judée. Arch Inst Paléont Humaine 24:179.

Ronen A, Kaufman D, Gophna R, Smith P, Bakler N (1975). Hef-
ziba-Hadera, An Epi-Palaeolithic site in the central coastal
plain of Israel. Quartar 26:53.

Roth H (1983) Comparaison statistique de la forme des arcades alveolaire et dentaire des mandibules des Hominidés
fossiles. Position de l'Homme de Tautavel dans l'évolution
humaine. Thèse Université Aix-Marseille I.

Smith P, Arensburg B (1977). A Mousterian skeleton from
Kebara cave. Eretz-Israel 13:164.

Stewart TD (1951). The problems of the earliest claimed
representatives of *Homo sapiens*. Cold Spring Harbor
Symposium Quantitative Biology 15:97.

Suzuki H, Takai F (1970). "The Amud Man and his Cave Site,"
Tokyo: Keigaku Publishing Co.

Tchernov E (1981). The biostratigraphy of the Middle East.
In Cauvin J, Sanlaville P (eds): "Prehistoire du Levant",
Paris: CNRS, pp 67-97.

Thoma A (1957-58). Métissage ou transformation? Essai sur

les hommes fossiles de Palestine. L'Anthropologie 61:
470, 62:30.
Tobias PV (1966) "A Member of the Genus *Homo* from Ubeidiya,"
Israel Academy of Sciences and Humanities, Jerusalem. p 1.
Vandermeersch B (1969). Les nouveaux squelettes mousteriens
découverts a Qafzeh (Israel): essai d'interpretation.
In Bordes F (ed): "The Origins of *Homo sapiens*", Paris:
UNESCO pp 49-54.
Vandermeersch B (1981). "Les Hommes Fossiles de Qafzeh
(Israel)", Paris: CNRS.

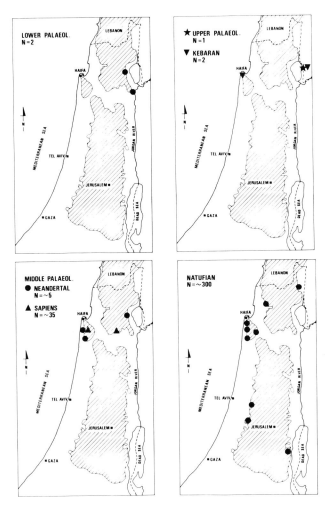

Fig. 1 Sites with human remains in Israel, by period.

Hominid Evolution: Past, Present and Future, pages 335–340
© *1985 Alan R. Liss, Inc.*

PRELIMINARY REPORT ON AN EARLY HUMAN BURIAL IN THE WADI
KUBBANIYA, EGYPT

T. DALE STEWART, M.D., ANTHROPOLOGIST EMERITUS,

United States National Museum of Natural
History, Smithsonian Institution,
Washington, D.C. 20560, USA

The Wadi Kubbaniya opens into the Nile Valley from the
west some 10-15 km north of Aswan (fig. 1). There in 1982
Fred Wendorf of Southern Methodist University, Dallas,
Texas, found eroding from a terrace what he judged to be a
Middle Paleolithic human skeleton (60-80,000 years old and
possibly Neandertal). The unique feature of this skeleton in Wendorf's experience of some 20 years of archeological field work in Egypt and the Sudan was that it had
been buried face down in a dug grave, the sandy fill of
which over time had become solidified by chemicals in
moisture infiltration. On account of wind erosion, by
1982 the part of the grave containing the skeleton was
standing about 30 cm (1 foot) above the present surface
level like a miniature mesa or butte (fig. 2).

Wendorf got word of all this to me from Egypt by
telephone in early April, 1982. He wanted to know whether
I would be willing to study the skeleton, if permission to
do so could be obtained from the Egyptian authorities. On
attempting to follow up on this matter during a visit to
Cairo in November of that year, I found the still embedded
skeleton boxed up and deposited in the Archeological Museum.
My request for a 1-year loan of the specimen for study was
not approved until nearly 5 months later (April 4, 1983)
when it was delivered to me in Washington.

My laboratory lacked the space and equipment needed for
extracting the bones from the block of solidified sand.
Therefore I sought permission from the Museum's Vertebrate
Paleontology Preparation Laboratory to carry out this phase

of the study there, using one of its Pneumatic Air Scribes
to loosen the sand. As received, the block was approximately 60 cm (24 inches) long, 50 cm (20 inches) wide, and
around 25 cm (10 inches) thick. Besides the block there
were bones and portions of bones which had become detached,
including parts of the skull, elements of the hands, and a
large number of elongated and polished fragments of long-
bone cortex.

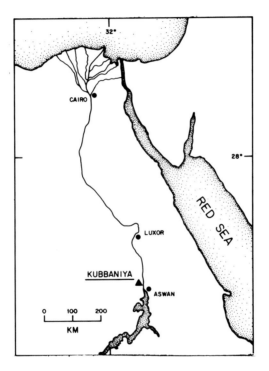

Fig. 1. Map of Egypt showing location of Wadi Kubbaniya.
(Courtesy Fred Wendorf.)

     One of the conditions imposed on me by the Egyptians
for the loan of the skeleton had been that by the end of a
year from the date of its arrival in Washington, all of the
bones found in and around the block, together with casts of
most of them, should be returned to Cairo. Added to this
was a plan to include also casts of the entire block at
different exposures of the contained bones. Although this

seemed like a good idea, I soon realized it was more than I
could manage alone. So the head of the Laboratory picked
one of his temporary assistants (Michael Tiffany) to take
over the laboratory work, leaving me only general supervis-
ion.

Since up to this stage I had concentrated almost entirely on the skull and the large block had not been touched, I
let Mike alternate work on these two parts by himself.
Although this slowed up the progress, the two operations
were finished by July 12. Pictures of the skull taken then
(fig. 3) show that only the facial parts are present. The
loss of the exposed back parts was due, of course, to sand-
blasting (ventifaction).

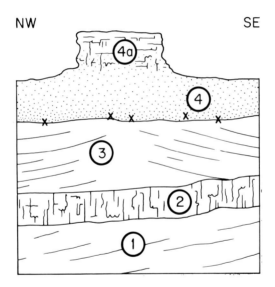

Fig. 2. Profile of burial(4a) and underlying strata (1-4).
(Courtesy Fred Wendorf.)

In front view (fig. 3, left) the face gives the best
indication of race: a somewhat archaic type of modern man
such as also represented by the skulls from site 117 near
Jebel Sahaba in Nubia described by JL Anderson (in
Wendorf 1968).   Evidently this type of man could not
have been as ancient as Wendorf had thought. Also, it was
not Neandertaloid.   Most likely it was not older than

20,000 years, which even then makes it unusual in Egypt.
Male sex is strongly suggested by the size of the jaws and
supraorbital ridges. Early adult age is suggested by the
moderate wear of the posterior teeth.

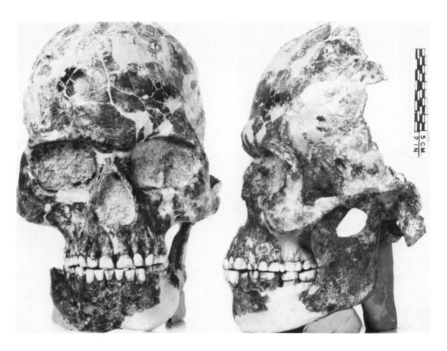

Fig. 3. (right) Loss of back parts of skull from ventifaction); (left) Racial characteristics visible in full face.
(Both courtesy Smithsonian Institution.)

    Turning now to the top surface of the block after remov-
ing the loose sand and increasing the bone exposure (fig. 4),
it is clear that, like the back of the skull, the back parts
of the trunk were highest in the grave and as a consequence
underwent extensive ventifaction. Gone were the spinous
processes and most of the neural arches from the vertebrae,
the vertebral ends of the ribs, and the bodies of the
scapulae.

    There being little more of significance to be learned
from the top surface of the block, on May 24, 1983 the block
was turned over(fig. 5). But the only evidence of leg bones

found there was a piece of the head of the left femur still
in its acetabulum. Although some other pieces of leg bone
had been found outside the block, there was no answer to the
question: What broke the leg bones into such small pieces?

Fig. 4. Ventifacted bones of the back of the skeleton indicating face-down burial. (Courtesy Smithsonian Institution.)

Further exposure of the arm bones and lumbar vertebrae
after this picture was taken showed 1) a second bladelet
lying against the second and third lumbar vertebrae, and 2)
injury to the lateral condylar ridge of the left humerus.
An X-ray of this part showed a stone chip within the line
of breakage. This appeared to be an inflicted injury at or
just before the time of death. This view is reinforced by
the presence of a healed parry fracture of the right ulna.

Removal of the two humeri from the block showed that the right one was larger throughout than the left one, thus indicating right-handedness and resulting arthritic enlargement of the sterno-clavicular joint on that side, a state that accords with the early adult age estimate.

Fig. 5. Front of skeleton after block was turned over and the bones exposed. Note absence of bone ventifaction and presence of bladelet no. 1. (Courtesy Smithsonian Institution.)

REFERENCE

Anderson, JE, Late paleolithic skeletal remains from Nubia. In Wendorf, F (ed): "The Prehistory of Nubia," Vol. 2. Dallas, Texas: Fort Bergwin Research Center and Southern Methodist University Press, 1968, 996.

Hominid Evolution: Past, Present and Future, pages 341–354
© *1985 Alan R. Liss, Inc.*

THE EUROPEAN, NEAR EAST AND NORTH AFRICAN FINDS AFTER
AUSTRALOPITHECUS AND THE PRINCIPAL CONSEQUENCES FOR THE
PICTURE OF HUMAN EVOLUTION

Jan Jelínek, Ph.D., D.Sc.

Head of the Anthropos Institute

nám. 25. února 7, 659 37 Brno - Czechoslovakia

The discovery of Australopithecus in 1924 was of great
importance for palaeoanthropology. Not only did it bring
new knowledge on Plio-Pleistocene hominids, but it was also
a turning point in our thinking, prompting further research,
resulting in new achievements. It started the dynamic advance
of palaeoanthropology, so characteristic of the second half
of this century. Directly or indirectly it influenced further
research and discoveries in South and East Africa; it brought
new ways of behavioural research in primatology, and the establishment of a new discipline taphonomy; with the introduction of new dating methods, it has led to new understanding of
the tempo and even mode of human evolution.

If we focus our attention on North Africa, the Near East
and Europe and, if we compare the individual finds of human
remains known before and after the discovery of Australopithecus, we see the tremendous difference in knowledge, quantitative and qualitative.

The most important post-1925 European discoveries include the fragmentary skull from Ehringsdorf (1925), the two
skulls from Saccopastore near Rome (1929, 1935), the parietal
bone found in Cova Negra (1933), the Steinheim skull (1933),
the Neandertal skull from Monte Circeo (1939), the Swanscombe
occipital bone and a parietal bone (1935, 1936), as well as
the second parietal found in 1955, modest remains from La
Chaise (1949), the calva from Fontéchevade (1947), the mandible from Montmaurin (1949). Between 1950 and 1985 followed
the Lazaret find (1964), the Petralona skull (1960), the
Azych mandible from Soviet Azerbaijan (1968) and the infant

skull from Gariguela (1954-55), the endocast from Gánovce
(1937-1956) and the Neandertal frontal bone from Šala,
Czechoslovakia (1961), the Atapuerca finds in Spain (1976),
the Hortus mandible (1960-64), the Orgnac teeth (1960), Arago remains (1964-71), Vergranne teeth (1968), the pelvis from
Grotte du Prince (1968), the famous finds of Vértesszölös
(1965), Bilzingsleben (1972), Biache (1976) and St. Cesaire
(1980). This list of the most important finds is certainly
not exhaustive.

In the Near East the first group covers the finds discovered before World War II: Tabun and Skhul (1929-1934),
Zuttiyeh (1925), and Qafzeh (1933-35) with new excavations
in the 1966-70 period. Then came the Amud skull (1961),
Shanidar (1953, 1957-60), a group of Crimean finds (Staro-
selie 1953 , Zaskalnaya 1970, 1972), and Teshik Tash (1938)
in southern Uzbekistan in the U.S.S.R. The latest find in
this region is the second skeleton from Kebarah (1983) not
yet published.

In North Africa let us mention the finds from Rabat
(1933), Tangier (1939, 1947), Sidi Abderrahman (1955), Haua
Fteah (1952, 1955), Ternifine (1954-55), Témara (1956),
Djebel Irhoud (1961, 1963, 1968), Thomas Quarry (1969, 1972)
and Salé (1971).

Such a wealth of new palaeoanthropological finds gives
us not only a better idea of human morphological variability
and evolutionary changes. With the help of new dating methods, specialists have been able to set up a more plausible
chronological sequence. Only thus are we able to understand
better human evolution, a historical process, and its complexity, both chronologically and geographically.

These facts testify that scientific progress is furthered
not only by new finds, but also by new methods. New dating
methods have been such an example. The multidisciplinary approach is another. Of course this is not a new idea: from
time to time and with varying intensity it was used by many
specialists long ago. But the widespread following of the
multidisciplinary approach appears to be a progressive method,
without which modern research is unthinkable. Archaeology and
palaeoanthropology widely use these methods; the methods of
the exact sciences have become an indispensable part of the
research.

If we focus on North Africa, the Near East and Europe and, if we study the palaeoanthropology of these regions separately, we find the local evolutionary tendencies documenting that the importance of the autochthonous population was much greater than the influence of immigrants. In anthropological research developments in recent decades thus place more emphasis on the local roots of the populations of large and well defined geographical-ecological areas. It seems that although some immigrations are indisputable, they were in most cases easily absorbed by well-adapted local populations. Only very seldom did such immigrations bring about fundamental changes in the principal characters of local peoples. Chamla (1968), Vallois (1969), Jaeger (1973), Ferembach (1979), and others stressed that a direct and uninterrupted evolutionary line can be traced in North Africa from the earliest finds of human remains (Ternifine), through later finds at Sidi Abderrahman, Thomas Quarry, Salé, Témara, Rabat, Tangier, Djebel Irhoud, Haua Fteah, to the Upper Palaeolithic Homo sapiens sapiens populations. It should not be forgotten that many years ago Biberson and others published rich archaeological materials, mostly pebble tools of considerable age found on the Moroccan sea-shore terraces. They document that the peopling of this part of North Africa is earlier than the hitherto known human remains and they pose the question, who made these tools? It is not clear whether it was an earlier type of Homo erectus or another hominid.

The ancient local roots of the core of the modern North African population are heavily supported by a similar situation in Europe. The African australopithecine finds were sometimes accompanied by the simplest of stone tools, and when finds were still scarce, they were considered to be products of those hominids. R.A. Dart believed he had discovered the earliest tool-producing evolutionary stage, which, from the bone-tooth-horn assemblage,he called osteodontokeratic culture; it preceded the invention and use of stone tools. Y. Coppens continues to consider the earliest flaked stone tools of Omo to be the product of Australopithecus. The whole situation is still far from clear. The important consequence is that archaeologists accepted the view that stone tools can be up to 2 m.y. old and they started to search for earlier stone tools also in Europe. Before the Taung find in 1924, the Abbevillian hand axes were known in Europe, but nobody knew their proper dating. Today, with the help of better dating methods, we have reached the 1 m.y. frontier of the earliest tools in

Europe and we are looking for even earlier finds. With the existence of Plio-Pleistocene stone tools in Africa this is no longer unexpected in Europe.

Very illustrative is the situation in Czechoslovakia. Up to the end of World War II there was a general consensus that the peopling of Czechoslovak territory reached only to the Upper Palaeolithic. Why was this so? Because the research was completely dominated by archaeologists of the older generation modelling themselves upon West European scientists. The other reason was that no Lower Palaeolithic hand-axes or typical Mousterian Middle Palaeolithic tools were known in this country. There was limited information also on the new ideas developing in other parts of the world. After World War II a new generation of archaeologists appeared and the latest three decades have brought not only rich Middle Palaeolithic finds of atypical Mousterian or other stone tools (atypical, due to the raw materials used or due to different cultural traditions), but also numerous human remains of Middle Palaeolithic provenience. The new generation of archaeologists did not limit their attention to the open-air Upper Palaeolithic loess sites or to the upper layers of cave deposits, but, with the help of advanced geological and palaeontological knowledge and equipped with better dating methods and excavation technology, they searched accessible Pleistocene deposits, wherever they found them.

Middle Palaeolithic occupation was recognized in Gánovce (East Slovakia) in an interglacial travertine spring layer; and in the Kůlna Cave, yielding also reliably dated human remains. In the 1970's a group of Early and Mid-Palaeolithic sites was found, recognized and excavated: Stránská Skála (Musil, Valoch 1968), with its stone and bone tools dated palaeomagnetically and accompanied by a very rich biostratigraphy, and Přezletice near Prague (Fridrich 1972), with stone tools, rich biostratigraphical material and good palaeomagnetic dating, have taken us to the limits of 700 000 y. BP. Bečov, another Mid-Pleistocene North Bohemian Locality (Fridrich 1979, 1980), yielded a multilayered Acheulean living site with stone tools, good stratigraphy and dating. A number of less important sites followed.

Now, in the 1984-85 season, new, even earlier stone tools have been found near Beroun in Central Bohemia in undisturbed stratigraphical position, superimposed by 20 soils of Interglacial type (Fridrich, personal communication and information by the Quaternary Research Committee). The palaeomagnetic dating of the site is put at over 0.7 m.y. BP. Thus, as in Southern France and the French Massif Central, Czechoslovakia in Central Europe also has crossed the 1 m.y. mark for the earliest stone tools. This fact illustrates well the relation between research and applied methodology. It shows, too, how psychology, in this case our preconceived models conditioned by fragmentary knowledge, can influence our approach to existing problems.

As we now have many more palaeoanthropological finds at our disposal, the continuity between Mid-Pleistocene hominids and populations of the modern type of man have been generally accepted. The Mid-Pleistocene finds from Vértesszölös, Petralona, Arago and Bilzingsleben demonstrate surprising variability and presumably also strong sexual dimorphism. The varying degree of brain development, as demonstrated by the Vértesszölös occipital, the Petralona skull and the Arago frontal and temporal bones, was one of the reasons why A. Thoma (1966) designated the Vértesszölös remains as H. erectus seu sapiens and why the Petralona endocast is recognized as of a modern type in several basic features. The clear link between the Mid-Pleistocene European finds and the Upper Pleistocene H. sapiens neanderthalensis influenced also the creation of the provisional group of "pre-Neanderthals" in Europe. In principle, the finds earlier than the classical West European Neandertals were recognized as direct ancestors of the latter. There are no other finds which could be ancestral to West European Neandertals.

Two taxonomic views are currently applied:
1. that the European pre-Neandertals are early H. sapiens finds, and
2. that they represent the European type of H. erectus.
If the difference from H. sapiens is not on the species but only the subspecies level, as some authors argue (Hennig 1966; Bilsborough 1969; J. Jelinek 1976, 1980), then the correct name should be H. sapiens erectus and not H. erectus. The question remains whether this taxonomic solution is to be applied for Europe only, or whether it has a more general, i.e. also extra-European value.

Anyhow, we see an increasing tendency to recognize and stress local evolutionary changes and continuity, not only in the European geographical area.

The theory of local evolutionary continuity is replacing the earlier idea of large-scale migrations traditionally used to explain evident morphological differences between finds belonging to different periods. This was so in most cases due to the scarcity of finds. Now, when more and more finds have been discovered, our evolutionary picture is more complex. Thus we find not only fragments (Bilzingsleben) which were described a few years ago without much hesitation as H. erectus (Vlček 1977, 1978), but also remains with transitional characters or with advanced morphological features usually characteristic of modern man (e.g. the dimensions of the occipital scale from Vértesszölös or the well developed brain size and morphology in the Petralona skull). Such transitional finds with mixtures of archaic and progressive characters are found not only in Europe, the Near East and North Africa, but also in East Africa and continental East Asia, indicating that it was due to insufficient material that the gradual mosaic changes in local populations had not been recognized earlier.

In Europe we face not only the problem of transition between the Mid-Pleistocene and Upper Pleistocene finds, but also the relationship between so-called classic Neandertals and Upper Palaeolithic H. sapiens sapiens. The new finds enable us to find a way out of the blind alley of the evolutionary line of Neandertal man. With the rapidly increasing number of Central and East European Neandertal finds, the idea of classic and progressive Neandertals was presented (Howell 1951). Today a degree of variability within the classic Neandertals is admitted (e.g. La Ferrassie) and finally the new find of St. Césaire has been accepted as an individual with transitional characters, such as small bone thickness, teeth morphology etc. (Léveque, Vandermeersch 1980, 1981). But, alas - more detailed morphological studies have shown that the find belonged to a female individual and that most cranial characters are similar to those of the classic Neandertals! Only the stone tools belonging to the Châtelperronian and not to the Mousterian, as one would expect, are exceptional. We should take into account this cultural situation, as we know that cultural and physical types can be, but need not be related. On the other hand we lack solid evidence that the Châtelperronian tools were manufactured by H. sapiens sapiens populations. It is quite possible that they were made by Homo sapiens neanderthalensis, or that they were not bound to a single type of Upper Pleistocene man. Here

again we await further new finds. The materials documenting
this Upper Pleistocene transitional period in Europe are still
too fragmentary.

The idea still survives that classic Neandertals were
not the forefathers of H. sapiens sapiens, but there is no
solid evidence for this. It was argued that chronologically
there was no time for the Neandertals to develop into the
later H. sapiens sapiens, almost their contemporary. This
assumption is based mainly on archaeological materials (on
late Mousterian, Châtelperronian, Szeletian and Aurignacian
tools), but again stone tools and physical types are not necessarily inseparable. Only in exceptional cases (e.g. the
Mladeč H. sapiens sapiens and the Šipka H. sapiens neanderthalensis (Jelinek 1965, 1983)) do human remains of these
two species appear side by side in the same region. In various regions they can have different dating and hence we
cannot exclude the participation of Neandertals in our pedi-
gree. In principle we have a transitional type between Mid-
Pleistocene hominids (H. erectus) and H. sapiens sapiens.
With more numerous finds this transition becomes evident
everywhere, not only in Europe, North Africa and the Near
East. Similar transitional types were found also in East
Africa (Ngaloba beds, Omo Shungura formation) and in continental East Asia (Dali, Mapa). Here we see again that if we have
enough well dated finds from the respective period, gradual
changes are more important than external (immigration) changes.

A classical example still open to discussion is the
transition from Middle Palaeolithic to Upper Palaeolithic in
the Near East. Although the dating of this transition was
discussed for a long time, the results of repeated excavations
have finally decided the matter (A. Jelínek 1973, 1982 ; Farrand 1979; Bar, Vandermeersch 1981). The two well known
sites of Mt. Carmel, Skhul and Tabun, yielded a number of
human remains, those from the Skhul Cave showing more numerous
H. sapiens sapiens features. Most anthropologists consider
this Skhul population early H. sapiens sapiens with a Mous-
tero-Levalloisian stone assemblage. The skeletal remains
vary in the frequency of robusticity, certain archaic features and sexual dimorphism. Most morphological features ally
these finds to H. sapiens sapiens. Most scholars involved in
Near Eastern anthropology consider the Skhul-Qafzeh group as
ancestral to Cro-Magnonoids or to proto-Cro-Magnonoids. We
consider the designation "archaic Homo sapiens" more suitable.
If we accept Neandertal man (the classic Neandertal) as limi-

ted to Western Europe during the first part of the last glaciation, then a similar solution should be accepted also for
the later Cro-Magnonoids - the West European population of
the ensuing Early Upper Palaeolithic. The use of the name
Cro-Magnonoid for other, sometimes even for extra-European
populations, is misleading. In these cases, as a rule, no
genetic links with true Cro-Magnonoids can be proved; in
fact direct genetic relationship between them is rather improbable. However, those who use this designation for the
Skhul and Qafzeh finds usually defend the idea that these
Near Eastern finds represent real ancestors of the Upper
Palaeolithic West European Cro-Magnonoids, and that they
immigrated to West Europe from the Near East, the cradle of
Upper Palaeolithic H. sapiens sapiens. This idea is far from
being proved. It seems very probable that the representatives of the archaic H. sapiens from Skhul and Qafzeh are the
forefathers of the local (i.e. of the Near Eastern) Upper
Palaeolithic H. sapiens sapiens.

For the European Upper Palaeolithic H. sapiens sapiens,
several other origins can be suggested. In Central Europe we
know a number of Early Upper Palaeolithic H. sapiens finds
with a varying number and various degrees of archaic characters.   Well known are the Gravettian (Pavlovian) finds of
Předmostí, Brno II, Pavlov and Dolni Věstonice. Another group
of early Aurignacian remains comes from the Mladeč caves
(Mladeč I, IV, V, VI), where another complex of plesiomorphous and apomorphous features have been preserved (J. Jelinek 1983). For Western Europe we can mention a group of other
populations (Cro-Magnon, Grimaldi, Chancelade). All these
Upper Palaeolithic groups have some specific character, but
they are definitely not homogeneous. Their characteristic
features cover a wide variety and various degrees of different characters. This is readily perceptible when we find
several individuals of the same physical and cultural type in
the respective period.

Naturally exact dating is the starting point for the
study of the evolutionary position of the South-West Asian
H. sapiens finds (Skhul and Qafzeh) and of their possible relation to the Early Upper Palaeolithic finds of Europe. Unfortunately we do not have at our disposal exact dating (c.f.
Vogel, Waterbolk 1963, for Skhul; and Bada, Helfman 1976 for
Qafzeh). The Skhul finds date probably to 41,000 y. BP or
even earlier (W.R. Farrand in A. Jelinek et al 1973). Morphologically both the Skhul and Qafzeh finds are simply primi-

tive forms of H. sapiens. The dating of the Qafzeh finds is
70-80,000 y. BP, according to stratigraphic, sedimentological
and microfaunal analyses. The absolute dating (through the
racemisation of amino-acids) of bones from layer No. XXII is
68,000-78,000 y. BP. This is of course inconsistent with
the early European Upper Palaeolithic finds of H. sapiens sapiens. Furthermore such a dating would mean that two different
populations, the primitive sapiens represented by Qafzeh, and
the Neandertal-like represented by Amud and Shanidar, were
living side by side somewhere in the Levant for a long time.

Looking for an easily comprehensible model, some authors
(e.g. Trinkaus 1983) plead for a younger absolute dating of
Qafzeh, as the Mousterian lithic material from levels VII-
XVII is similar to the Skhul stone assemblage from layer B.
Trinkaus questions also the other features supporting an earlier dating, such as microfaunal, stratigraphic and sedimentological evidence, and proposes a later date "somewhat less
than 50,000 years." The only reasons for this can be only
the similarity of stone tool assemblages and the tendency to
fit a simpler evolutionary model. A later dating of Qafzeh
would make it possible to divide the Near Eastern finds into
two groups, an earlier (Amud, Tabun, Shanidar), corresponding
more to West European Neandertals, and a later one (Skhul,
Qafzeh), resembling a primitive type of modern man. But this
is not valid and the situation is much more complex. First
there is an interglacial (therefore early) find from Zuttiyeh,
which certainly does not fully accord with the other Neandertal-like finds (Amud, Tabun, Shanidar). But the above-
mentioned three finds have no fully reliable dating. The C2
Tabun mandible comes from layer C, the same as the partial
female skeleton C1, of a probable age of 50-60,000 y. BP.
Trinkaus mentions the possibility that skeleton C1 might come
from layer D, and could be dated to 70-80,000 y.BP. This
possibility was mentioned by Trinkaus (1983) as a personal
communication to him by A. Jelinek in 1980. In such a detailed and delicate discussion, it is certainly better to avoid
mentioning unpublished personal communications, to avoid
possible misunderstandings. It seems to be more important
that the two finds, Tabun C1 and Tabun C2, have appreciable
morphological differences. This may be due to variability
within one population (Vandermeersch 1981) or to the existence of two different populations, either living side by
side or chronologically different. To make the situation
even more complex, we should bear in mind the Kebara 1

infant skeleton, now followed by the Kebara 2 adult skeleton,
accompanied by a lithic assemblage comparable to the stone
tools from the Tabun B layer. These stone tools are supposed
to be slightly less than 50,000 y. BP. (A. Jelinek 1982).
The Amud skull is unfortunately not reliably dated. The
Tabun B layer, dated to less than 50,000 y. BP (Trinkaus
1983), seems to be slightly earlier than the Skhul B layer
with primitive <u>sapiens</u> remains - as documented by the lithic
assemblage (A. Jelinek 1982). Here we can see how labile
the chronology of the Near Eastern finds is, often based on
nothing more than the similarity of the lithic assemblage or
on certain similar (or different) morphological characters.

The morphological comparisons are logically based on the
characters to be compared, on characters that are more important than others, characters whose function is well understood
and on finds that are well dated. Their changes continue
through a certain period and we know that the tempo of these
changes can vary. Even periods of stasis of a functional
system can appear.

In the West European classic Neandertals, facial prognathism is often stressed. Unfortunately the facial skeleton
of the Zuttiyeh remains has not been preserved, but the high
nasofrontal angle does not signal much midfacial prognathism.
In this respect it may be compared with Shanidar 2 and 4, but
it differs in the interorbital breadth, and certainly the
frontal vaults of Zuttiyeh and Shanidar 4 are different. On
considering the Shanidar sample we see that Shanidar 2 and 4
have less midfacial prognathism than Shanidar 1 and 5. Possible explanations are :
1. we have to do with a degree of variability between these
individuals,
2. they are of different age inside the Levallois-Mousterian
layer. As for the Amud skull, its midfacial prognathism is
less evident than is the case with the Shanidar sample.

As for the nasal height-orbital breadth index differentiating the classic Neandertal from other, both earlier and
later hominids, Shanidar 1 has the lowest value of any of the
classic Neandertals. As regards the nasal area Shanidar 1 has
a value within the range of modern man, while Shanidar 5 boasts
the largest nasal area of any known hominid. We see therefore
that some fundamental facial differences between the Shanidar
individuals cannot be explained by different geological ages,
as the Shanidar 1 and 5 skeletons were found in the same

level. Stringer and Trinkaus (1981) clearly show the variability of some metrical and morphological characters for
certain Shanidar skulls. There is marked variation in the
occipitomastoid region, which again cannot be explained
through chronological differences between the two Shanidar
subsamples, Shanidar 1 and 5 as later and Shanidar 2 and 4 as
earlier individuals, with a minimum difference of 15,000 y.
between them (according to Solecki, 1960).

   As for the cranial vault the Amud skull is quite exceptional with its high sagittal vault (Suzuki, Takai 1970),
cranial length and breadth, while the height above the
Frankfurt plane is greater than in Neandertals. Its cranial
capacity reaches 1740 cm^3, so far the biggest in fossil men.
The Amud skull has several other features quite exceptional
for Neandertals, such as a slightly more progressive form of
supra-orbital torus, less pronounced than those of the classic Neandertals and Shanidar. The frontal sinus is wide and
does not extend to the frontal squama, as in most classic
Neandertals. The orbital margin is sharp as in modern man
and not rounded as in most Neandertals and the inclination of
the frontal squama is steeper than in Neandertals, but less
steep than in modern man. The upper outline of the temporal
squama is well curved, resembling the situation in modern
man and its mastoid process is better developed than in the
Neandertals. The fossa mandibularis is not so deep as in
modern man, but is deeper than in Neandertals. The tympanic
plate also resembles that of modern man and the petromastoid
angle is greater than in the Neandertals.

   I mentioned these characters of cranial morphology to
demonstrate that, although in the principal features there is
evident resemblance between European Neandertals and West
Asian Neandertaloids, there are also marked differences between them.

   Without doubt there is a similar situation in the dental
characters, in mandibular morphology, endocranial features
and in the post-cranial skeleton. As in North Africa and in
Europe - when the finds are numerous enough and are reliably
dated - the whole situation reveals mosaic changes in smaller
or greater functional systems, differing in time sequence and
in quantity. These mosaic changes are characteristic of the
evolutionary process and they represent the most suitable
model of gradual transition. The recognition of periods of
morphological stasis or of rapid change depends on the

features we select and on their complexity. The more complex the selected feature or system of features we are following, the more gradual are the changes in the model. If we choose a simple isolated character, we have a better chance to discover its periods of stasis followed by periods of rapid change.

Let us summarize our short discussion:

The discovery of <u>Australopithecus</u> forms an integral and inseparable part of our palaeoanthropological knowledge. It is extremely difficult to define the most important impact of such a discovery, since once incorporated into our knowledge, it forms an integral part of the ensuing scientific progress. This has certainly contributed to our understanding of brain evolution, dental evolution, sexual dimorphism, bipedal locomotion, systematics of hominids and higher primates in general, and many other anthropological questions. It also gave impetus to the birth of a new discipline, taphonomy, to the study of the origin of stone tool production, to the study of primate behaviour, etc.

The most important consequences of the above development for Europe, North Africa and the Near East, are:

1. Recognition of the important rôle of local evolution within divers large geographical regions - with the same evolutionary result - <u>H. sapiens sapiens</u>.
2. Dynamic development of palaeoanthropology as a science.
3. Recognition of transitional evolutionary stages within a gradual stream of unequal mosaic changes.
4. Shifting of the focus of our search for earlier human remains and tools in Europe and their actual discovery in Middle and Early Pleistocene deposits, hence the proof of Early Pleistocene peopling of Europe.

The importance of a discovery should be measured by the consequences it directly or indirectly creates.

REFERENCES

Bada J. Helfman P (1976). Application of amino-acid racemization in palaeoanthropology and archaeology. Colloque du Union Internationale des Sciences Préhistoriques et Protohistoriques. 18e Congres 1:39.

Bar Y, Vandermeersch B (1981). Notes concerning the possible age of the Mousterian layers in Qafzeh Cave. In Cauvin et Saulaville (eds.): "Prehistoire du Levant", Paris: CNRS, p 281.

Bilsborough A (1969). Rates of evolutionary change in the hominid dentition. Nature 223:146.

Chamla MC (1968). L'évolution du type de Mechta-Afalou en Algérie occidentale. CR Acad Sc Paris 267:1849.

Farrand WR (1979). Chronology and paleoenvironment of Levantine prehistoric sites as seen from sediment studies. J Arch Sci 6:369.

Ferembach D (1979). L'émergence du genre Homo et de l'éspéce Homo sapiens. Les Faits, les incertitudes. Biométrie humaine. Paris 11:18.

Fridrich J (1972). Altpaleolithische Industrie aus dem Altpleistozän in Prezletice, Kr. Prag-Ost. Archeologické rozhledy 24:241.

Fridrich J (1979). Stredopaleolitické osídlení Čech. Unpubl. thesis. Praha, Arch. ústav CSAV.

Fridrich J (1980). Bečov IV., District of Most - an Acheulian site in Bohemia. Anthropologie 2-3:291.

Hennig W (1966). Phylogenetic Systematics. Urbana. Univ. of Illinois Press.

Howell FC (1951). The place of Neanderthal man in human evolution. Am J Phys Anthropol 9:379.

Jaeger JJ (1973). Découverte d'un crâne d'hominidé dans le Pleistocene moyen du Maroc. Problémes actuels de Paléontologie. Évolution des Vertébrés. Paris.

Jelinek A (1982). The Tabun cave and paleolithic man in the Levant. Science 216:1369.

Jelinek A, Farrand WR, Haas G, Horowitz A, and Goldberg P (1973). New excavations at the Tabun Cave. Mount Carmel. Israel: a preliminary report. Paléorient 1, 2,: 151.

Jelínek J (1965). Der Kiefer aus der Sipka Höhle. Anthropos XVII:135.

Jelínek J (1976). A contribution to the origin of Homo sapiens sapiens. J Hum Evol 5:497

Jelínek J (1978). Homo erectus or Homo sapiens? In "Recent advances in Primatology. Vol. 3:419.

Jelínek J (1983). The Mladeč finds and their evolutionary importance. Anthropologie 1:57.

Léveque F, Vandermeersch B (1980). Découverte des restes humaines dans un niveau castelperronien à Saint-Césaire (Charente-Maritime) Compte r. de 1 'A.

Léveque F, Vandermeersch B (1981). Le Neandertalien de Saint Césaire. La Recherche 12:242.

Musil R, Valoch K (1968). Stránská skála: its meaning for Pleistocene studies. Curr Anthropol 9 (5), part II:534.

Solecki R (1960). Three adult Neanderthal skeletons from Shanidar Cave, Northern Iraq. The Smithsonian Report for 1959. Washington Publ. 4414:503.

Stringer CB, Trinkaus E (1981). The Shanidar Neanderthal Crania. In Stringer CB (ed) "Aspects of Human Evolution", p 124.

Suzuki H, Takai F (1970). The Amud Man and his cave site. Tokyo. Univ. of Tokyo: p 439.

Thoma A (1966). L´occipital de l´homme mindelien de Vértes-szölös. L´Anthropologie 70: 495.

Trinkaus E (1983). The Shanidar Neanderthals. London. Academic Press. p 502.

Vallois HV (1969). Les Hommes de Cro Magnon et les Guanches: les faits acquis et les hypothéses. Anuario de Estudios Atlanticos 15: 97.

Vandermeersch B (1981). A Neanderthal skeleton from a Chatelperronian level at Saint Césaire (France). Am J Phys Anthropol 54:286. (abstract).

Vlček E (1978). A new discovery of Homo erectus in Central Europe. J Hum Evol 7:239.

Vlček E, Mania D (1977). Ein neuer Fund von Homo erectus in Europe, Bilzingsleben (GDR). Anthropologie 15:159.

Vogel JC, Waterbolk MT (1963). Groningen Radiocarbon Dates IV. Radiocarbon 5:163.

Hominid Evolution: Past, Present and Future, pages 355–365
© *1985 Alan R. Liss, Inc.*

HUMAN EVOLUTION AT THE PERIPHERIES:
THE PATTERN AT THE EASTERN EDGE

Milford H. Wolpoff

Department of Anthropology
University of Michigan
Ann Arbor, Michigan 48109 USA

Just as the basis for the modern understanding of African australopithecines lies firmly in the insights and works of Raymond Dart, the basis for the worldwide perspective of Middle and Upper Pleistocene human evolution outside of Africa is to be found in the papers and monographs of Franz Weidenreich. The parallels between Dart (Tobias 1984) and Weidenreich are surprising. Both were trained as anatomists, but in fact became preeminent evolutionary morphologists. Both left their native lands and established careers as paleoanthropologists tens of thousands of miles away. Both were convinced that they were dealing with the earliest hominids. Both developed viewpoints and interpretations that have remained valid for a half century or more, and both produced publications that remain insightful and refreshing to this day in spite of the explosive growth of the fossil record. Ironically, however, each virtually completely ignored the other; this was because of the most fundamental difference between them — apart from a common focus on *Gigantopithecus* which they treated quite differently, they dealt with completely non-overlapping parts of the fossil record.

Another parallel between Dart and Weidenreich is found in Grafton Elliot Smith. It was he who sent Dart to South Africa, recommending that he apply for the Chair of Anatomy at the University of the Witwatersrand (and serving on the committee that recommended the appointment). Weidenreich was recommended by Grafton Elliot Smith to replace Davidson Black at the Cenozoic Research Laboratory of the Beijing Union Medical College. Moving to Beijing in 1935, he also became director of the ongoing excavations at Zhoukoudian. The move placed him in the position to describe the human fossils from the Zhoukoudian cave. This he accomplished in a series of monographs that to this day have remained unparalleled in their detail and insight.

By the end of the 1930's Weidenreich had accumulated firsthand experience with the human fossil record of three regions. These were: Europe, where he was trained and did his early work; north Asia, as the result of the rapidly accumulating Zhoukoudian finds from both the lower and the upper cave; and southeast Asia, as a consequence of communications and an exchange of casts with G.H.R. von Koenigswald, followed finally by an exchange of visits. Weidenreich was the only paleoanthropologist of his time (and one of the very few of any time) with a detailed knowledge of three regions. As a scholar well educated in the morphological and evolutionary traditions of central Europe, this put him in a unique position to appreciate the evidence for a worldwide pattern of human evolution. Weidenreich found such a pattern, and developed his theory of polycentric evolution.

According to Weidenreich's polycentric interpretation (1939, 1943) all fossil hominids belong to a single species, *Homo sapiens* (this original "single species hypothesis" first appears clearly stated in his 1943 cranial monograph). He contended that there was no single center of evolution from which new hominid types appeared from time to time to replace older ones. Instead, in his interpretation there were at least four centers of origin: Asia Minor, east or south Africa, north China, and the Sunda islands (perhaps today better referred to as the Sunda subcontinent because of the extensive land areas exposed during the glaciations). He argued that no fossil group or type could be excluded from the paternity of recent hominids, and that racial differences (which he regarded as "minor details") were as old as human evolution.

Weidenreich's separate geographic lines shared a common ancestry and evolved in the same direction. The crux of the problem he faced was how to explain this. Here he failed. In the mid-40's he proposed an explanation based on orthogenesis (1947), just at the time when the foundations for the new evolutionary synthesis were being laid by Huxley, Mayr, Simpson, and others — a synthesis which denied any role for orthogenesis.

With Weidenreich's death, polycentrism was gratefully retired. The grateful were led by Hooton and later by his student Howells, by Leakey, and by Vallois. What these diverse scholars had in common was the belief that human evolution was characterized by replacement, and in particular that what was called "modern man" at the time, meaning modern populations of Europe, came to Europe from somewhere else and replaced the Neandertals who lived there. Suitably modified, this view still predominates in western Europe, and in other areas influenced by western European ideas. Punctuationalists find great solace in it, as the replacements seen in earlier interpretations of human evolution became the punctuational events predicted by their theory.

Where this replacement view does *not* predominate is in central Europe, where the evolutionary traditions of Gorjanović-Kramberger and Schwalbe that strongly influenced Weidenreich still prevail, and in certain peripheral regions where the fossil record is good enough to show that a replacement explanation for human evolution cannot be correct. Two peripheral regions with excellent fossil records are at the north and south ends of what was the eastern most extent of the human range for the Middle and most of the Upper Pleistocene, with scattered remains known from the areas between them. These regions are the north China and the Sunda subcontinent centers of human origins described by Weidenreich; even a half-century ago, they were known for the excellence of their fossil records.

Today, however, the fossil records are much better than Weidenreich could have imagined. Within these two areas, the interpretations of human fossil remains have been taken up by local scholars, intimately acquainted with the paleontological materials of their regions and particularly concerned with the origins of the populations that inhabit them now.

For instance, from the Indonesian perspective, links between the Sangiran *Homo erectus* specimens, the later hominids from Ngandong, and finally the recent inhabitants of Kow Swamp in Australia have been recognized (Sartono 1982). A new reconstruction of Sangiran 17 (Thorne and Wolpoff 1981), the most complete *Homo erectus* cranium, allowed comparisons of the face to be made for the first time. This proved to be the anatomical area with the strongest evidence for regional continuity when the Sundaland and the latest Pleistocene and Holocene samples from Australia were compared. Facial features supporting the contention of a special relation between these samples primarily involve the maintenance of a very large face with pronounced alveolar prognathism. Thus, for instance, posterior tooth size and facial heights show little reduction until well into the Holocene. Other facial features include the marked ridge paralleling the zygomaxillary suture, the eversion of the lower border of the zygomatic, the rounding of the inferolateral orbital border, the lack of a distinct line dividing the nasal floor from the subnasal face of the maxilla, the curvature of the posterior alveolar plane of the maxilla that corresponds to the mandibular "curve of Spee," the posterior position of the minimum frontal breadth, and the relatively horizontal orientation of the inferior border of the supraorbital torus. It is clear that the inhabitants of this southern region have been morphologically distinct from the populations of its more northerly neighbors for the entire time east Asia has been inhabited (Thorne and Wolpoff 1981; Jacob 1981; Jelínek 1982; Wolpoff, Wu, and Thorne 1984).

The case for specific ancestral-descendant relations between hominid samples more than a half million years apart has improved dramatically since Weidenreich (1943) first proposed it for Sundaland (Thorne and Wolpoff 1981). Sambungmachan is an excellent intermediary between the Sangiran *Homo erectus* remains (*including* Sangiran 17 which, contrary to the claims of a number of authors, very clearly is a male *Homo erectus* specimen and not a member of what could be loosely called the "Solo group") and the Ngandong hominids (Jacob 1976). While the Sambungmachan male is somewhat smaller than the Ngandong males, it resembles them in the development of an angular trigone at the lateral corner of the supraorbital torus, the reduction in basal pneumatization and the position of the maximum cranial breadth, the flattening of the occipital plane of the occiput (and its vertical orientation), the doubled digastric sulci, and the tall vertical posterior border of the temporal squama. On the other hand, although the cranial capacity is larger the specimen is morphologically closer to the Kabuh *Homo erectus* remains than it is to the Ngandong hominids.

Even without Sambungmachan, Weidenreich (1951) was able to firmly link the Kabuh and Ngandong remains. Moreover, Weidenreich (1943, pp. 248–250) was also able to relate the Ngandong specimens to the Australian Aborigines, thus arguing for the interpretation of an unbroken line of descent from the earliest hominids of Sundaland to the Australian natives of today which Howells (1976) denies. Weidenreich recognized evidence for continuity in the occasional appearance among Australian Aborigines of a morphological complex that included a well-developed supraorbital torus which combines a discontinuity at glabella with the lack of a supratoral sulcus and a long, flat, receding forehead. Additional evidence was found in prelambdoidal depressions, sharp angulations between the occipital and nuchal planes, and short or even nonexistent sphenoparietal articulations in the region of pterion.

Thus, from the Australian perspective, an Indonesian link (in the words of Macintosh, "the mark of Java") has long been recognized for the Aboriginal populations (Larnach and Macintosh 1974; Thorne and Wolpoff 1981), but with increasing evidence of additional links with the Asian mainland proposed particularly by A.G. Thorne (1977). Weidenreich's contention was quantified through the implications of a study conducted by Larnach and Macintosh over a decade ago (1974). These authors compared a number of Australian and New Guinea crania with Europeans and Africans, scoring them for the eighteen characters that Weidenreich (1951) claimed were unique for the Ngandong hominids. Six of these were absent in all modern samples, while nine of the twelve other features were found to attain their highest frequencies in the Australian and New Guinea natives.

The comparisons with the modern Aborigines are further borne out by Australian fossil remains uncovered since Weidenreich and Larnach and Macintosh published their contentions. Beginning with the temporally earliest of these, the WLH 50 hominid from the Willandra Lakes of Australia is a very convincing intermediary between the Ngandong specimens and the recent and modern Aboriginals of the continent because of its robustness, the thickness of its vault, the position of the maximum cranial breadth, and a number of other distinct morphological features (Thorne 1984). WLH 50 is only the most recent addition to this evidence linking hominid populations across the Pleistocene of the Sunda subcontinent. The Indonesian erectus connection is strengthened when the earlier discoveries of subfossil remains from Kow Swamp and Coobool Crossing are taken into account, because so many of the evolutionary changes in Australia accelerated during the Holocene. For this reason, these late Upper Pleistocene/Holocene specimens retain a much higher frequency of features that are archaic in Sundaland than do the modern populations of the continent, and thus evince features that reinforce the notion of morphological continuity.

In sum, if one firmly believed that modern populations had a single recent origin and replaced their predecessors throughout the world, the evidence discussed above would strongly suggest that Sundaland if not greater Australia itself comprised the region of origin, and indeed somebody has suggested exactly that (Gribbin and Cherfas 1982).

Of all those who would disagree with such a contention, the foremost would probably be among the scholars of the IVPP in Beijing, because they know that the evidence for continuity between fossil and living populations is at least as good in North China (Wu and Zhang 1978). Since liberation there has been a continuous increase in the fossil record, with new discoveries of *Homo erectus* from Zhoukoudian, Lantian, Longgudong, and Hexian. New early (or archaic) *Homo sapiens* specimens include major remains from Dali, Maba, Yingkou, and Xujiayao, and a number of more fragmentary specimens. New individuals that are terminal Pleistocene or perhaps even Holocene in age include Chilinshan, Huanglong, Liujiang, Muchienchiao, Tzeyang, and a palate from South China found in a Hong Kong drugstore. These specimens confirm Weidenreich's interpretations, showing both morphological continuity within China and regional distinctions of the Chinese fossils from other areas with good fossil records such as Sundaland and greater Australia (Jacob 1981; Wolpoff, Wu, and Thorne 1984) and Europe (Wu and Wu 1982).

Working as he was in North China, Weidenreich (1939, 1943) was much more detailed in his description of morphological continuity for that region. He described 12 features that he felt had particular importance in showing continuity, and over the decades since his death a number of additional observations have been added (Aigner 1976; Wu 1981; Wolpoff, Wu and Thorne 1984). Wu, in his description of the Dali cranium, was able to add a number of facial features which have great importance because the face seems to show more regional features than does the cranial vault in north China, just as is the case in the Sundaland hominids (Thorne and Wolpoff 1981).

The Dali cranium (Wu 1981), and evidently also the morphologically similar, newly discovered and largely complete Yingkou specimen (Wei 1984), have added much to an already very convincing case established by Weidenreich and elaborated on the basis of more fragmentary remains recovered through the end of the last decade (Wolpoff, Wu and Thorne 1984). Regionally distinct features of these early *Homo sapiens* vaults include sagittal keeling, the forehead profile (in particular the development of the frontal boss), and the Inca bone development. There is also a suite of regionally distinct facial features such as facial flatness (including the associated distinct angulation in the zygomatic process of the maxilla and anterior orientation of the frontal process of the zygomatic), lack of anterior facial projection and low degree of prognathism, minimal nasal projection (the frontonasal and frontomaxillary sutures are virtually level), a very low nasal profile and the shape of the orbits. Other fragmentary remains of early *Homo sapiens*, especially those from Maba (with important confirmations of the nasal profile and projection, and frontal process of the zygomatic orientation in the upper face), Xujiayao, and Dingcun, also show these details when the appropriate parts are preserved.

Among the more recent remains, Liujiang from south China and the Upper Cave specimens from Zhoukoudian have a number of features that are characteristically regional (Woo 1959; Wu 1961; Wolpoff, Wu and Thorne 1984). Wu argued strongly against Weidenreich's "Eskimo, Chinese, Melanesoid" trichotomy for the Upper Cave crania, instead showing that they all fall within the expected range of variation for the area. The *early Homo sapiens* remains have been particularly important in providing firm evidence for a morphological link between the late Pleistocene/early Holocene specimens and the Zhoukoudian *Homo erectus* sample — a link that was lacking in Weidenreich's time, hampering his interpretation (1939) of the Zhoukoudian Upper Cave specimens.

In sum, the basis for claiming regional continuity in China, especially northern China, is at least as good as in Sundaland. The differences between features showing continuity in these two ends of the eastern periphery were recognized as long ago as in Weidenreich's time and have continued to be recognized by scholars familiar with the fossil remains from both areas (Jacob 1981; Wu and Zhang 1978; Wu and Wu 1982; Wolpoff, Wu and Thorne 1984). It was this evidence that led to Coon's chapter headings "Pithecanthropus and the Australoids" and "Sinanthropus and the Mongoloids".

The *explanation* of regional continuity should be distinguished from the *observation* of the phenomenon. Weidenreich's (1947) orthogenesis explanation was unacceptable and Coon's (1962) explanation involving separate geographic subspecies evolving at different rates was perhaps even more so. The most recent explanation has been proposed by myself in collaboration with colleagues from Australia (A.G. Thorne) and China (Wu Xinzhi). We have suggested a theory of multiregional evolution that should apply to long-term evolutionary processes in any widespread polytypic species. The multiregional evolution theory (Wolpoff, Wu and Thorne 1984) proposes that regional variations should first become distinct at the species' peripheries because of the "centre and edge" pattern of habitation (Thorne 1981), and that these regional differences are subsequently maintained because of a balance between gene flow and selection that develops to form long-lasting morphological clines. This dynamic way of viewing the evolutionary process accounts for how human populations (especially the peripheral ones) initially develop and maintain regional differences, while at the same time they continue to exchange genetic material so that selectively advantageous allele frequency changes (and the occasional new allele) can rapidly spread throughout the human range. The hypothesis connects seemingly contradictory processes and explains how initial differences are maintained in the face of gene flow between regions.

Paradoxically, then, the multiregional evolution theory rests on two lines of evidence that could be considered contradictory if considered individually — data concerned with regional continuity for different peripheral areas, and data showing gene flow between regions and from the center towards the edge that could account for the spread of genetic information (allele frequency changes) throughout the range of the species. Indeed, it was the apparent contradiction between these two very different lines of evidence that created so many difficulties in both Weidenreich's and Coon's thinking, but that also underlies the dynamic approach of considering the evolutionary importance of standing clines created by various gene flow — selection balances in the multiregional evolution model.

The evidence for regional continuity in two areas of the eastern periphery has been reviewed, and it is as good as probably could be expected given the nature of the fossil record. But what of the evidence for gene flow? To some extent the presence of common major evolutionary trends constitutes a valid source of evidence for gene flow. At the north and south ends of the eastern periphery there are similar timings for brain size expansion, as well as for reductions in muscularity, in sexual dimorphism, and in anterior tooth size. These and a number of additional common evolutionary trends result in many of the synapomorphic features shared by the modern populations of these areas and could be taken as evidence of gene flow. This is especially the case given the lack of technological and other forms of adaptive similarities. However, this evidence is not enough because the data the hypothesis was created to explain can hardly also be taken as independent support of the hypothesis itself. Moreover, it is likely that there are some common elements of selection promoting these changes across the human range.

There is a growing body of direct evidence for gene flow along this eastern end of the human range. In the Upper Pleistocene this evidence was first recognized by Weidenreich during his comparisons of the Wadjak remains with Australian fossils such as Keilor (1945). He argued that these specimens are virtually identical, and my observations also support the notion that there is a marked degree of similarity between them in size, proportions, and facial breadth and flatness. Moreover, the south China specimen from

Liujiang shares many, perhaps most, of these features (Wolpoff, Wu and Thorne 1984). These and more fragmentary materials seem to provide a link between the Asian mainland, Sundaland, and greater Australia that indicates an early, important presence of persistent gene flow.

The evidence for gene flow begins much earlier, however. The Hexian *Homo erectus* cranium (Wu and Dong 1982) resembles the Zhoukoudian remains in many details of the forehead profile as well as in the moderate frontal boss and the rounding of the superior orbital border. However, the weak expression of the supratoral sulcus on the frontal and in contrast the marked expression of the occipital's supratoral sulcus, the form of the basal pneumatization as seen in the posterior contour, the parietal angulation, and the expanded cranial breadths all much more closely resemble the Indonesian hominids. Indeed, a case could be made that Hexian is a morphological as well as a geographical intermediary. Yet it is interesting that to the west, the newly discovered Narmada cranium (Sonakia 1984) from Madhya Pradesh, central India, closely resembles the Chinese *Homo erectus* remains. Especially in the shape and configuration of the supraorbitals and the curvature of the frontal squama, its frontal is much like Hexian while the round orbits resemble Maba. The temporal and occipital are similar to Zhoukoudian specimens (especially the H5 vault) in the form of the temporal squama, the expression and position of the angular torus, the curvature of the occipital in the sagittal plane, and the morphology of the nuchal torus. Evidently on the Asian mainland the features first recognized in China predominate, even though intermediate specimens may be found. While in its totality the Asian evidence supports the interpretation of significant gene flow in more than one direction along the eastern periphery, it would appear that at least on the continent there was some directionality.

At present the data from the eastern periphery seem to support the multiregional evolution hypothesis, and no data from these regions stand in contradiction to it. Of equal importance, there is another explanatory hypothesis affected by these data that seeks to provide an alternative explanation for human evolutionary changes in east Asia — punctuated equilibrium. The pattern of human evolution along this eastern periphery provides the opportunity for what I believe is an absolutely convincing test of the punctuational model in the hominids. According to punctuational theory, speciation can only happen in allopatric populations and is most likely to occur at the periphery of the range. If *Homo sapiens* originated at either end of the north-south cline at the eastern end of the Pleistocene human range, the apparent (but under these circumstances incorrect) evidence for morphological continuity at the *other* end is a coincidence that defies the imagination. If the replacement came from somewhere else, the coincidences are even more striking. The point is that under a model of evolution by peripheral speciation and replacement it is very difficult to determine which periphery is the source of *Homo sapiens*. The evidence from different regions is equally good, and the assumption that any one peripheral region is the source area raises very difficult problems of homoplasy in a number of important features during any attempt to account for the continuity that cannot actually be evolutionary continuity in the other peripheral areas. Thus the source area for *Homo sapiens* is unknown, and perhaps unknowable. Put another way, the data for this region do not support a replacement model for *Homo sapiens* origins.

Yet if there were no replacement, then a great deal of evolutionary change (from the earliest inhabitants of east Asia to the modern populations of the region) can be documented that involves *no* speciation *events*. Considerable magnitudes of change without a speciation event is hardly a prediction of the punctuational model. To the contrary, it is a disproof. Thus, without the replacement interpretation, it seems to me that the punctuational equilibrium model fails the test. One must conclude that the hominid data as understood from east Asia provides good reasons for concern among supporters of the punctuated equilibrium model.

At the moment, the pattern of hominid evolution at the eastern periphery appears to be much as Weidenreich described. It involves a morphological cline spread over the eastern periphery with ends that are unique and distinguishable as far back into the past as the earliest evidence for habitation can be found, but also involves intermediate specimens that show there never was populational isolation. This pattern fits the predictions of the multiregional evolution model and seems to discount punctuational theory as a valid explanation of Pleistocene hominid evolution.

I am very grateful to Alan Thorne and Wu Xinzhi, my coworkers in developing the multiregional evolution hypothesis, for permission to examine the fossil human remains in their care, and for allowing me to join them in the development of these ideas about the pattern of human evolution in the region they know so well. I thank my many other friends and colleagues in Australia and at the IVPP in Beijing, and my Indonesian friends T. Jacob and D. Kadar for the many kindnesses and courtesies they extended to me during my visits to their laboratories. F.B. Livingstone, A. Clark, K. Rosenberg, and L. Schepartz very kindly helped in correcting this manuscript and made many useful suggestions for its improvement. Finally, I thank P.V. Tobias for his gracious invitation to this symposium, allowing me to present the results of a good deal of work.

REFERENCES

Aigner JS (1976). Chinese Pleistocene cultural and hominid remains: a consideration of their significance in reconstructing the pattern of human bio-cultural development. In Ghosh AK (ed): "Le Paléolithique Inférieur et Moyen en Inde, en Asie Centrale, en Chine et dans le sud-est Asiatique," UISPP Colloque VII, Paris: Centre Nationale de la Recherche Sciéntifique, pp 65–90.

Coon, CS (1962). "The Origin of Races." New York: Knopf.

Gribbin J, Cherfas J (1982). "The Monkey Puzzle: Reshaping the Evolutionary Tree." London: Bodley Head.

Howells WW (1976). Physical variation and prehistory in Melanesia and Australia. Am J Phys Anthropol 45:641.

Howells WW (1980). Homo erectus — who, when and where: a survey. Yearb Phys Anthrop 23:1–23.

Jacob T (1976). Early populations in the Indonesian region. In Kirk RL, and Thorne AG (eds) "The Origins of the Australians," Canberra: Australian Institute of Aboriginal Studies, pp 81–93.

Jacob T (1981). Solo man and Peking man. In Sigmon BA and Cybulski JS (eds): "Homo erectus. Papers in Honor of Davidson Black," Toronto: University of Toronto Press, pp 87–104.

Jelínek J (1982). The east and southeast Asian way of regional evolution. Anthropos(Brno) 21:195

Larnach SL, Macintosh NWG (1974). A comparative study of Solo and Australian Aboriginal crania. In Elkin AP, and Macintosh NWG (eds): "Grafton Elliot Smith: The Man and his Work," Sydney: Sydney University Press, pp 95–102.

Sartono S (1982). Characteristics and chronology of early men in Java. In "L'Homo erectus et la Place de L'Homme de Tautavel parmi les Hominidés Fossiles," Volume 2, Nice: Jean-Louis Scientific and Literary Publications, pp 491–541.

Sonakia A (1984). The skull-cap of early man and associated mammalian fauna from Narmada Valley alluvium, Hoshangabad area, Madhya Pradesh (India). Rec Geol Surv India 113(6):159.

Thorne AG (1977). Separation or reconciliation? Biological clues to the development of Australian society. In Allen J, Golson J, and Jones R (eds): "Sunda and Sahul: Prehistoric Studies in Southeast Asia, Melanesia, and Australia," London: Academic Press, pp 187–204.

Thorne AG (1981). The centre and the edge: The significance of Australian hominids to African paleoanthropology. In Leakey REF and Ogot BA (eds): "Proceedings of the 8th Panafrican Congress of Prehistory and Quaternary Studies, Nairobi, September 1977," Nairobi: TILLMIAP, pp 180–181.

Thorne AG (1984). Australia's human origins — how many sources? Am J Phys Anthropol 63(2):227.

Thorne AG, Wolpoff MH (1981). Regional continuity in Australasian Pleistocene hominid evolution. Am J Phys Anthropol 55:337.

Tobias PV (1984). "Dart, Taung and the 'Missing Link'." Special Publication, Institute for the Study of Man in Africa. Johannesburg: University of the Witwatersrand Press,

Wei Liming (1984). 200,000 year-old skeleton unearthed. Beij Rev (Dec 3), No.49:33

Weidenreich F (1939). Six lectures on *Sinanthropus pekinensis* and related problems. Bul Geol Soc China 19:1–110

Weidenreich F (1943). The skull of *Sinanthropus pekinensis*: A comparative study of a primitive hominid skull. Palaeont Sinica, n.s. D, No. 10 (whole series No. 127).

Weidenreich F (1945). The Keilor skull. A Wadjak skull from southeast Australia. Am J Phys Anthropol 3:21.

Weidenreich F (1947). The trend of human evolution. Evolution 1:221.

Weidenreich F (1951). Morphology of Solo man. Anthrop Pap Am Mus Nat Hist 43(3):205–290.

Wolpoff MH, Wu Xinzhi, Thorne AG (1984). Modern *Homo sapiens* origins: a general theory of hominid evolution involving the fossil evidence from east Asia. In Smith FH, and Spencer F (eds): "The Origins of Modern Humans: A World Survey of the Fossil Evidence," New York: Liss, pp 411–483.

Woo Jukang (Wu Rukang) (1959). Human fossils found in Liujiang, Kwangsi, China. Vertebr PalAsiat 3:109.

Wu Rukang, Dong Xingren (1982). Preliminary study of *Homo erectus* remains from Hexian, Anhui. Acta Anthrop Sinica 1:2

Wu Rukang, Wu Xinzhi (1982). Comparison of Tautavel man with *Homo erectus* and early *Homo sapiens* in China. In: "L'*Homo erectus* et la Place de l'homme de Tautavel parmi les Hominidés Fossiles," Volume 2, Nice: Louis-Jean Scientific and Literary Publications, pp 605–616.

Wu Xinzhi (1961). Study on the Upper Cave man of Choukoutien. Vertebr PalAsiat 3:202.

Wu Xinzhi (1981). A well-preserved cranium of an archaic type of early *Homo sapiens* from Dali, China. Sc Sinica 24(4):530.

Wu Xinzhi, Zhang Yinyun (1978). Fossil man in China. In: "Symposium on the Origin of Man," Beijing: Science Press, pp 28–42.

Hominid Evolution: Past, Present and Future, pages 367–380
© *1985 Alan R. Liss, Inc.*

THE ORIGIN OF THE AUSTRALIAN ABORIGINES: AN ALTERNATIVE
APPROACH AND VIEW

Phillip J. Habgood

Department of Anthropology,
University of Sydney,
N.S.W. 2006, Australia

People have been intrigued by the Australian Aborigines
since their unique morphology and culture were first observed
in the late 18th century. Many early scholars saw the modern
aboriginal cranial morphology as 'primitive' and theorised
that they were a relic population that had become isolated in
Australia and changed little over thousands of years (Hrdlička
1930). Subsequently, many dated human fossils have been discovered in Australia (Oakley et al 1975). Based on this additional material two main explanations of aboriginal origins
have been proposed: one, that they result from several migrations at different periods by morphologically and geographically distinct groups, and the second, that they result from a
migration to Australia by a group from a single homeland (see
Kirk, Thorne 1976, for references).

A.G. Thorne, an exponent of the dual source hypothesis,
has distinguished two groups of late Pleistocene/earlier
Holocene Australian skulls (1971a, 1971b, 1976, 1977). A
'robust' type is exemplified by crania from Kow Swamp, Cohuna,
Mossgiel and Lake Nitchie, and a 'gracile' type by crania from
Lake Mungo and Keilor (Fig. 1). The 'robust' type has relatively low, rugged crania, flat receding frontal bones that are
broad anteriorly and have a prebregmatic eminence, postorbital
constriction, large supra-orbital tori, broad prognathic faces,
moderate gabling of the vault, the bones of which are quite
thick, occipital tori and large palates, mandibles and teeth.
The 'gracile' type has high, rounded and more modern-looking
crania with expanded frontal and temporal regions, slight
brow-ridge development, thin vault bones, short delicate
faces and relatively small palates, mandibles and teeth. The

Fig 1:   Map of Sahul with Australian late Pleistocene sites.

'gracile' type is earlier, being dated to 30,000 y. BP at Lake
Mungo, while the earliest 'robust' forms are dated to
14,000 y. BP at Kow Swamp (Table 3).

     Thorne believes the differences between these groups are
'too great to be ascribed to adaptation and genetic effects
alone' (1981); so 'the case for the two morphologies being the
result of migration from outside is stronger' (1977). He
follows Weidenreich (1947) in highlighting similarities
between some early Australian skulls and Indonesian fossils
from Sangiran and Ngandong. The 'mark of ancient Java' which
Thorne sees on the 'robust' crania includes: flat receding
frontal bones, the prebregmatic eminence, marked facial prognathism and large teeth, mandibles and palates (Thorne,

Wolpoff 1981, Wolpoff et al 1984). Thorne (1980) sees the 'gracile' crania as displaying the 'stamp of ancient China'. He especially stresses the similarities between the Keilor and Liujiang skulls (Wolpoff et al 1984).

The two differing types within the late Pleistocene/ earlier Holocene Australian sample are explained as the result of two distinct groups of migrants entering Australia, a robust, archaic group from Indonesia and a more gracile group from China. These subsequently interbred within Australia to produce the Holocene aboriginal morphology, which in general has more Homo erectus cranial traits than any other modern group (Macintosh, Larnach 1972).

This hypothesis is testable. If the dual source hypothesis is correct, 'gracile' crania such as Keilor which are said to be the result of migration from China, should be more similar to Liujiang and the Zhoukoudian Upper Cave skulls than to either the 'robust' Australian crania or late Pleistocene/ earlier Holocene crania from other regions. The present investigation was designed to test this deduction. Multivariate techniques were used because they define what is being compared, allow procedures to be repeated and remove subjective interpretations of features and affinities (Habgood 1985).

MATERIAL

The data consisted of 20 sets of 24 cranial measurements (Table 1) taken from 19 crania of early, anatomically modern Homo sapiens from Asia, Africa and Europe (Table 2). Measurements from a new reconstruction of Wajak 1 were included (C.B. Stringer pers comm). The Asian material included most crania which Thorne considers similar to the Australian 'gracile' type. The measurements were taken on either original crania or casts, or obtained from published data (Table 2).

The early Australian crania were added to the data set, but due to their fragmentary nature it was impossible to place them, individually, in the analyses. For this reason each available measurement for members of either 'gracile' or 'robust' groups as defined by Thorne (1977:Fig 1), with the addition of the Lake Tandou cranium to the 'gracile' group and the Cossack cranium to the 'robust' group, were used to obtain the mean values for the individual measurements for each group. This has given data sets for composite, hypothetically typical 'robust' and 'gracile' crania. Keilor, the only Australian cranium for which all 24 measurements were available, was not only included in the 'gracile' group mean values, but also was added to the overall data set, to establish how a late Pleistocene Australian cranium would behave within the analyses (Table 3).

TABLE 1: VARIABLES - CRANIAL MEASUREMENTS*

Glabella - Occipital Length	Orbital Breadth
Basion - Nasion Length	Nasal Breadth
Basion - Bregma Height	Mastoid Length x
Auricular - Bregma Height +	Mastoid Breadth x
Maximum Cranial Breadth	Nasion - Bregma Chord
Maximum Frontal Breadth	Nasion - Bregma Arc +
Bistephanic Breadth	Bregma - Lambda Chord
Biauricular Breadth	Bregma - Lambda Arc +
Biasterionic Breadth	Lambda - Opisthion Chord +
Nasion - Prosthion Height	Glabella - Inion Length z
Nasal Height	Lambda - Inion Chord +
Orbital Height	Minimum Frontal Breadth

* For definitions see Howells 1973, + Morant 1930, x Smith 1976, z Martin, Saller 1958.

METHOD

The multivariate techniques used were a standardization procedure, a principal components analysis and a sum of squares cluster analysis. The standardization procedure consisted in, first, standardization by variable, which gives equal weighting to all variables and stops large ones from swamping smaller ones. This was followed with standardization by object, which removes absolute size differences between objects and leaves only shape information in the data matrix (Corruccini 1973).

The double standardized data set was then fed into a principal components analysis (PCA) which shows similarity or dissimilarity between objects in a sample (Orton 1980). It attempts to describe the variation within the data matrix by calculating new sets of orthogonal axes which include progressively less of the variation within the sample. That is, the first new axis or principal component accounts for the maximum possible variance in the sample, the second for the maximum remaining variance and so on. The discontinuity method

TABLE 2: THE CRANIA USED AND SOURCES OF MEASUREMENTS

CRANIA	ABBREVIATION	MEASUREMENT SOURCE
AUSTRALIAN		
Robust mean	R	cf. Table 3
Gracile mean	G	cf. Table 3
Keilor	K	Freedman, Lofgren 1979, Cast
EAST AND SOUTHEAST ASIAN		
Liujiang	Lj	Coon 1963, Cast
Upper Cave 101	UC101	Coon 1963, Cast
Upper Cave 102	UC102	Coon 1963, Cast
Upper Cave 103	UC103	Coon 1963, Cast
Wajak 1 old reconstruction	W1o	Coon 1963, Cast
Wajak 1 new reconstruction	W1n	Cast
AFRICAN		
Fish Hoek	FH	Coon 1963, Cast
Afalou 32	A32	Stringer pers comm, Suzuki, Takai 1970
Taforalt 18	T18	Stringer pers comm, Suzuki, Takai 1970
WESTERN EUROPE		
Chancelade	Ch	Morant 1930, Cast
Solutré 5	So5	Morant 1930, Cast
Abri Pataud 1	AP1	Original
Le Placard 5	LP5	Original
CENTRAL EUROPE		
Oberkassel 1	O1	Original
Oberkassel 2	O2	Original
EASTERN EUROPE		
Dolni Vestonicé	DV3	Jelinek 1953, Cast
Brno 3	B3	Stringer pers comm, Cast
Mladec 1	M1	Original
SOUTHWEST ASIAN		
Hotu 2	H2	Cast

TABLE 3: AUSTRALIAN CRANIA

Crania	Approximate Dates and References	No. of Measurements	Source of Measurements
Gracile			
Lake Mungo 1	24,500-26,500 BP, Thorne 1981	13	Thorne 1976, Cast
Lake Mungo 3	30,000 BP, Thorne 1981	6	Cast
Keilor	12,900 $\pm$120 BP, Macintosh, Larnach 1976	24	Freedman, Lofgren 1979, Cast
Lake Tandou	15,000 $\pm$ 160 BP Freedman, Lofgren 1983	8	Freedman, Lofgren 1983
Robust			
Kow Swamp 1	9,500-14,000 BP, Thorne, Wolpoff 1981	21	Thorne, Wolpoff 1981, Cast
Kow Swamp 5	9,500-14,000 BP, Thorne, Wolpoff 1981	21	Thorne, Wolpoff 1981, Brown 1981, Cast
Cohuna	9,500-14,000 BP, Thorne, Wolpoff 1981	19	Thorne, Wolpoff 1981, Brown 1981, Cast
Mossgiel	6010$\pm$125 BP, Macintosh 1971	19	Freedman, Lofgren 1979, Cast
Cossack	6,500 BP, Freedman, Lofgren 1979	13	Freedman, Lofgren 1979
Lake Nitchie	6,820$\pm$200 BP, Macintosh 1971	13	Freedman, Lofgren 1979, 1983
Talgai	11,650$\pm$100 BP, Macintosh 1971	Not used in analyses because the individual is an adolescent	

(Rummel 1970) was used to remove from further consideration principal components that incorporate 'noise' or redundant data, such as measurement error, and which impair the sensitivity of the analysis (Kowalski 1972). The discontinuity occurred at the fifth principal component where 69% of the total within-group variance was accounted for.

The component scores, which document how much each of the crania contributed to the five principal components, were subjected to a sum of squares cluster analysis (SSA). This

method is a polythetic, hierarchial and agglomerative technique, based on Euclidean distance, which probes for structure
within the matrix of component scores by combining similar
crania into clusters in a dendrogram (Orlocci 1978).

RESULTS

Within three major groupings on the dendrogram (Fig. 2),
eight smaller clusters are discernible. These reflect geographical distribution quite closely, Asian crania forming three
groups, European crania three groups and African crania their
own group. The 'robust' and 'gracile' types and Keilor form
their own cluster. These results suggest a high degree of geographical identity of cranial morphology of early anatomically
modern H.sapiens. This is significant for Australia, because
it shows that late Pleistocene/earlier Holocene crania are more
similar to one another than to early anatomically modern crania
from other regions. The 'gracile' Australian crania are not
more similar to Wajak 1, Liujiang or Zhoukoudian Upper Cave
than they are to 'robust' Australian crania.

The crania which are most similar to the Australian
material come from Asia: Wajak 1 and Zhoukoudian Upper Cave
101. This suggests that these Asian and Australian fossils
are similar not only in grade but also in clade, which may reflect regional, Australasian   continuity (Thorne, Wolpoff
1981; Wolpoff et al 1984).

These results dispense with the need to recognise two
groups within late Pleistocene/earlier Holocene Australian
skeletal material. Therefore, the dual source hypothesis would
appear negated, because its basic premise is the presence of
two morphological types. What the results do suggest is that
Australian skeletal material is part of a single, variable population. There are no grounds for assuming more than one homeland for this population.

Others have arrived at similar conclusions by different
methods (cf Kirk, Thorne 1976, for references). Abbie studying
modern Australian Aborigines suggested that 'Aborigines widely
dispersed through ... the continent are practically uniformly
homogeneous' (Abbie 1963:102) and that 'all the evidence points
to the fact that the Aborigines, everywhere, are one people'
(Abbie 1966:42).

Larnach and Macintosh (1966, 1970), using features identified as diagnostic of crania from coastal New South Wales and
Queensland, found that the crania of Cohuna, Talgai, Mossgiel
and Keilor, fitted within a normal distribution of modern
aboriginal skulls. The 'robust' types (e.g. Cohuna) tended
towards the higher, and the 'gracile' forms (e.g. Keilor)
towards the lower index range (Macintosh 1971). They concluded
that 'the subjective impression of two separate types is not
confirmed' (Macintosh, Larnach 1976:115) and that the prehistoric crania are 'equal representatives ranged towards ...
either end of a continuum of a single population' (Macintosh,
Larnach 1976:114).

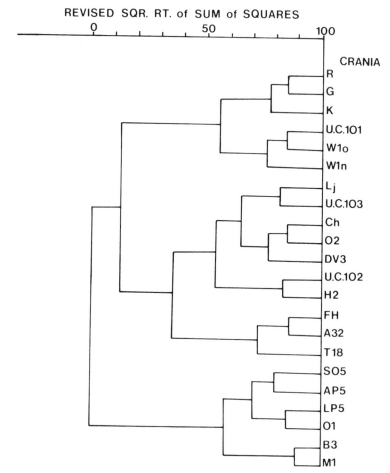

Figure 2:   Dendrogram from the SSA cluster analysis.

Some of Thorne's data suggest that the late Pleistocene/ earlier Holocene Australian crania are from a single, variable population. He demonstrated that a number of Kow Swamp crania (especially nos. 8, 9, 14) do not significantly differ from modern Victorian crania, except for greater overall size (Thorne, Wilson 1977). There is similarity also in the frontal and supra-orbital regions between Kow Swamp 14 and 15 and Lake Mungo 3, while Lake Mungo 1 resembles female crania from Kow Swamp such as nos. 4 and 16. Thorne (1976) suggests that differences in tooth and mandibular size also separate the two groups; yet only the large Kow Swamp molar breadths fall out- side of a modern aboriginal dental range (Freedman, Lofgren 1979), while of 12 published jaw measurements from Kow Swamp only two breadth dimensions exceed those of a modern Victorian sample (Thorne 1976: Table 4).

Cranial deformation appears to contribute to the large variation within early Australian crania. Brown (1981) has shown that some features of the 'robust' group, such as the flat receding frontal bone, prebregmatic eminence and low position of maximum biparietal breadth, are likely to be the result of cranial deformation. The probable method was re- peated pressure to the front and back of the infant's cranium, a practice known to have occurred in the Cape York area during the 19th century. Cranial pressing, unlike binding, would allow for highly variable deformation, from markedly deformed crania, such as Kow Swamp 5 and Cohuna, to those with little or no deformation, like Kow Swamp 1 and Keilor.

There is then, strong evidence from different types of studies that early Australian skulls represent a single var- iable population. Is it possible to reconstruct a scenario that accounts for the large variation in the early Australian sample?

The earliest conclusive evidence of human occupation of Australia is from a terrace of the upper Swan River, Western Australia (Fig.1), where a date of $39,500 \pm 2,300/1,800$ BP was obtained (Pearce, Barbetti 1981). Stratigraphically equiva- lent to the carbonized wood which produced the date were stone artefacts. Another site in Western Australia is Devil's Lair where a date of $35,160 \pm 1,800$ BP has been obtained (Dortch 1979). On the other side of the continent, sites around Lake Mungo (Fig. 1) have dates in excess of 30,000 y.BP, associated with hearths, faunal remains and stone tools (Bowler et al 1970). At Keilor (Fig. 1), quartzite artefacts have been

recovered from a deposit thought to date between 25,000 BP and
at least 36,000 BP (Bowler 1976). The initial colonization
must therefore have occurred prior to 40,000 y.BP. The most
likely entry point into Sahul (the combined landmass of Australia, Tasmania and New Guinea) was N.W. Western Australia (Fig.
1). The migration from Sunda, the landmass of S.E. Asia, has,
for more than 30 million years, involved sea voyages. Even at
times of lowest sea levels the shortest route across Wallacea,
the island zone between the two landmasses, involved at least
eight voyages, the last, from Timor to the N.W. Sahul coast,
being almost 90 km (Birdsell 1977).

A probable date for the first migration to Greater Australia, which might mark the earliest use of sea craft, is 53,000
y.BP when the sea level was 120-150 m lower than at present
(Chappell 1976). The first migratory group could have been
small, possibly a family of a male, two females and children
of both sexes. Such a group could, within a thousand years,
increase to a few hundred individuals (White, O'Connell 1982).

The most probable ancestors are represented by the
Ngandong fossils from Java which 'are, on the whole, surprisingly alike' (Weidenreich 1951:239). If the concept of regional continuity (Thorne, Wolpoff 1981) is correct, the earliest Australians should be advanced Ngandong type Homo sapiens.
The recent discovery of Willandra Lakes hominid 50 (WLH50)
near Lake Mungo (Flood 1983; Thorne 1984) supports this deduction in that it has many similarities to Ngandong. The calvaria
of WLH50 is 'more robust and archaic than any Australian hominid found previously' (Thorne 1984) but unfortunately it is, at
present, undated. A date in excess of 30,000 y.BP is a possibility and would be in keeping with the morphology of WLH50.
If so, we could assume that WLH50 was, morphologically, representative of the earliest inhabitants of Australia.

At the time of the earliest migration the N.W. coast of
Greater Australia would have been cooler, with a mean annual
temperature up to 5°C lower, and drier, with a possible 50%
lower rainfall, than at present. It was probably covered by
open woodland and savannah. As small groups began to colonize
the rest of the continent (Habgood, White 1985), they moved into diverse terrains and climates. By at least 15,000 y.BP most
parts of Australia show evidence of habitation. Due to the
enormous size of the continent and small population numbers,
these colonizing groups would probably have become isolated
from one another. Then, selection, mutation, founder effect

and differing rates of genetic drift could have increased the variation of the migratory group to the level found in late Pleistocene/earlier Holocene Australian skulls. With an increasing population during the Holocene (Ross 1981; Lourandos 1983, 1984; Beaton 1983), closer interaction through such mechanisms as mate exchange between groups and increases in gene flow would have reduced the range of variation (Thorne 1977: Fig 2).

A scenario such as the above can explain the morphology of the skeletal material from both late Pleistocene and Holocene periods in Australia. The late Pleistocene/earlier Holocene stone tools also suggest a single group originally settled the continent. The stone tools, marked by the Australian 'core-tool-and-scraper tradition', are remarkably similar for over 20,000 years, even though made from varying raw materials and used in different environments (Habgood, White 1985).

The most economical interpretation of the morphological and archaeological data, is that the late Pleistocene/earlier Holocene skulls from Australia represent a variable population derived from a single homeland.

Sincere thanks are offered to the following for allowing me to study material in their care: Dr. H.E. Joachin, Mlle B. Senut, Prof Y. Coppens, Dr. J-L Heim, Dr. J. Szilvassy, Dr. C.B. Stringer and R.G. Kruszynski. Prof R.V.S. Wright and E. Roper assisted with the use of the computer and in the interpretation of results. Great help was offered by Dr. A.G. Thorne, who not only showed me material in his care, including some that is unpublished, but willingly discussed the origin of the Australians. Finally, I thank Dr. R.F. Fletcher for useful comments on an earlier draft and Michelle Cranston for the illustrations.

REFERENCES

Abbie AA (1963). Physical characteristics of Australian Aborigines. In Shiels H (ed): "Australian Aboriginal Studies", Oxford: Oxford Univ Press, p 89.

Abbie AA (1966). Physical characteristics. In Cotton BC (ed): "Aboriginal Man in South and Central Australia", part 1, Adelaide: Government Printer, p 9.

Beaton JM (1983). Does intensification account for changes in the Australian Holocene archaeological record? Archaeol Oceania 18:94.

Birdsell JH (1977). The recalibration of a paradigm for the first peopling of Greater Australia. In Allen J, Golson J, Jones R (eds): "Sunda and Sahul: Prehistoric Studies in Southeast Asia, Melanesia and Australia", London: Academic Press, p 111.

Bowler JM (1976). Recent developments in reconstructing late Quaternary environments in Australia. In Kirk RL, Thorne AG (eds): "The Origin of the Australians", Canberra, Australian Inst Aboriginal Stud, p 55.

Bowler JM, Jones R, Allen H, Thorne AG (1970). Pleistocene human remains from Australia: a living site and human cremation from Lake Mungo, western New South Wales. World Archaeol 1:39.

Brown P (1981). Artificial cranial deformation: a component in the variation in Pleistocene Australian aboriginal crania. Archaeol Oceania 16:156.

Chappell JMA (1976). Aspects of late Quaternary palaeogeography of the Australian-East Indonesian region. In Kirk RL, Thorne AG (eds): "The Origin of the Australians", Canberra: Australian Inst Aboriginal Stud, p 11.

Corruccini RS (1973). Size and shape in similarity coefficients based on metric characters. Am J Phys Anthropol 38: 743.

Dortch C (1979). Devil's Lair: an example of prolonged cave use in south-western Australia. World Archaeol 10:258.

Flood J (1983). "Archaeology of the Dreamtime." Sydney: Collins.

Freedman L, Lofgren M (1979). Human skeletal remains from Cossack, Western Australia. J Hum Evol 8:283.

Freedman L, Lofgren M (1983). Human skeletal remains from Lake Tandou, New South Wales. Archaeol Oceania 18:98.

Gordon AD (1981). "Classification Methods for the Exploratory Analysis of Multivariate Data." London: Chapman and Hall.

Habgood PJ (1985). The origin of the Australians: a multivariate approach. Archaeol Oceania 20 (in press).

Habgood PJ, White JP (1985). La prehistoire de l'Australie. La Recherche (in press).

Howells WW (1973). "Cranial Variation in Man." Papers of the Peabody Museum 67.

Hrdlička A (1930). "The Skeletal Remains of Early Man." Smithsonian Misc Coll Vol. 83.

Jelinek J (1953). The fossil man of Dolni Vestanicé III (Czechoslovakia). Anthropockum 3:37.

Kirk RL, Thorne AG (1976). Introduction. In Kirk RL, Thorne AG (eds): "The Origin of the Australians", Canberra, Australian Inst Aboriginal Stud, p 1.

Kowalski CL (1972). A commentary on the use of multivariate statistical methods in anthropometric research. Am J Phys Anthropol 36:119.

Larnach SL, Macintosh NWG (1966), "The Craniology of the Aborigines of Coastal New South Wales." Oceania Monographs 13.

Larnach SL, Macintosh NWG (1970). "The Craniology of the Aborigines of Queensland." Oceania Monographs 15.

Lourandos H (1983). Intensification: a late Pleistocene-Holocene archaeological sequence from Southwestern Victoria. Archaeol Oceania 18:81.

Lourandos H (1984). Changing perspectives in Australian prehistory: a reply to Beaton. Archaeol Oceania 19:29.

Macintosh NWG (1963). Origin and physical differentiation of the Australian Aborigines. Australian Nat Hist 14:248.

Macintosh NWG (1965). The physical aspects of man in Australia. In Berndt RM, Berndt CH (eds), "Aboriginal Man in Australia", Sydney: Angus and Robertson, p 29.

Macintosh NWG (1971). Analysis of an aboriginal skeleton and a pierced tooth necklace from Lake Nitchie, Australia. Anthropologie 9:49.

Macintosh NWG, Larnach SL (1972). The persistence of Homo erectus traits in Australian aboriginal crania. Archaeol Phys Anthropol Oceania 7:1.

Macintosh NWG, Larnach S (1976). Aboriginal affinities looked at in world context. In Kirk RL, Thorne AG (eds): "The Origin of the Australians", Canberra: Australian Inst Aboriginal Stud, p 113.

Martin R, Saller K (1958). "Lehrbuch der Anthropologie." 3rd ed. Stuttgart: Fischer.

Morant GM (1930). Studies of Palaeolithic man 4. A biometric study of the Upper Palaeolithic skulls of Europe and of their relationships to earlier and later types. Ann Eugen 4:109.

Orlocci L (1978). "Multivariate Analysis in Vegetation Research." 2nd ed. The Hague: W. Junk.

Orton C (1980). "Mathematics in Archaeology." Cambridge: Univ Press.

Pearce RH, Barbetti M (1981). A 38,000 year old archaeological site at Upper Swan, Western Australia. Archaeol Oceania 16:173.

Ross A (1981). Holocene environments and prehistoric site patterning in the Victorian Mallee. Archaeol Oceania 16:145.

Rummel RJ (1970). "Applied Factor Analysis." Evanston: North-western Univ Press.

Smith FH (1976). "The Neandertal Remains from Krapina: a Descriptive and Comparative Study." Univ Tennessee Rep Invest 15.

Suzuki H, Takai F (eds) (1970). "The Amud Man and his Cave Site." Tokyo: Yugaku Printing Co.

Thorne AG (1971a). Mungo and Kow Swamp morphological variation in Pleistocene Australians. Mankind 8:85.

Thorne AG (1971b). The racial affinities and origins of the Australian Aborigines. In Mulvaney DJ, Golson J (eds): "Aboriginal Man and Environment in Australia", Canberra: Australian National Univ Press, p 316.

Thorne AG (1976). Morphological contrasts in Pleistocene Australians. In Kirk RL, Thorne AG (eds): "The Origin of the Australians", Canberra: Austral Inst Aborig Stud, p 95.

Thorne AG (1977). Separation or reconciliation? Biological clues to the development of Australian society. In Allen J, Golson J, Jones R (eds): "Sunda and Sahul: Prehistoric Studies in Southeast Asia, Melanesia and Australia", London: Academic Press, p 187.

Thorne AG (1980). The longest link: human evolution in Southeast Asia. In Fox JJ, Granaut RG, McCawley PJ, Mackie JAC (eds): "Indonesia: Australian Perspective." Canberra: Australian National Univ Press, p 35.

Thorne AG (1984). Australia's human origins - how many sources? Am J Phys Anthropol 63:227.

Thorne AG, Macumber PG (1972). Discoveries of late Pleistocene man at Kow Swamp, Australia. Nature 238:316.

Thorne AG, Wilson SR (1977). Pleistocene and recent Australians: a multivariate comparison. J Hum Evol 6:393.

Thorne AG, Wolpoff MH (1981). Regional continuity in Australasian Pleistocene hominid evolution. Am J Phys Anthropol 55:337.

Weidenreich F (1947). Facts and speculations concerning the origin of Homo sapiens. Amer Anthropol 49:187.

White JP, O'Connell F (1982). "A Prehistory of Australia, New Guinea and Sahul." Sydney: Academic Press.

Wolpoff MH, Wu Xin Zhi, Thorne AG (1984). Modern Homo sapiens origins: a general theory of hominid evolution involving the fossil evidence from East Asia. In Smith FH, Spencer F (eds): "The Origins of Modern Humans," New York: Alan Liss, p 411.

Hominid Evolution: Past, Present and Future, pages 381–386
© *1985 Alan R. Liss, Inc.*

RAPID MORPHOMETRIC CHANGE IN HUMAN SKELETAL TRAITS: AN
EXAMPLE FROM THE ANDEAN HIGHLANDS

Robert B. Eckhardt

Department of Anthropology and Graduate Program in
Genetics, The Pennsylvania State University,
University Park, PA 16802

Our area of inquiry, evolutionary biology, has one
great lesson; it is a lesson taught by both Darwin and Dart:
that scientific insights, no matter how clearly and accurate-
ly perceived, are in themselves inadequate to persuade skeptical colleagues. Even when the evidence in favor of a new
viewpoint was substantial, even when the traditional alternative was dubious, received wisdom in anthropology has been
very slow to change. This phenomenon can be illustrated no
more forcefully than with the reception accorded to Taung
itself, even after it was augmented by the discovery of hundreds of additional specimens representing the same taxon.
In their influential book, Fossil Men, Marcellin Boule and
Henri Vallois dismissed the australopithecines as "far from
being our ancestors". This opinion was rendered in 1957,
four years after Piltdown had been proved conclusively to be
a fraud. The detailed discussion by Boule and Vallois of the
place of australopithecines in hominid phylogeny, although
phrased in language which fairly could be called Delphic in
its ambiguity, leaves little doubt that they believed Taung
and its congeners to be aside from the direct ancestry of
later hominids (to be "less evolved cousins" in their exact
words). Further, despite the pervasive uncertainties about
the geologic ages of these South African hominids, Boule and
Vallois left little doubt that the humanity of the australopithecines was too little and too late. They had plenty of
company, of which full citation here could still cause dis-
comfort, while the only useful purpose served would be to
indicate the dubious value of widespread consensus. To in-
voke the American humorist James Thurber, there is no safety
in numbers, or in anything else.

There is a reason, beyond the valuable one of rendering to Professor Dart the credit that was so long overdue, for bringing these half-forgotten matters back to the surface: the old issues are still with us today. It is still rather easy to give examples from the recent literature of paleontology of unhedged statements about evolutionary rates and fates that are held to be unacceptable or impossible. Most cases incorporate a version of one of the central paleontological paradigms: a given specimen or taxon is too different from a later specimen or taxon for the earlier to have been ancestral to the latter (Eldredge and Tattersall 1982; Kurtén 1980; Stanley 1979; Stanley 1981; Trinkaus and Howells 1979). These paradigmatic models incorporate sometimes implicit, but often rather explicit, genetic models. I'll explore in detail just the last-cited case of this sort because it is notable for its especial clarity. After a detailed review of the anatomical contrasts between neandertals and modern populations, Trinkaus and Howells went on in an attempt to account for what they saw as the salient discontinuities, thereby making some very straightforward statements about the quantitative and causal limits to rates of human evolution.

My analysis combines current evidence and theory on how population structure and the genetic architecture of traits interact to limit or otherwise influence evolutionary rates. I will begin by noting the sobering fact that a set of conclusions similar to that now widely held--chiefly that hominid populations of more modern aspect replace neandertals too rapidly in Europe and the Near East to be explained by forces of evolution acting in situ, but also involving the contention that a widespread neandertal phase never existed --was published over a quarter of a century ago by another respected scholar, Thoma (1958). This congruence is sobering because Thoma's conclusions were presented as if they followed from his genetic analysis of neandertal anatomy. In his study morphological features such as the degree of development of the torus supraorbitalis, the shape of the malar region--27 characters in all, including stature--were treated as single-gene traits. It is difficult for someone with any degree of sophistication in genetics to deal seriously with work of this sort, which is contradicted by a wealth of empirical knowledge, some of which antedates Thoma's paper by several decades.

There is rather little basis for reasoning from a trait's genetic architecture to the magnitude of evolutionary change or the rates at which such shifts in frequency might occur over time. Polygenic traits can change very rapidly, not only on a geologic timescale but in real time. This conclusion too is grounded in both theory and observation. From among many papers on the subject I'll select one of central importance (Robertson 1960) because of its generality: the theory is developed for single genes, then extended to individual phenotypic measurements influenced by polygenes and environment. For a single gene with selective advantage s, expected gene frequency at the limit of selection is set by the chance of fixation, therefore being a function of only Ns, where N is the effective population size. For polygenic traits in which genes at different loci are additive in effect (commonly a reasonable approximation to reality), the limit to selection is given by the function $N\bar{i}$, where $\bar{i}$ is the selection differential expressed in standard deviations. For low values of $N\bar{i}$ (as in small, closed populations) the total advance from selection for additive genes is 2N times the gain in the first generation; for all practical purposes the ultimate limit to selection is set by the initial effective population size. This makes intuitive sense because selection depletes additive genetic variation, the stock of which was set by the few genomes present at the outset. As a corollary, in natural species populations, genetic limits to selection are probably rarely approached.

How long does it take a population responding to selection to reach its limit? Because an ultimate selection limit is approached as an asymptotic function, one gets closer and closer but may never reach it; this may be a real-world example of Zeno's paradox. Problems of this sort are often encountered in physics, with the decay of radioactive elements and the like. For the corresponding genetic situation Robertson introduced the concept of a "half-life of the selection process". For additive genes this half-life is estimated to be about 1.4N generations, so effective population size again takes on critical importance. I stress here that the population referred to is not the local group but the species, and can do no better than to quote from Sir Ronald Fisher's correspondence regarding

> ...the magnitude of the population number n. To reduce it to the number in a district requires that there shall be <u>no</u> diffusions even over the number

of generations considered. For the relevant purpose I believe n must usually be the total population on the planet, enumerated at sexual maturity,
and at the minimum of the annual or other periodic
fluctuation (Bennett 1983:273).

Since values of n for hominid species populations have
hardly ever dropped below hundreds of thousands, and commonly have exceeded such minima by several orders of magnitude,
genetic limits to selection must rarely be approached--even
halfway. As reported by Falconer (1967), before these limits
are reached, major phenotypic shifts can occur. In the face
of efficient selection programs response lasts for 20 to 30
generations, producing a mean shift of 5 to 10 phenotypic
standard deviations--and this when population size is small.

Over the last few years much of my energy has been devoted to exploring such possibilities in a field study that
uses statistical genetic approaches developed first for experimental purposes but transferrable to living populations
of our own species. By using these methods we can understand the interacting genetic and environmental bases of
complex morphological traits, and the rates at which such
traits can change on an evolutionary timescale.

The setting for the fieldwork is highland Peru, an extreme environment where the greatest stress--hypoxia--was
not ameliorable by culture. Human populations have lived on
the altiplano between 10 and 20 millennia. Against the overall background of hominid evolution this span is both relatively brief and known to an unusual level of precision.

Andean highlanders have some notable physical distinctions, especially thoraxes of high volume, both absolute and
relative to their short stature. These distinctions reflect
the underlying skeleton, so their development can be traced
over time. One of the best studies of the contrasts between
Andean highlanders and their lowland counterparts was that
of Vellard (1965), who measured 10 anthropometric traits on
838 subjects from 10 different Andean sites. The most notable contrast is between a Bolivian group from 3600 m and a
sea level sample of Peruvian natives. Chests were larger
overall in the highland group, though marginally in transverse thoracic diameter (17.73 cm versus 17.13 cm) and an-
terior-posterior thoracic diameter (13.53 cm against 11.96
cm), with a much greater contrast in sternal height (12.47

cm at altitude compared to 10.90 cm at sea level). Reflecting chiefly the sternal difference, calculated thoracic volume (Vellard 1959) is 21.54% greater at high altitude. Almost identical results are obtained when we use as the
lowland population an Indian group from the Amazon region
inland of the Andes, so what we are seeing here is undoubtedly a highland-lowland contrast. It remained to be resolved, at the outset of my own research in 1978, however,
whether there was any genetic component in these thoracic
variations. A negative answer to this question had been
given quite consistently, including as recently as the previous year by Schull and Rothhammer (1977).

Our original study population was the village of Camacani, situated on the shores of Lake Titicaca at 4000 m. In
the first field season we measured 20 anthropometric traits
on 845 people, of whom 480 were children and 365 were adults.
From these we were able to match up 265 father-offspring
pairs, 290 mother-offspring pairs, and 205 midparent-off-
spring pairs. To attain the maximum power in our heritability analysis, we employed a complex least squares technique,
differentially weighting families according to numbers of
children (Falconer 1963). Only a few key features of our
results can be reported here. In the Camacani population
stature shows the highest heritability ($h^2 = .51\pm.14$). All
of the linear chest measurements had lower heritabilities,
of which the highest was for sternal length ($h^2 = .34\pm.12$),
others ranging down to below .10. These rather closely-
clustered results indicate that thoracic measurements in the
high altitude Camacani population exhibit very low levels of
additive genetic variation, which would be quite consistent
with the past action of selection on these traits.

Although anthropometric measurements have been gathered
in abundance for a century, relatively few studies have collected these data in a family setting permitting estimation
of heritabilities; almost none have data comparable to ours
for thoracic dimensions. However, Susanne (1978) described
a Belgian population in which stature had a heritability be-
between .5 and .6 (about the same as we have found), while
sternal length had a heritability of .79, over twice that in
our population. In both groups stature seems to be a trait
that has been under little directional selection, since it
retains a relatively high level of additive genetic variation. The same is true for sternal length at low altitude,

which is in sharp contrast to our finding of low heritability for other thoracic dimensions at high altitude.

I have described here a case in which, over a known
span of time, a set of substantial morphological variations
of documented heritability and demonstrated adaptive value
have come to characterize a human population numbering tens
of millions, in the face of a high level of gene flow from
surrounding groups and in the absence of any speciation
event. The rate at which these distinctions have accrued
easily exceeds by an order of magnitude the most rapid instances I am aware of in the hominid fossil record.

Bennett JH (1983). "Natural Selection, Heredity, and Eu-
    genics." Oxford: Clarendon, p 273.
Boule M, Vallois, H (1957). "Fossil Men." New York:
    Dryden.
Eldredge N, Tattersall, I (1982). "Myths of Human Evolu-
    tion." New York: Columbia.
Falconer DS (1963). Quantitative inheritance. In Burdette
    WJ (ed): "Methodology in Mammalian Genetics," San Fran-
    cisco: Holden-Day.
Falconer DS (1967). "Quantitative Genetics." New York:
    Ronald Press.
Kurtén B (1980). "Dance of the Tiger." New York: Pantheon.
Robertson A (1960). A theory of limits in artificial select-
    ion. Proc Roy Soc (Lond) Ser B 153:234.
Schull WJ, Rothhammer, F (1977). Analytic methods for gene-
    tic and adaptational studies. In Laughlin WS (ed): "Ori-
    gins and Affinities of the First Americans," New York:
    Gustav Fischer, p 241.
Stanley S (1979). "Macroevolution--Pattern and Process."
    San Francisco: Freeman.
Stanley S (1981). "The New Evolutionary Timetable." New
    York: Basic Books.
Susanne C (1978). Heritability of anthropological charact-
    ers. Hum Biol 49:573.
Thoma A (1958). Métissage ou transformation? Essai sur les
    hommes fossiles de Palestine. L'Anth 61-62:470.
Trinkaus E, Howells, WW (1979). The neanderthals. Sci Amer
    241:118.
Vellard J (1959). "Manual de Antropologia Fisica." Lima:
    Instituto Riva-Agüero.
Vellard J (1965). Le thorax des Indiens des Andes. CR Acad
    Sci Paris 261:529.

Hominid Evolution: Past, Present and Future, pages 387–394
© *1985 Alan R. Liss, Inc.*

EVIDENCE AND INTERPRETATION OF RED OCHRE IN THE EARLY PRE-
HISTORIC SEQUENCES

Ernst E. Wreschner

Department of Anthropology
University of Haifa
Haifa, Israel

In the course of excavations at the Ge'ula Cave on Mount
Carmel in the early sixties, we retrieved from a Middle
Paleolithic layer 52 grams of yellow and red ochre (Wreschner
1967: 69-89). It was at that time a rather unique discovery
in an Israeli Mousterian context. Some years later Vandermeersch (1969: 157) reported the find of red ochre in a
Mousterian level at the Qafze Cave near Nazareth. Almost at
the same time R.A. Dart published a stimulating paper on the
role of red color pigments in prehistoric ritual and symbolism
(Dart 1968:15-27). Intrigued and encouraged by Vandermeersch's
discovery and by Dart's publication, I began to familiarize
myself with ochre occurrences outside Israel. To my surprise
I found that the phenomenon of prehistoric ochre had so far
not been dealt with in a comprehensive framework, either by
prehistorians or by anthropologists. Reports relating to the
subject were scattered in a wide range of publications.

Scholars had first and foremost focused their attention
on the aspects and implications of tool making. True, tool
making is older than the human manipulation of red coloring
matter; however, ochre phenomena in prehistory are a unique
socio-cultural parallel to tool making; there are claims for
finds of red color pigments already in Lower Paleolithic
contexts. But at least since Middle Paleolithic periods
fairly abundant finds of red ochre are attested, and the use
of red color pigments persists well into historical time.

Red and yellow ochre have been found in the early
excavations of living sites, as well as of burial sites,
which prompted Gordon Childe (1942) to remark that "it is

symptomatic of the tenacity of tradition that the practice of
sprinkling the dead with ochre persisted for 20,000 years,
long after the experience should have convinced everyone of
its futility". The pigments in question, generally referred
to under the collective term "ochre", are either iron oxides
or iron hydroxides and are usually well preserved through the
ages. The bearing of archeological data on ochre occurrence
stems from its inclusion in specific biological and cultural
stages of the human past. Because red ochre plays a prominent
role in behavior, especially in symbolic activities of prehistoric hunters, we assume that valid information on red
ochre manipulation is apt to contribute to the discussion on
symbolism and the development of language. I should like to
suggest that in the evolutionary perspective, relationships
and interactions have existed between symbolic actions and
articulate speech; it is, therefore, imperative to attempt
to clarify the question: at what cultural stage and with
what type of hominid does evidence for true symbolic action
appear? According to presently available data, claims for
finds of red and yellow ochre show the following distribution:

Iron Age	3 sites
Bronze Age	1 site
Neolithic 2,500 BC - 4,500 BC	12 sites
Neolithic 4,500 BC - 6,000 BC	11 sites
Neolithic 6,000 BC - 8,000 BC	12 sites
Mesolithic (Late Paleolithic)	91 sites
Upper Paleolithic	90 sites
Middle Paleolithic	17 sites
Lower Paleolithic	4 sites

This review may not be complete and additional reports on
ochre finds in Upper Paleolithic and later sites may exist.

It is generally acknowledged that red ochre plays an
important role in iconism and in symbolic activities of
Homo sapiens sapiens ; however it is the occurrence of red
ochre in the Middle and especially in Lower Paleolithic
contexts that must be carefully examined. These finds raise
the question whether they are valid evidence for ideational
activities of *pre-Neandertal* and *Homo erectus* populations.
We have already stressed the persistence of red ochre in the
Upper Paleolithic. Although a fair number of red color pigments have been found in Mousterian contexts and a few are
claimed for the Lower Paleolithic, one cannot speak of the
persistence of the ochre phenomenon since the early Acheulean.
In order to avoid such an unwarranted assumption, a brief

survey may be given of claims for the finding of ochre in
early archeological contexts.

For the time being, early occurrences of red color pigments in Lower Paleolithic habitation sites are scarce and
sporadic. There is L.S.B. Leakey's claim of two lumps of
ochre at Olduvai BK (1958) in a context probably older than
500,000 years. Chronological assignments claimed for three
further locations are approximately 380,000 BP for Terra Amata,
France (de Lumley 1966, 1976), 250,000 BP for Ambrona in Spain
(Howell 1966) and 250,000 BP for Becov, Czechoslovakia (Fridrich, 1975). The affiliations suggested by the excavators are
Homo erectus at Terra Amata and Ambrona and *pre-Neandertal* at
Becov.

Although we have no datings, the next finds of red ochre
appear, significantly, with *Homo sapiens neanderthalensis* of
the early and middle Mousterian, roughly between 100,000 and
50,000 years ago. Finds of red ochre lumps have been reported
from France, Israel and Russia, which geographical distribution has roughly the character of nuclear diffusion areas
(Wreschner 1980). Whereas in the earlier Mousterian stages
no indication for red ochre application could be found, clear
evidence for red ochre use appears in the late Mousterian.
The pigment is associated with human burials and with ritual
animal deposits. It is noteworthy that, basically, the ochre
practices in the Upper Paleolithic and in later stages retain
these two late Mousterian innovations.

The use and spread of ochre were accompanied by the development of ochre technology. Its use in art, besides that of
other mineral pigments, and its probable application in the
working of hides can be seen as incentives to develop the
methods and the tools for production and storage of the much
desired mineral pigments. The well documented use of red
ochre as a symbolic vehicle in belief systems of *Homo sapiens
sapiens* may explain why the claims for the Lower Paleolithic
finds of ochre have lately become so attractive for the
theorizing on developments of human cognitive capacities and
on the genesis of color symbolism. Thus the Terra Amata ochre
claims led Marshack (1981) to argue: "Leaving out any discussion of possible meaning or usage it is clear that what we
have at the simplest level of discussion are cognitive skills
not much beyond Piaget stages four, five and six of the
human child, but which were used here in an adult, nonsubsistence
context, one that involved an adult semantic. One part of
the meaning and viability of this symbolic activity, one part

of its deep structure, would have involved an affective limbic
component" (1981: 188). Marshack even cites the Lower Paleolithic ochre claims as evidence for the existence at that early
period of a widespread cultural tradition.

These inferences provoke the crucial question: are the
Lower Paleolithic ochre finds, especially those at Terra Amata,
sufficiently substantiated culturally and chronologically and
is their validity sufficiently established to permit the construction of theories on cognitive capacities and symbolic
actions of *Homo erectus*?

The only way to deal with this question is to make a solid
investigation of the finds and their circumstances. In previous publications, I must confess, I have unhesitatingly
accepted the validity of the Terra Amata claim (Wreschner 1976).
However, in the course of conducting experimental studies with
ochres in archeological contexts and of becoming acquainted
with various aspects of ochre behavior and technology, I have
been obliged to revise my approach to the early ochre finds
and especially to those of Terra Amata. It should be noted
again that the reports of ochre from the Lower Paleolithic,
that is, from Early to Middle Acheulean context, are very
scarce. Moreover,   they are unique in their chronological
setting and their hominid affiliation.

It is remarkable that so far we have no ochre finds on
record from all the succeeding Acheulean stages in Europe
and elsewhere, although a few pieces of red color pigments
are claimed to have been found in some sub-Saharan contexts
preceding the South African Middle Stone Age. It follows that
presently a significant hiatus exists. We appear to be confronted with the problem that about 5000 generations of
Acheulean populations in Europe and elsewhere have left no
evidence for red color activities between Terra Amata times,
on one hand, and Ambrona and Becov times on the other. A
similar hiatus of about 120,000 years would exist between
Ambrona-Becov and the earliest Mousterian ochre occurrences.
This rather difficult situation implies that we should concentrate our investigation on the ochre finds in the Lower
and Middle Paleolithic stages when evaluating them as indicators for formative symbolic processes.

At Terra Amata, 73 pieces of ochre with different hue
and chroma values varying from yellows to reds to browns,
have been recovered. According to the inventory they were
found in the beds D, J, M, P, S, in the "Dune"; some came from

unidentified loci; 22 fragments came from zone G of beds D
and M, the others from 7 zones of three other beds from the
"Dune" and the unmarked spots. The significant feature of
the ochre pieces is their small size, in all find localities
of Terra Amata. Seventy-one pieces are between 2mm and 18mm
in diameter, the remaining two pieces are 19mm - 22mm in length,
of brown color, probably the color of the original raw material.
It seems improbable that the fragmentary character of the 71
pieces is the result of use or that they are the remains of
rubbed pieces. Among the Terra Amata ochre pieces, 20 are
multicolored specimens with hue and chroma sequence changing
from the original brown to red and dark-red; this sequence is
typical of clayey ochreous compounds.

I have observed that clayey ochres, after having been
exposed to thermal influence tend, when still hot, to break
and fragment under the influence of rain or water action
(Wreschner 1983). It may be assumed that the relationship
between the measurements of ochre pieces at Terra Amata and
their hue and chroma values, may help us to evaluate whether
we are dealing with the results of human activity or of natural
causes. This assumption is based on results derived from
experimental firing tests with similar clayey ochres. Experiments with controlled hearth roasting of yellow and brown
clayey ochres, and the exposing of similar ochres to incidental brush fires, show this relationship between the sizes
of single pieces and the degree of color change towards reds.
It was observed that brush and grass fire can produce thorough
uniform color change to reds in pieces with maximal measurements of 36mm length, 18mm breadth and 16mm thickness.

On the other hand, intentional roasting can produce consistent uniform color change to reds in pieces with maximal
measurements of 49mm length, 38mm breadth and 20mm thickness.
In repeated tests, it could be observed that brush and grass
fire, sweeping over an area where yellow or brown clayey
ochres are exposed on the topsoil, can cause similar hue and
chroma changes as those in the Terra Amata material. It
seems that this is what could have happened at Terra Amata,
where short intensive heat could produce the different hue and
chroma values and the different shades could be the result of
varying thermal influence, dependent on the position of the
pieces. On the strength of these experiments it seems highly
improbable that the ochres found in the seven locations were
intentionally roasted. Therefore it may be reasonable to
assume that the Terra Amata ochres were modified by natural
agencies and were not processed by man. Villa's study of the

Terra Amata industries (1978) mentions thermoluminescence
dating of burnt flint and burnt silicified limestone from
layers P and M. However, information is not given on the
amount and location of this burnt material and whether it
could be related to hearths. Although human collecting of
the Terra Amata ochres cannot be ruled out, their dominant
features and find circumstances suggest that they were present on the soil surface during prehistoric occupation periods, and that they were exposed to incidental conflagrations
and subsequent rain or water action leading to fragmentation.
Such conditions and processes would have occurred more than
once and are reflected in the stratigraphical record in the
respective find circumstances.

The red ochre found on the floor of an artificial pit
habitation at Becov (Fridrich 1975) seems to be sufficiently
documented to serve as valid archeological evidence. Marshack's (1981) study of a crystalline stone from the ochre-
bearing layers shows that this stone had been used for grinding. However, it has not been established by experiment that
the material ground was ochre, as maintained by that author.
Fridrich's discovery of ochre lumps together with unused
similar crystalline stones provides an unique opportunity
for such an experiment. He discerns two habitation phases
in the pit dwelling and one after the destruction of the pit.
All three phases yielded similar industries which, according
to Fridrich, "are of Middle Paleolithic character". The
scant typological information in the publication does not
allow for a critical appraisal. The date of 250,000 years
BP was obtained by a combination of micro-morphological soil
analysis of the profile and archeological typological criteria.
From the brief mention of the applied methods in the publication we must conclude that the claimed date is that of the
profile but not that of the ochre-bearing layer. It is, therefore, hard to say whether the claimed date reflects the date
of occupation by the ochre-users.

At Ambrona Howell claims "a slab of ochre, apparently deliberately shaped by trimming" (Howell 1966:129). Says Butzer
(1980:635), "The 'ochre slab' at Ambrona was a reddish siltstone
which, because of natural fractures along the laminate structure, appeared to be flaked: to prove human modification would
be next to impossible." That the Ambrona hunters were well
acquainted with the requirements and properties of stones for
tool-making is demonstrated by the Acheulean industry in the
layer; siltstone, however, is soft and not suitable for Acheulean tools but it has a reddish color and can be easily

ground into powder. Most probably it was its color that prompted its transportation to the site. Howell assumes a date of 250,000 BP.

Thermoluminescence dating of burnt stones from layers M and P at Terra Amata resulted in a provisional estimate of c. 230,000 years. Notwithstanding the well founded doubts as to the intentional production of red pigment, the time sequence would place Terra Amata ochre roughly in the same chronological contexts as Becov and Ambrona, that is, in the later stage of the Lower Paleolithic.

It may be concluded that the earliest use of ochre, according to up-to-date finds, cannot be linked with *Homo erectus*. In Neandertaloid sites as well the finds are sporadic and inconclusive.

Evidence for the symbolic use of red color pigments is found in Middle Paleolithic sites, associated with *Homo sapiens neanderthalensis* of La Chapelle-aux-Saints, Molodova and Nahr Ibrahim. In La Chapelle, ochre might have been a personal belonging of the deceased or been offered as a funeral gift to be used in the netherworld; just as the Nahr Ibrahim ochre in the ritual inhumation of the Fallow Deer might have been conceived as symbolizing the procreation of these beasts, on whose meat the population of hunters depended.

In the course of time, the red color that is red ochre spread in clothing, in medicine, and in sympathetic magic that is involved in a great range of behavior.

I end with the words of Professor Dart written to me in a personal letter dated 8 March 1980 "... That may lead to the realisation that the red color was everywhere: in paintings, in pottery and in fire to brighten everything ... So the story is as endless as man and woman themselves."

Butzer KW (1980). Comments in Wreschner EE, Red ochre and human evolution. A case for discussion. Curr Anthropol 21(5):631.
Childe VG (1942). "What Happened in History" Pelican Books.
Dart RA (1968). The birth of symbology. African Studies 27:15.
Fridrich J (1975). Ein Beitrag zur Frage nach den Anfaengen des kuensterlischen und aesthetischen Sinnes des Urmenschen

(Vor-Neanderthal). Pamatky archeologicko 68(2):5.   (In Czech and German) Praha.

Howell FC (1966). Observations on the earlier phases of the European Lower Paleolithic. American Anthropologist 68(2): 88.

Klein R (1973). "Ice Age Hunters of the Ukraine",Chicago: University Press

Leakey LSB (1958). Recent discoveries at Olduvai Gorge, Tanganyika. Nature 19(4):1099.

de Lumley H (1966). Les Fouilles de Terra Amata a Nice. Bull Mus d'Anthrop; Monaco 13.

de Lumley H (1976). "Le Cadre Chronologique et Páleoclimatique du Quaternaire in La Prehistoire Francaise." Tome I pp 5-23.

Marshack A (1981). Paleolithic ochre and the early uses of color and symbol. Curr Anthrop 22 (2): 188.

Solecki R (1975). The Middle Paleolithic site of Nahr Ibrahim (Asfureah Cave) in Lebanon. In Wendorf F, Marks R (eds) "Problems of Prehistory in North Africa and the Levant", Dallas, SMU Press.

Villa P (1978). The stone artifact assemblage from Terra Amata, a comparative study of Acheulian industries in Southwestern Europe. PhD thesis (unpublished) University of California, Berkeley.

Wreschner E (1976). The red hunters. Curr Anthrop 17:717.

Wreschner E (1980). Red ochre and human evolution. A case for discussion. Curr Anthrop 21(5):631.

Wreschner E (1983). Studies in prehistoric ochre technology. PhD thesis (unpublished),Jerusalem, Hebrew University.

PART III
TOMORROW AND TOMORROW AND TOMORROW

Winding the Biological Clocks

Hominid Evolution: Past, Present and Future, pages 397–400
© *1985 Alan R. Liss, Inc.*

CHROMOSOMES AND THE ORIGIN OF MAN

Brunetto Chiarelli

Institute of Anthropology
University of Florence
50122 Firenze, Italy

In 1968, on the invitation of the promoter of the Taung Diamond Jubilee Symposium, Phillip Tobias, in a special issue of the South African Journal of Science I stated: "The most obvious difference between the karyotype of man and that of the apes consists in the chromosome number. Man has 23 pairs of chromosomes, the apes 24. The most likely hypothesis to explain this difference in the chromosome number is that a centric fusion between two acrocentric chromosomes might have occurred in an ancestral pre-hominid. In 1962 (Chiarelli 1962) we made a list of some of the indirect evidences which support this hypothesis. The chromosome which might have arisen from such a centric fusion could have been chromosome n.5 or chromosome n.2 of the Denver classification. Such chromosomes in fact are not present in the karyotype of the apes and their sizes and morphology are in good agreement with our hypothesis." (Chiarelli, 1968:78).

Today, after the introduction of the chromosomal banding techniques and gene mapping, this hypothesis has received strong support. It is now certain that such an event occurred.

The comparison between chromosomes n.12 and n.13 of chimpanzee and the arms p and q of chromosome n.2 of man, shows a banding pattern surprisingly similar. Thus, a translocation between similar chromosomes of a common ancestor of chimpanzee and man can be proposed. In support of this interpretation it has been found with detailed banding techniques that chromosomes homologous to 2q and 2p occur in almost all the Old World Primates.

Even more impressive are the data of gene mapping. Around 20 genes have been assigned to chromosome 2 of man. For the apes and in general for the Cercopithecinae there is limited information. At present, however, 4 of these genes, ACP1, IDH1, GALT and MDH1, which are located on chromosome 2 of man, have been found also on chromosome 12 or 13 of the chimpanzee. Three of these genes have been found in homologous chromosomes of three species of Cercopithecinae (Cercopithecus aethiops, Macaca mulatta, and Papio papio).

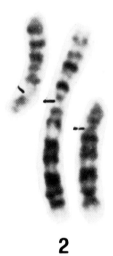

2

Fig. 1. Human chromosome n.2 compared with chromosomes 12 and 13 of the chimpanzee.

The origin of chromosome n.2 of man by a translocation mechanism, from two chromosomes similar to n.12 and n.13 of chimpanzee, in an early human ancestor, seems therefore acceptable. Obviously, other chromosomes also underwent chan-

ges such as inversions, translocations and duplications, each
of which had its own evolutionary history during human phylogeny.

The study of banded chromosomes with different techniques, which now allows more than 600 bands per karyotype,
and the simultaneous use of the BrdU incorporation technique,
which allows us to show the replication sequences of every
major band in the chromosomes, has allowed my colleague (Ro-
scoe Stanyon) and myself to reply to another question I posed
in the same paper in 1968: how and when such an "hominizing"
transformation occurred.

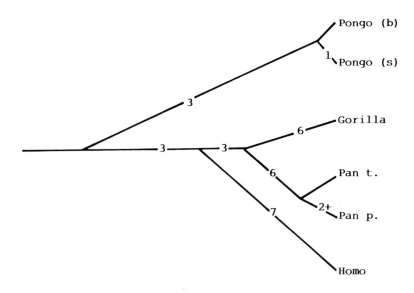

Fig. 2. A possible chromosome phylogeny for the great apes
and humans. The numbers refers to number of major chromosome rearrangements that occurred in a particular evolutionary line.

Figure 2 shows the number and types of chromosome muta-
tions we could detect in the karyotype of the different apes
compared to man.

The type and number of changes, up to now detected, demonstrate that the orangutan is the most conservative and the most unrelated to man, among apes, while the African apes (especially the chimpanzee) share a number of derived changes with the human karyotype.

In producing the phyletic tree, based on karyological data, we have however to evaluate the probability that each mutation has to pass through the meiotic sieve, and establish itself in small and scattered populations, with peculiar social structure as the one of early man.

Following such an accurate evaluation we now hypothesize that the separation of African apes and the human lineage occurred around 8 to 6 million years ago. We are now beginning the gene mapping of hominoid chromosomes, trying to locate human genes which have already been mapped.

A second approach we are developing is to detect the replication sequences in the chromosomes of the different primate species, using the BrdU technique. In this way a new type of dynamic karyology can be developed to establish the replication sequence homology between primate taxa. When such a data are available, and we hope soon, the contribution of cytogenetics to the understanding of human origins and primate evolution will be greatly enhanced.

References

Chiarelli B (1962). Comparative morphometric analysis of primate chromosomes. I. The chromosomes of the anthropoid Apes and Man. Caryologia 15:99.
Chiarelli B (1968). From the karyotype of the apes to the human karyotype. S Afr J Sci 64:72.
Stanyon R, Chiarelli B (1982). Phylogeny of the Hominoidea: the chromosome evidence. J. Hum Evol 11:493.
Stanyon R, Chiarelli B, Romagno D (1983). Origine del cromosoma 2 del cariotipo umano. Antrop. Contemp. 6:225.

Hominid Evolution: Past, Present and Future, pages 401–410
© *1985 Alan R. Liss, Inc.*

RADIOIMMUNOASSAY OF EXTINCT AND EXTANT SPECIES

Jerold M. Lowenstein, M.D.

Clinical Professor of Medicine
University of California, San Francisco
San Francisco, CA 94143, USA

All life on earth apparently had a single origin and shares a single genetic code. Therefore, the primary relational difference between life forms lies not in their gross morphology, which can be reduplicated in the course of evolution by adaptation to similar lifeways, but in the DNA of their genomes, which determines all the biochemical, developmental and morphological characteristics of each organism. DNA encodes RNA which encodes proteins, so that comparison of organisms on the most fundamental level involves comparisons of their DNA, RNA or proteins, not their bones or teeth. During the past two decades, these kinds of investigations of living organisms have given rise to the new field of molecular evolution. But the study of extinct species has continued to rely almost exclusively on morphometric classifications.

Several years ago I began to apply the technique of radioimmunoassay (RIA) to phylogenetic problems of extinct as well as extant species (Lowenstein 1980b). The persistence of amino acids as well as protein material had been demonstrated in some fossils (Wyckoff 1973), but the techniques used had not provided useful genetic information. RIA has the advantages of being able to detect extremely small amounts of proteins or polypeptides (Luft and Yalow 1974),to distinguish those of closely related species, like the African and Indian elephants (Lowenstein, Sarich and Richardson 1981) or to compare those of widely divergent groups, like mammals, birds, and reptiles (Lowenstein 1981). Although I developed this RIA method to investigate relationships among fossil species, it has also found application in the identification of forensic and archeological materials.

The RIA I use is a solid-phase double-antibody technique (Lowenstein 1980a). First, extracts of the fossil or other material are placed in the cups of plastic microtiter plates, where some of the protein binds to the plastic. Second, rabbit antisera to various proteins are placed in the cups, where some of the antibodies bind to the already-bound protein. Third, radioactive (I-125 labeled) goat anti-rabbit gamma-globulin (GARGG) is added to the cups and binds to the rabbit antibody. Finally, the radioactivity in each cup, which reflects the affinity of the antisera for the fossil protein, is measured. If the fossil is human, for example, antisera to human proteins will bind better than those to monkey proteins (Lowenstein 1980b). In this way, the relations of the fossil proteins to those of living species can be estimated.

FOSSIL PHYLOGENIES BY RIA

The first success of the RIA method was in detecting albumin and collagen in a 40,000 year old frozen Siberian mammoth and showing the albumin to be nearly identical with, and equally different from, albumins of the living African and Indian elephants (Lowenstein, Sarich and Richardson 1981). Recently, bones of an American mastodon 10,000 years old have been studied by RIA and found also to contain identifiable elephant-like albumin and collagen (Shoshani in press).

The closest living relatives of the elephants are sea cows and hyraxes. Steller's sea cow (Hydrodamalis gigas) was discovered on the Commander Islands near Kamchatka in 1741 and by 1768 the species was extinct (Haley 1980). By RIA, proteins in the bones of Steller's sea cow were compared with proteins in the bones and sera of the four living species of sea cows. As expected, Steller's sea cow is genetically closer to the living dugong than to the manatees, but the degree of closeness is suprising. Palaeontologists have generally assumed that Steller's sea cow and the dugong diverged from a common ancestor about 20 million years ago, whereas the immunological similarities indicate a much more recent branching of about 5 million years (Rainey, Lowenstein and Sarich 1984). Thus to a

molecular phylogeny of the small surviving family of two
elephants and four sea cows we have now been able to add 3
extinct species, the woolly mammoth, the American mastodon,
and Steller's sea cow (Fig. 1).

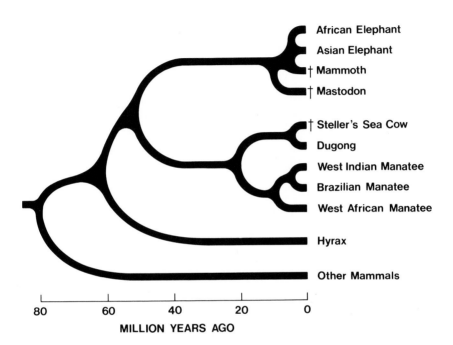

Fig. 1. RIA phylogeny of elephants and sea cows, including
3 extinct species-- mammoth, mastodon and Steller's sea
cow. Mammoth is immunologically equidistant from the two
extant elephant species. The affinity of Steller's with
the dugong is expected on anatomical grounds but the
divergence time shown here is much more recent than
estimates derived palaeontologically. Likewise, the
divergence of the dugongids and the 3 manateids appears
much more recent than was believed on the basis of
reconstructions from the fossil record. RIA confirms the
association of elephants, sea cows and hyraxes posited by
most comparative anatomists.

Another recently vanished species of particular interest to
those who live in southern Africa is the quagga (<u>Equus
quagga quagga</u>). There used to be huge herds of these zebra
-like animals in Cape Province before the growing human
population destroyed their habitat and slaughtered them to
extinction (Rau 1978). There has been controversy about
the relations of the quagga to the three other species of
African zebras, some contending that it was a distinct
fourth species, others maintaining that it was merely the
southernmost variant of the Plains zebra (<u>E. burchelli</u>).
These systematic analyses have been based largely on ranges
and striping patterns. RIA of quagga skins from museums
shows that the residual proteins are much more similar to
those of the Plains zebra than to those of the Mountain
zebra (<u>E. zebra</u>) or Grevy's zebra (<u>E. grevyi</u>). The results,
shown in Fig. 2, suggest that the quagga was probably a
variant of the Plains zebra rather than a separate species
(Lowenstein and Ryder 1985).

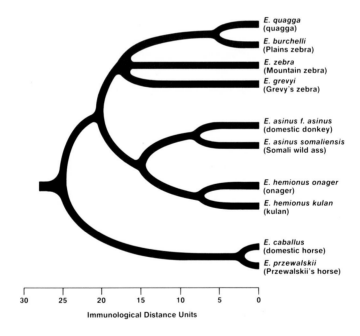

Fig. 2. RIA phylogeny of the genus <u>Equus</u>, including the
extinct quagga. Immunologically, the quagga is much closer
to the Plains than to the Mountain or Grevy's zebras.

It would obviously be desirable to apply this approach to gathering data about the relations of the various species of African hominids to each other and to living hominoids. Problems hindering this pursuit are the small amounts of hominid material available and the very low protein content of fossils more than a million years old. The only hominoid fossil group from which I have thus far succeeded in extracting useful phylogenetic information is the Miocene Ramapithecus/Sivapithecus. Fragments of 5 different specimens, combined and injected into a single rabbit, yielded antisera which reacted strongly with sera of orangutan, gibbon and gorilla, less strongly with sera of humans and chimpanzees, and weakly or not at all with sera of monkeys and other mammals (Lowenstein 1983). These results suggest that Ramapithecus/Sivapithecus was hominoid but not hominid and more or less equally distant from living Asian and African hominoids.

So far, I have been unable to detect proteins in hominid material from Taung, Makapan and Sterkfontein, generously provided me by Dr. Phillip Tobias. I am working on methods for refining the RIA technique so as to further increase its sensitivity.

ARCHEOLOGICAL AND FORENSIC USES OF RIA

It is frequently important to identify the species of a bone, tooth or tissue specimen which lacks diagnostic landmarks. A spectacular example is the notorious Piltdown hoax (Weiner 1955), a combination of a human skull, a fragmentary ape jaw with the teeth filed down, and an isolated canine tooth. For 40 years it was debated whether this chimaera was a forgery or "the first Englishman", until in 1953 it was shown by bone fluorine analysis that the skull and jaw were both recent, not ancient, and that they did not go together (Weiner, Oakley and Le Gros Clark 1953). The question still remained whether the jaw and canine tooth were the same species of ape and whether they were orangutan or chimpanzee. RIA analysis of very small amounts of the jaw and canine tooth proved that they were both orangutan (Lowenstein, Molleson and Washburn 1982).

Until recently, the Jivaro of Ecuador and Brazil commonly cut off the heads of their enemies, removed the skin from the skull and shrank it with hot sand. These shrunken heads have been avidly collected by individuals and museums. Lately, though, missionaries have discouraged head-hunting, and the supply has dwindled. Some enterprising Jivaro, however, have manufactured shrunken "human heads" from monkeys and from horse hide. This has left many curators uncertain whether or not their collections are the real thing. The British Museum (Natural History) had me test their two shrunken heads, one of which had been pronounced "probably horse". By RIA, both were unequivocally human (Molleson 1982).

It would be of great interest to make species identification of much fragmentary archeological material, which might thus provide clues to the lifeways of early peoples. Not infrequently blood-stains are found on ancient weapons. Thomas Loy, of the Provincial Museum in Victoria, British Columbia, provided scrapings from several arrowheads found during a road excavation in northern Canada. RIA testing with antisera to species-specific albumins identified two bloodstains as bison, two as caribou, and one as human.

Forensic applications of RIA come unsolicited from time to time. A man was suspected of being a hit-and-run driver, and examination of his automobile revealed a fragment of flesh on one hubcap. RIA identified this as rat, not human. An individual in Puerto Rico was arrested for selling what he claimed were pieces of sea turtle penis, believed by some men to have aphrodisiac powers. Sea turtles are endangered and protected species in the Caribbean. But sometimes goat meat, which is more easily obtained, is advertised as sea turtle. I was asked to identify 9 specimens, all of which RIA showed to be sea turtle (Lowenstein 1982). This information was used in court to convict the poacher.

RIA PHYLOGENIES OF LIVING GROUPS

In Africa, and especially in South African cave sites, which cannot be dated radiometrically, faunal dating is often used. Bovids, which have radiated profusely here during the past few million years, provide a favorite palaeontological calendar (Vrba 1975). I began to do some work on fossil bovids and then realized that there is no agreed-upon phylogeny for living bovids-- that there are,

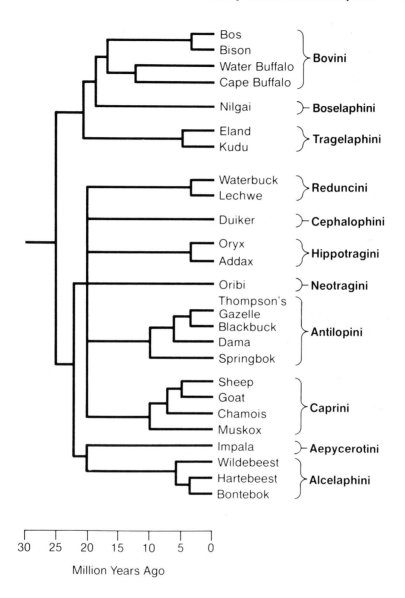

Fig. 3. RIA phylogeny of the bovids. The tribal clusters (boldface) agree generally with traditional morphometric analysis but many of the branching orders have been disputed.

in fact, many disputes: for example, whether the water
buffalo (Bubalus) is more closely related to the Cape
buffalo (Syncerus caffer) or domestic cattle (Bos taurus)
(Groves 1980); or the branching order among the Alcelaphini
-- hartebeest (Alcelaphus buselaphus), wildebeest
(Connochaetes) and bontebok (Damaliscus dorcas)
(Vrba 1980).

For several years, I have been working on an immune
phylogeny of the bovids (Fig. 3). The sera were provided by
Elisabeth Vrba, of the Transvaal Museum, by Vincent Sarich
of the University of California, Berkeley, and by the San
Diego Zoo. The family tree gives the answers to the two
disputes mentioned above. Water buffalo is closer to Cape
buffalo than to domestic cattle. Hartebeest and bontebok
are close to each other, and wildebeest is the outgroup.
The association of the Bovini, Boselaphini, and
Tragelaphini is not unexpected, but the apparent grouping
of Cephalophini (duikers) and Hippotragini (oryx-addax)
with the Antilopini (antelopes) and Caprini (sheep and
goats) will come as a surprise to most systematists, as it
fails to conform with the basic division into two tooth
types, Boodontia and Aegodontia (Vrba 1980).

Bovid systematics have largely been worked out in
the past on the basis of tooth and horn anatomy. With the
profuse radiation of this group during the past 20 million
years or so, there have doubtless been many cases of
parallelism and convergence. The molecular approach, as
exemplified here by RIA, avoids the problems of anatomical
convergence in deriving a family tree of this complex and
important group of large mammals.

SUMMARY

Radioimmunoassay is a sensitive and powerful
immunological technique which can be applied to studying
the genetic affinities of extinct and living organisms.

REFERENCES

Groves CP (1980). Systematic relationships in the Bovini (Artiodactyla, Bovidae). Zeit Zool Systematik Evol 15:264.

Haley D (1980). The great northern sea cow: Steller's gentle siren. Oceans 13(5):7.

Lowenstein JM (1980a). Immunospecificity of fossil proteins.In Hare PE (ed): "Biogeochemistry of Amino Acids." New York: John Wiley & Sons.

Lowenstein JM (1980b). Species-specific proteins in fossils. Naturwissenschaften 67:343.

Lowenstein JM (1981). Immunological reactions from fossil material. Phil Trans Royal Soc London B292:143.

Lowenstein JM, Sarich VM, Richardson BJ (1981). Albumin systematics of the extinct mammoth and Tasmanian wolf. Nature 291:409.

Lowenstein JM, Molleson T, Washburn SL (1982). Piltdown jaw confirmed as orang. Nature 299:294.

Lowenstein JM (1982). Sea turtles, aphrodisiacs and machismo. Oceans 15(4):72.

Lowenstein JM (1983). Fossil proteins and evolutionary time. Pontif Acad Sci Scripta Varia 50:151.

Lowenstein JM, Ryder OA (1985). Immunological systematics of the extinct quagga. Experientia, in press.

Luft R and Yalow RS (eds)(1974). "Radioimmunoassay: Methodology and Applications in Physiology and in Clinical Studies." Stuttgart: Georg Thieme.

Molleson T (1982). Shrunken heads-- but whose? New Scientist 96:3/8.

Rainey WE, Lowenstein JM, Sarich VM, Magor DM (1985). Sirenian molecular systematics-- including the extinct Steller's sea cow (Hydrodamalis gigas). Naturwissenschaften, in press.

Rau RE (1978). Additions to the revised list of preserved material of the extinct Cape Colony quagga and notes on the relationship and distribution of southern Plains zebras. Ann S Afr Mus 77:27.

Shoshani J, Walz DA, Goodman M, Lowenstein JM (1984). Protein and anatomical evidence on the phylogenetic position of Mammuthus primigenius within the Elephantinae. Acta Zoologica Fennica, in press.

Vrba ES (1975). Some evidence of chronology and palaeoecology of Sterkfontein, Swartkrans and Kromdraai from the fossil Bovidae. Nature 254:301.

Vrba ES (1980). The significance of bovid remains as
   indicators of environment and predation patterns. In:
   Behrensmeyer AK & Hill AP (eds) "Fossils in the Making."
   Univ of Chicago.
Weiner JS, Oakley KP, Le Gros Clark WE (1953).
   The solution of the Piltdown problem.
   Bull Br Mus Nat Hist A2(3):141.
Weiner JS (1955). "The Piltdown Forgery."
   London: Oxford University Press.
Wyckoff RWG (1972). "The Biochemistry of Animal Fossils."
   Baltimore: Williams and Wilkins.

Hominid Evolution: Past, Present and Future, pages 411–416
© *1985 Alan R. Liss, Inc.*

NONLINEAR MOLECULAR CLOCKS AND APE-HUMAN DIVERGENCE TIMES

Philip D. Gingerich

Museum of Paleontology
The University of Michigan
Ann Arbor, Michigan 48109

Preference for simple rather than complex hypotheses is axiomatic in science. "Occam's razor" and "parsimony" both express this principle. It is important to note in framing and comparing evolutionary hypotheses that an operational preference for simple explanations does not imply that evolution itself is simple. We study evolution, however complex, by building, testing, and sometimes rejecting progressively more complicated models to account for new observations earlier and simpler models could not anticipate (let alone explain). This process of learning requires inductive hypothesis formation based on a collected body of past observation, followed by deductive testing using newly acquired observations. New observations may corroborate and strengthen the hypothesis under which they were collected, or new observations may require modification of a given working hypothesis. In either case, once made, new observations become a part of the collected body of past observation and the process of learning continues.

This approach to learning, with its continual interplay of induction and deduction, is nowhere more apparent than in the tortuous development of our present understanding of human evolution. A "missing link" is a paleontologist's hypothesis about a stage of evolution not represented by fossils. It is a prediction about what one expects to find. Some missing links are found, corroborating and strengthening models predicting their existence (in which case the "links" are, of course, no longer "missing"). Sometimes the search for missing links yields unexpected results, requiring modification of one's working hypothesis.

Many missing links have been found in sixty years of human paleontology in Africa, beginning with Dart's recognition of the significance of Australopithecus africanus. Other predicted links have been abandoned as new finds rendered their discovery improbable.

My purpose here is not to review the human fossil record, but rather to discuss interpretation of new evidence, originating outside the realm of paleontology, that bears directly on the relationships of humans and extant apes, our closest living relatives. Living African apes are virtually unknown as fossils, and any estimate of the closeness or distance of their genealogical relationship to humans requires comparison based on visible morphology, which is difficult to quantify and interpret, or comparison of structure at a molecular level, which appears easier to measure and relatively straightforward in interpretation.

The current consensus model of molecular evolution, as befits hypotheses in young sciences, is among the simplest of possible models. It is based on a Poisson metric (Zuckerkandl, Pauling 1965) transformed (augmented or corrected) to make it linear. The "neutralist hypothesis" of molecular evolution follows from the Zuckerkandl-Pauling model, indeed it is assumed implicitly in this model. The Zuckerkandl-Pauling model was constructed for blood proteins (globins), and initially calibrated using a single paleontological estimate of the time of human (primate) versus other mammal divergence. Zuckerkandl and Pauling demonstrated at once the great value of paleontological calibration for placing molecular difference in evolutionary context and the potential of a calibrated molecular clock for clarifying unknown molecular or organismal genealogies and divergence times.

Recent studies recognize variation in molecular difference/divergence time ratios, but this is usually assumed to reflect random variation about a linear trend. Reed and Lestrel (1970), Radinsky (1978), and Corruccini et al. (1980) questioned the appropriateness of linear models. Here I shall briefly outline preliminary results of an ongoing effort to test the linear model using multiple calibration points derived from the primate fossil record. These results are important for understanding evolution at the molecular level, and they have an important bearing on molecular clocks used to predict ape-human divergence times.

Simply stated, the Zuckerkandl-Pauling model predicts that augmented or corrected molecular difference (MD) should be a linear function of divergence time (DT). Algebraically:

$$MD = \underline{a}(DT) \tag{1}$$

where $\underline{a}$ is a constant. A linear model like that in Equation 1 is a special case of the general power function:

$$MD = \underline{a}(DT)^{\underline{b}} \tag{2}$$

where the exponent $\underline{b} = 1$. Linearity can be tested by examining whether an empirically derived $\underline{b}$ differs significantly from unity. Relative rate tests (Sarich, Wilson 1967) measure internal consistency in a set of molecular data, but they do not bear on the problem of how rates behave in real time. Tests in real time require molecular difference values and geological divergence times for a minimum of two nonzero calibration points.

Primates are the most intensively studied mammals. The fossil record of primate evolution is among the best known for any group, offering the greatest potential for testing the linearity assumption inherent in the Zuckerkandl-Pauling model and the neutralist hypothesis. Three primate divergence intervals are sufficiently well established paleontologically to be of use here. These are Hominoidea-Cercopithecoidea at the Oligocene-Miocene transition (ca. 20-30 myBP), Catarrhini-Platyrrhini at the Eocene-Oligocene transition (ca. 35-45 myBP), and Anthropoidea-Prosimii at the Paleocene-Eocene transition (ca. 50-60 myBP) (Radinsky 1978; Szalay, Delson 1979; Pilbeam 1984; Gingerich 1984). These divergence intervals are bounded on the left by the appearance of both descendants in the fossil record and on the right by complete absence of primates of the grade in question known in the fossil record. Central values for each major divergence (25, 40, and 55 myBP) are used in the following analysis. Ape-human divergences, the subject of this study, are open to interpretation paleontologically and thus here considered unknowns to be predicted.

The most widely cited molecular studies predicting ape-human divergence times are those of Sarich (1968, 1970), based on immunological comparisons of primate albumins (Fig. 1). Other studies providing additional measures of molecular difference that can be scaled against the primate

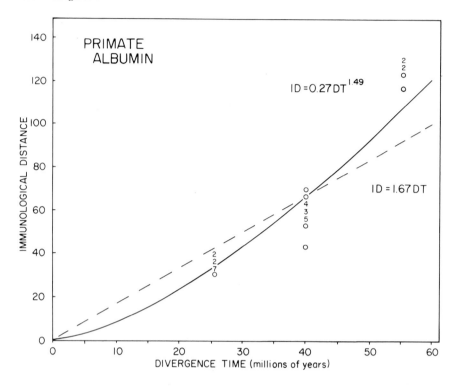

Fig. 1.   Comparison of Hominoidea-Cercopithecoidea,
Catarrhini-Platyrrhini, and Anthropoidea-Prosimii albumin
immunological distances (ID; from Sarich 1968, 1970) with
paleontologically constrained divergence times (DT) of 25,
40, and 55 myBP (see text).   Exponent 1.49 of power function
fit to data has 95% confidence interval of 1.31-1.67,
significantly greater than unity and precluding linear model
ID = 1.67(DT) employed by Sarich (1970: 194).   Anthropoid-
prosimian divergence at 75 myBP is predicted by the Sarich
model, which would require a last common ancestor for
anthropoids and prosimians 20 m.y. older than any known
primate of prosimian grade (and fully 10 m.y. older than any
known mammal of remotely primate grade).   Linear model
predicts recent divergence times that are too young and
ancient divergence times that are too old.   Power function
is more consistent with fossil record over entire range of
primate history and Cenozoic time.   Albumin ID of 7 units
for Homo-Pan yields a predicted DT of 4.2 myBP using Sarich
model and a predicted DT of 8.9 myBP, more than twice as
great, using power function fit to same data.

Table 1.   Predicted divergence times (myBP) for apes and humans, based on temporal scaling of molecular evolution.

Source	Homo vs.:	Pan	Gorilla	Pongo	Hylobates
Immunology					
Albumin (ID)		8.9	10.5	12.8	14.8
(Sarich 1968,1970)					
Transferrin		10.1	-	-	-
(Sarich, Cronin 1976)					
Summed proteins (AD)		7.6-11.2	9.5-10.3	18.9	18.2-21.3
(Dene et al. 1976)					
Nucleic acid hybridization					
DNA (del TS C)		11.2	-	-	-
(Kohne et al. 1972)					
DNA (del TmR C)		7.0-7.3	7.3-7.9	(13.1)	18.1-19.4
(Benveniste, Todaro 1976)					
DNA (del mode C)		10.0	13.3	19.1	20.9
(Sibley in Pilbeam 1983)					
Means		**9.2**	**9.8**	**16.0**	**18.7**
Standard deviations		(1.7)	(2.1)	(3.5)	(2.2)

fossil record to yield ape-human divergence times are listed in Table 1.   All have exponents b greater than 1.0, indicating similar nonlinear scaling of MD and DT (sequence data show this pattern but exhibit few ape-human differences).   Molecular estimates of ape-human divergence times based on paleontologically calibrated nonlinear clocks suggest humans and chimpanzees diverged 9.2 myBP, humans and gorillas diverged 9.8 myBP, humans and orangutans diverged 16.0 myBP, and humans and gibbons diverged 18.7 myBP.

A linear model and neutralist hypothesis of molecular change may have been justifiable on the basis of simplicity a decade ago, but molecular differences are now sufficiently well documented to permit scaling against the fossil record and geological time.   Available evidence supports Goodman's (1963) early suggestion of a slowdown of molecular change during primate evolution.   Power functions appear adequate to describe this slowdown at present, but new evidence may show such curves to be oversimplified as well.

Benveniste RE, Todaro GJ (1976). Evolution of type C viral genes. Nature 261:101.

Corruccini RS, Baba M, Goodman M, Ciochon RL, Cronin JE (1980). Non-linear macromolecular evolution and the molecular clock. Evolution 34:1216.

Dene HT, Goodman M, Prychodko W (1976). Immunodiffusion evidence on the phylogeny of the primates. In Goodman M, Tashian RE (eds): "Molecular Anthropology", New York: Plenum Press, p 171.

Gingerich PD (1984). Primate evolution: evidence from the fossil record, comparative morphology, and molecular biology. Yearb Phys Anthropol 27:57.

Goodman M (1963). Man's place in the phylogeny of primates as reflected in serum proteins. In Washburn SL (ed): "Classification and Human Evolution", New York: Wenner Gren Foundation, p 204.

Kohne DE, Chiscon JA, Hoyer BH (1972). Evolution of primate DNA sequences. J Hum Evol 1:627.

Pilbeam D (1983). Hominoid evolution and hominid origins. In Chagas C (ed): "Recent Advances in the Evolution of Primates", Rome: Pont Acad Sci Scripta Varia 50:43.

Pilbeam D (1984). The descent of hominoids and hominids. Sci Am 250(3):84.

Radinsky L (1978). Do albumin clocks run on time? Science 200:1182.

Read DW, Lestrel PE (1970). Hominid phylogeny and immunology: a critical appraisal. Science 168:578.

Sarich V (1968). The origin of the hominids: an immunological approach. In: Washburn SL, Jay JC (eds): "Perspectives on Human Evolution", New York, Holt Reinhart and Winston, p 94.

Sarich V (1970). Primate systematics with special reference to Old World monkeys, a protein perspective. In Napier JR, Napier PH (eds): "Old World Monkeys", New York: Academic Press, p 175.

Sarich V, Cronin J (1976). Molecular systematics of the primates. In Goodman M, Tashian RE (eds): "Molecular Anthropology", New York: Plenum Press, p 141.

Sarich V, Wilson AC (1967). Rates of albumin evolution in primates. Proc Nat Acad Sci 58:142.

Szalay FS, Delson E (1979). "Evolutionary History of the Primates", New York: Academic Press, 580 pp.

Zuckerkandl E, Pauling L (1965). Evolutionary divergence and convergence in proteins. In Bryson V, Vogel HJ (eds): "Evolving Genes and Proteins", New York: Academic Press, p 97.

*New Shadow Pictures Beyond
Röntgen's Wildest Visions*

Hominid Evolution: Past, Present and Future, pages 419–426
© *1985 Alan R. Liss, Inc.*

ENDOCRANIAL VOLUME DETERMINATION OF MATRIX-FILLED FOSSIL
SKULLS USING HIGH-RESOLUTION COMPUTED TOMOGRAPHY

Glenn C. Conroy, Ph.D.
Michael W. Vannier, M.D.
Dept. of Anatomy&Neurobiology/Anthropology (GCC)
Dept. of Radiology (MWV)
Washington University Medical Center
St. Louis, Missouri 63110 USA

Endocranial volume is one of the most important parameters derived from fossil skulls (Jerison 1973,1982,1983;
Radinsky 1973,1975,1977,1980,1981; Tobias 1971). The evolution of brain size and shape is a particularly significant
(and sometimes controversial) theme in paleoanthropology
(Holloway 1972,1973,1975,1983; Tobias 1967,1971; Pilbeam
1969,1970; Wolpoff 1969,1981).
     Unfortunately, intact primate skulls are relatively
rare in the fossil record. Indeed, the actual fossil evidence of 70 million years of pre-hominid primate brain evolution is based upon only a dozen or so specimens in which
relatively accurate measures of endocranial volume have been
determined (Radinsky 1977b,1979; Gurche 1982; Gingerich
1977; Walker et al 1983). Because of their rarity, it becomes imperative that paleoanthropologists extract as much
information as possible from each precious specimen that is
discovered. Scientists are sometimes hampered in this goal,
however, by the presence of stone matrix within the cranial
cavity of fossil skulls which may preclude direct and accurate measurement of important intracranial parameters
(Conroy, Vannier 1984; Holloway 1981; Leakey et al 1971;
Leakey, Walker 1976).
     The difficulty in calculating accurate endocranial
volume in matrix-filled skulls presently affords paleontol-
ogists only one option for directly measuring this param-
eter, namely, physically "invading" the original fossil in
order to manually remove the stone matrix. Otherwise, only
indirect estimation of endocranial volume can be obtained
(Holloway 1973; Mackinnon et al 1956). It is understandable
that many fossil curators are reluctant to have their rare

fossils directly examined in such a manner! For this reason, some invaluable fossil specimens still remain unpre-
pared on museum shelves with important morphological information seemingly lost forever inside their matrix-filled
cranial vaults. Until now, this dilemma has seemed insurmountable.

In order to circumvent this problem, we have developed
a research strategy utilizing recent advances in high-
resolution computed tomography to produce accurate, non-
invasive intracranial volume measurements of matrix-filled
fossil skulls. This study derives from recent advances in
computer imaging techniques that allow us to produce geo-
metrically accurate three-dimensional images of fossil
skulls from two-dimensional CT data (Conroy, Vannier 1984;
Vannier et al 1983a-e,1984; Marsh, Vannier 1983a,b). These
techniques can make various portions of the fossil skull
"transparent" on the computer screen allowing one to view intracranial structures noninvasively. In effect, this allows
the investigator to "dissect" electronically the
fossil skull in any plane desired. In addition, the CT
densities of matrix and mineralized bone can be distinguished from one another on our CT scanner (Siemens DR3).
This allows the computer to "remove" the stone matrix from
the picture leaving behind only the true osseous contours of
the fossil (Conroy, Vannier 1984).

While volumetric determinations using computed tomography have proven their value in certain clinical evaluations (Brenner et al 1983; Forbes et al 1983), no one, to
our knowledge, has ever successfully applied these techniques to the fossil record.

In order to illustrate the potential paleontological
value of these methods, we have examined two 30-million year
old fossil mammal skulls in which the cranial cavities were
completely filled with a hard sandstone matrix (Stenopsochoerus sp.-AMNH 45192 and Merycoidodon culbertsonii-AMNH
39430) (see Conroy, Vannier 1984). A contiguous sequence of
high-resolution, thinly collimated (2mm), CT scans were
taken through each skull. These images were produced from
third generation (Siemens Somatom 2 and Somatom DR3) CT
scanners in the Mallinckrodt Institute of Radiology,
Washington University Medical Center. All scans were taken
at 125kv with a scanning time of 10 seconds to produce 256x
256 resolution slice images. Voxel (volume element) size
equals pixel (picture element) edge length squared times
slice thickness. Depending on the zoom or magnification
factor, pixel edge length ranged between 0.5 and 1.0mm.

Further details of the scanning procedure are documented elsewhere (Vannier et al 1983a-e; Marsh, Vannier 1983a,b; Conroy, Vannier 1984).

It was necessary to employ extended bone density measurement range technology in our CT analyses of the fossils since the CT attenuation of mineralized bone and sandstone matrix is well beyond +1000 Hounsfield units, the upper limit for normal CT scanning. By manipulating display window width and level controls, we were able to distinguish the mineralized bone from the underlying sandstone matrix (+1600 Hounsfield units and higher in these fossils). The sandstone matrix was then "removed" from the image by setting the window level to the CT threshold value separating mineralized bone from stone matrix. This level slicing technique displayed the intracranial skull contours for each CT slice with the matrix "removed" (see Fig. 1 in Conroy, Vannier 1984). The procedure resulted in a sequence of non-overlapping, 2mm thick CT sections of the complete fossil skull with the matrix "removed".

As each 2mm skull section was displayed on the computer screen, region-of-interest (ROI) outlines of "matrix free" endocranial cavity were manually traced with a marker stylus on a resistor pad x-y digitizer built into our CT viewing console (Siemens Evaluscope B or RC). The computer (DEC PDP-11) automatically computed the volume of the cranial cavity section outlined by the operator (Fig. 1).

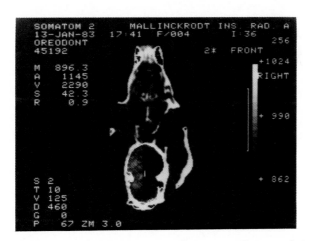

Fig. 1. 2mm CT slice of <u>Stenopsochoerus</u>. The sandstone matrix is distinguishable from the mineralized bone. The volume (V) of the cranial cavity for this slice is 2290 mm^3.

Besides volume, other data automatically generated on the
computer screen includes: arithmetic mean of CT attenuation
coefficients of all points in the region of interest (ROI),
the area of the ROI, and the number of pixels in the ROI.
By suming the volumes generated at each CT slice, the total
endocranial volume was determined.

$$V= \sum_{i=1}^{N} \sum_{i=1}^{Mi} TP$$

> N=number of slices
> Mi=number of voxels in ROI for slice i
> T=slice thickness in mm
> P=pixel edge length in mm

In addition, the endocranial outlines could be processed through a three-dimensional computer reconstruction
algorithm to produce a three-dimensional model of the
matrix-filled cranial cavity (Fig. 2).

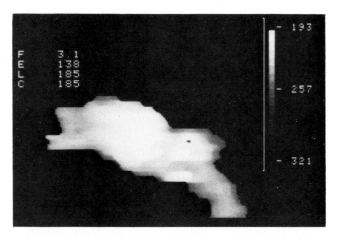

Fig. 2. Computer-generated three-dimensional lateral view of
the "endocast" of the matrix-filled fossil skull.

This figure is the first computer generated three-
dimensional image of a fossil "endocast" from a matrix-
filled skull ever produced. Frontal lobes are at the top
left and the spinal cord is found at the bottom right of the
image. Cerebral hemisphere and cerebellum are clearly dis-

tinguishable in this lateral view. In this image, depth information is encoded in gray scale. Thus, the brain surface is lighter the closer it is to the observer.

Using the procedures outlined above, the endocranial volumes for Stenopsochoerus and Merycoidodon were calculated as 29.3 and 48.5 cubic centimeters respectively (see Thorpe 1937 and Moody 1922 for endocranial volume determinations of other closely related genera from the fossil record).

The accuracy of the fossil volume determinations was then assessed by duplicating these procedures on several modern skulls of known endocranial volume (two H. sapiens, one Gorilla, and one Papio). These specimens were chosen to provide a wide range of cranial size and shape. A direct measure of endocranial volume in these four skulls was made by the traditional method of filling each with seed and measuring the seed volume in a graduated cylinder. Five determinations were made for each skull (these determinations were made without prior knowledge of the volume determinations made by the computer on these same specimens).

The results of the five trials for each of the four modern skulls are summarized in Table 1. Endocranial volume determinations derived from the seed method and from the CT scanner are compared in Table 2. Values derived from the CT scanner are all within 1-3% of those values directly determined by filling the cranial cavity with seed. Thus, we are confident that the CT-derived endocranial volumes for the matrix-filled fossil skulls are extremely accurate and provide previously unattainable data for paleontological investigations.

Table 1. Endocranial Volume (seed trials)

Statistics	Gorilla	H. sapiens	H. sapiens	Papio
Count (n)	5	5	5	5
mean	560.00	1361.00	1271.00	143.40
s.e.	5.24	2.92	7.81	1.44
st.dev.	11.73	6.52	17.46	3.21
maximum	570.00	1365.00	1280.00	147.00
minimum	545.00	1350.00	1240.00	140.00
range	25.00	15.00	40.00	7.00

Table 2. Comparison of Endocranial Volume Determinations

Taxa	Vol. mean (seed)	CT volume	% difference
H. sapiens	1361	1348	1%
H. sapiens	1271	1245	2%
Gorilla	560	566	1%
Papio	143	139	3%

Previously, accurate measurements of endocranial volume in such matrix-filled fossils would have been impossible to obtain noninvasively. We believe these new non-invasive computer imaging techniques can have important applications in future paleoanthropological investigations by allowing anthropologists to view and analyze fossil primate material in ways never before possible. Anthropologists will now be able to easily generate precise data on intracranial volume, area, angles, or linear distances within matrix-filled fossil skulls. Such data can also be easily obtained in osteological specimens or in living subjects (e.g., craniofacial growth studies in primates) and thus could be applicable to many studies in archeology and comparative primate biology as well. Most importantly, these analyses can all be accomplished in a totally safe way, non-destructive to the original fossil, osteology specimen, or living subject.

We gratefully acknowledge the technical assistance of R. Knapp in this research and financial support from the National Science Foundation. We thank M. Novacek of the American Museum of Natural History for the loan of the fossil specimens.

References

Brenner D, Whitley N, Houk T, Aisner J, Wiernik P, Whitley J (1982). Volume determinations in computed tomography. J Am Med Assoc 247:1299.

Conroy G, Vannier M (1984). Noninvasive three-dimensional computer imaging of matrix-filled fossil skulls by high-resolution computed tomography. Science 226:456.

Forbes G, Gorman C, Gehring D, Baker H (1983). Computer analysis of orbital fat and muscle volumes in Graves Ophthalmopathy. Am J Neuroradiol 4:737

Gingerich P (1977). Correlation of tooth size and body size in living hominoid primates, with a note on relative brain size in Aegyptopithecus and Proconsul. AJPA 47:395.

Gurche J (1982). Early primate brain evolution. In Armstrong E, Falk D (eds): "Primate Brain Evolution," New York: Plenum, p 227.

Holloway R (1972). Australopithecine endocasts, brain evolution in the Hominoidea, and a model of hominid evolution. In Tuttle R (ed): "Functional and Evolutionary Biology of the Primates," Chicago: Aldine, p 185.

Holloway R (1973). Endocranial volumes of early African hominids and the role of the brain in human mosaic evolution. J Hum Evol 2:449.

Holloway R (1975). The role of human social behavior in the evolution of the brain. "43rd James Arthur Lecture on the Evolution of the Human Brain," New York: AMNH, p 1.

Holloway R (1981). Exploring the dorsal surface of hominoid brain endocasts by stereoplotter and discriminant analysis. Philos Trans R Soc Lond B 292:155.

Holloway R (1983). The O.H.7 parietal fragments and their reconstruction. Am J Phys Anthropol 60:505

Jerison H (1973). "Evolution of the Human Brain and Intelligence." New York: Academic Press, p 482.

Jerison H (1982). The evolution of biological intelligence. In Sternberg R (ed): "Handbook of Human Intelligence," Cambridge: Cambridge University Press, p 723.

Jerison H (1983). The evolution of the mammalian brain as an information-processing system. In Eisenberg J, Kleiman D (eds): "Advances in the Study of Mammalian Behavior," Amer Soc Mammal Spec Publ 7:113.

Leakey R, Mungai J, Walker A (1971). New australopithecines from East Rudolf, Kenya. Am J Phys Anthropol 35:175

Leakey R, Walker A (1976). Australopithecus, Homo erectus and the single species hypothesis. Nature 261:572.

MacKinnon I, Kennedy J, Davies T (1956). The estimation of skull capacity from roentgenologic measurements. Am J Roent Rad Ther Nucl Med 76:303.

Marsh J, Vannier M (1983a). The "third" dimension in craniofacial surgery. Plas Reconstr Surg 71:759.

Marsh J, Vannier M (1983b). Surface imaging from CT scans. Surgery 94:159.

Moody R (1922). On the endocranial anatomy of some Oligocene and Pleistocene mammals. J. Comp Neur 34:343.

Pilbeam D (1969). Early Hominidae and cranial capacity. Nature 224:386.

Pilbeam D (1970). Early Hominidae and cranial capacity (continued). Nature 227:747.

Radinsky L (1973). Evolution of the canid brain. Brain Behav

Evol 7:169.

Radinsky L (1975). Evolution of the felid brain. Brain Behav Evol 11:214.

Radinsky L (1977a). Brains of early carnivores. Paleobiology 3:333.

Radinsky L (1977b). Early primate brains: facts and fictions. J Hum Evol 6:79.

Radinsky L (1979). The fossil record of primate brain evolution. "49th James Arthur Lecture on the Evolution of the Human Brain," New York: Amer Mus Nat Hist, p 1.

Radinsky L (1980). Endocasts of amphicyonid carnivorans. Am Mus Novitates 2694:1.

Radinsky L (1981). Evolution of skull shape in carnivores 1. Biol J Linn Soc 15:369.

Thorpe M (1937). The Merycoidodontidae. Mem Peabody Mus Nat Hist 3:1.

Tobias P (1967). "Olduvai Gorge vol 2." Cambridge: Cambridge University Press, p 264.

Tobias P (1971). "The Brain in Hominid Evolution." New York: Columbia University Press, p 170.

Vannier M, Marsh J, Warren J (1983a). Three-dimensional computer graphics for craniofacial surgical planning and evaluation. Comput Graph 17:263.

Vannier M, Marsh J, Warren J, Barbier J (1983b). Three-dimensional computer aided design of craniofacial surgical procedures. Diagn Imaging 5:36.

Vannier M, Gado, Marsh J (1983c). Three-dimensional display of intracranial soft tissue structures. Am J Neuroradiol 4:520.

Vannier M, Marsh J, Warren J, Barbier J (1983d). Three-dimensional CAD for craniofacial surgery. Elect Imaging 2:48.

Vannier M, Marsh J, Gado M, Totty W, Gilula L, Evens R (1983e).Clinical applications of three-dimensional surface reconstruction from CT scans. Electromedica 51:122.

Vannier M, Marsh J, Warren J (1984). Three-dimensional CT reconstruction images for craniofacial surgical planning and evaluation. Radiology 150:179.

Wolpoff M (1969). Cranial capacity and taxonomy of Olduvai hominid 7. Nature 223:182.

Wolpoff M (1981). Cranial capacity estimates for Olduvai hominid 7. Am J Phys Anthropol 56:297.

Hominid Evolution: Past, Present and Future, pages 427–436
© *1985 Alan R. Liss, Inc.*

HIGH-RESOLUTION COMPUTED TOMOGRAPHY OF FOSSIL HOMINID SKULLS: A NEW METHOD AND SOME RESULTS

Frans W. Zonneveld, M.Sc.
Radiology Dept., Utrecht University Hospital,
P.O. Box 16250, Utrecht, The Netherlands

Jan Wind, M.D.
Institute of Human Genetics, Free University,
P.O. Box 7161, Amsterdam, The Netherlands

INTRODUCTION

Computed tomography (CT) is a diagnostic tomographic imaging modality (Hounsfield 1973), providing cross-sectional images which represent the X-ray attenuation coefficient of the tissues in the imaged plane. CT does not suffer from the superimposition of structures outside the plane of interest which is the case in traditional projection radiography and polytomography (Wind, Zonneveld 1985). This quality allows CT to visualize small density differences. The CT image is reconstructed by a computer from X-ray transmission data obtained during a scan (Zonneveld 1980a).

Since the technique's introduction, spatial resolution has improved from 3 mm to 0.5 mm. Also, the minimum slice thickness has been reduced from 8 mm to 1.5 mm. This, in turn, has reduced blurring resulting from the averaging of all details at one location but distributed along the thickness of the slice (partial volume averaging). CT appears to be a promising tool to study fossils (Wind 1984) since their internal structures, inaccessible to other techniques, can be assessed with sufficient detail in a non-destructive manner. And obviously, knowledge of intracranial morphology is of paleoanthropological interest. However, two problems remain: (i) the fossils' mineralization causes their X-ray attenuation coefficient to rise beyond the maximum value that is accounted for; and (ii) CT scanners are usually calibrated for patients, i.e. for soft tissue and not for fossils.

MATERIALS AND METHODS

All CT scans were performed on a Philips Tomoscan 310 or
350. A spatial resolution of 0.6 mm was used while the sharp-
ness of the image obtained was achieved by using a slice
thickness of 1.5 mm. Overall (24x24 cm) and zoom reconstructions (8x8 cm) were made. The scans were carried out in three
planes: sagittal, transverse (parallel to the nasion-biauricular plane), and coronal (perpendicular to the Frankfurt Horizontal). The temporal bones were completely imaged by a series of consecutive scans with a slice index of 1.5 mm.

The standard CT-numbers, which represent the linear attenuation coefficient, normally range between -1000 (air) and
+3095 (human tooth enamel) with intermediate values of 0 (water) and +1000 (cortical bone). However, when fossils are
mineralized to such a degree that the CT number rises above +
3095, the number drops back completely to -1000, and then
starts to rise again from thereon. This results in an image
which is difficult to interpret (Fig. 1a). We have, therefore,
modified the CT-number scale in such a way that the numbers'
range between -1000 and +7191 is transposed on the normal scale
(Fig. 2) thus avoiding discontinuity in the CT-number representation.

The linear attenuation coefficient is partially determined by the X-ray energy used. Consequently, as the X-rays pass
through the fossil, the low-energy rays are filtered out more
rapidly which raises the effective energy of the beam and thus

a   b

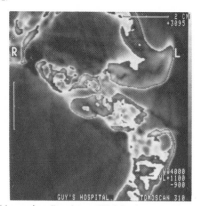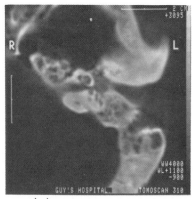

Fig. 1. Broken Hill 1 scan prior to (**a**) and after adaptation
(**b**) of the CT-number scale (see also Fig. 2).

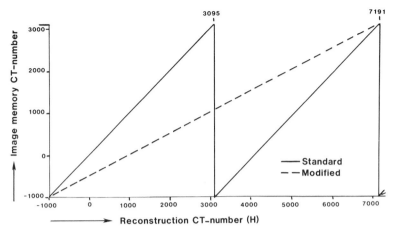

Fig. 2. Standard and modified representation of the CT-number scale which effectively results in a twofold range.

lowers the attenuation coefficient. In order to eliminate this effect, which is known as 'beam hardening', the scanner is usually calibrated for the type of tissue which is normally examined, such as muscle or fat. Plexiglass has been found suitable for representing these soft tissues and is, hence, normally used for calibration. However, when this material is used in calibration for scanning mineralized fossils, the image's centre appears to have a lower density than its edges due to a more severe beam hardening than was allowed for during calibration. This non-homogeneity is called 'cupping' and can be prevented by using calibration material having an atomic number roughly equivalent to that of the fossil itself. We found aluminium to be very suitable for this purpose.

The CT scanner allows the taking of normal projection radiographs (Scanogram) (Zonneveld 1980b) which facilitates positioning in the correct plane and also, subsequently, permits registration of the scan positions (Fig. 3).

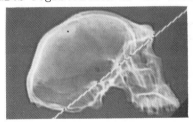

Fig. 3.
Lateral scanogram of Broken Hill 1 with a line indicating the plane of a transverse scan.

Ten fossil skulls have been examined in this way: Wadjak 1, Gibraltar 1, Broken Hill 1, Olduvai Hominid 9, Taung 1 (endo-cranium partly including the right petrous pyramid), STW 53, MLD 37/38, STS 5, SK 48, and TM 1517.

ASSESSMENT OF INTRACRANIAL MORPHOLOGY

Labyrinthine structures, such as the cochlea, vestibule, and semicircular canals, can usually be recognized in fossil hominid skulls as is the case in modern man (Zonneveld 1983). There are, however, a few basic differences, even if we compare the fossil with a modern dry skull. In the latter, bony struct-ures are displayed as white, while air-filled cavities are shown as black. In fossils, of course, these cavities were originally also filled with air, and for some this is still true (Broken Hill 1, STW 53) (Fig. 4). During fossilization in many others, however, a layer of mineral-rich material, such as calcite, has become deposited on the surface of these cavities. As a result, the walls of the vestibule, for example, are seen lined with a layer of calcite but contain air in the centre, while the semicircular canals appear completely filled with calcite (MLD 37/38, STS 5, Taung 1) (Fig. 5). The calcite is more dense than the bone and, therefore, appears as lighter on the CT-images. In some cases, a very narrow air space remains in the centre of somewhat wider canals such as the cochlear spiral (MLD 37/38, Taung 1) or the common crus; in others, the cochlea is also completely calcite filled (STS 5). By completely imaging the labyrinth with a series of con-secutive scans its lateral and superior views can be recon-structed (Fig. 6).

a   b

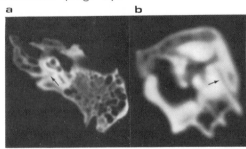
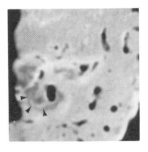

Fig. 4. STW 53. **a.** Transverse scan with the cochlear aqueduct (arrow). **b.** Sagittal scan demonstrating the vestibular aqueduct (arrow).

Fig. 5. Calcite-filled lateral semicircular canal in Taung 1 (ar-rowheads).

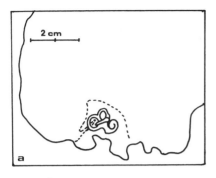
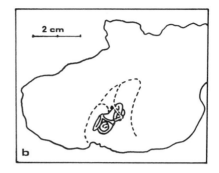

Fig. 6. Reconstruction of the labyrinth in the Taung 1 endocranium. **a.** Lateral view with the vestibular aqueduct. **b.** Superior view with the facial nerve canal.

In modern man, as in fossils, the labyrinthine capsule cannot readily be recognized in CT scans. However, in Gibraltar 1 we found a dark zone surrounding the cochlea which is also found in patients with cochlear otospongiosis (Damsma et al. 1984) (Fig. 7).

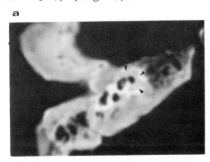
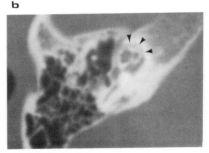

Fig. 7. Radiolucent zone around the cochlea (arrowheads). **a.** Gibraltar 1. **b.** Cochlear otospongiosis in modern man.

Of the other intratemporal structures, we found the vestibular and cochlear aqueducts in STW 53 (Fig. 4) and even, in spite of their endocranial matrix, which, until now, had prevented visualization, in Taung and in MLD 37/38 (Dart 1962).

Of the **middle ear structures,** ossicles have, until now, rarely been described in fossil hominids (Rak, Clarke 1979). We feel that, in the left epitympanic recess of Broken Hill, a malleus is present, not in its original position but rotated, as a result of which the head, neck and manubrium can simultaneously be seen (Fig. 8). The structure's length is 8 mm,

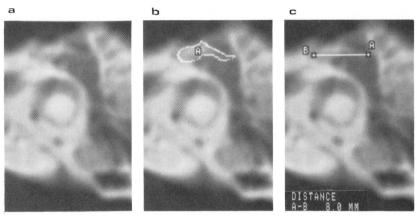

Fig. 8. Transverse scan of Broken Hill 1 demonstrating the dislocated malleus. **a.** Original. **b.** Outline. **c.** Size.

which corresponds with the modern malleus' size (Heron 1923). In Wadjak 1 we even recognized the tiny cavity of the stapedius muscle (Fig. 9).

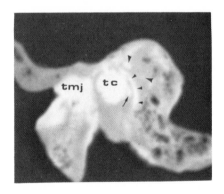

Fig. 9.
Sagittal scan of right temporal of Wadjak 1; mineral filling of some cavities but not of the mastoid air cells. Cavity of stapedius muscle (arrow); facial nerve canal (small arrowheads); lateral semicircular canal (large arrowheads); tympanic cavity (tc); temporomandibular joint (tmj).

The **frontal sinuses** can be readily recognized in OH9. The scans clearly demonstrate here the difference between bone and matrix density, thereby visualizing the actual outlines of the frontal sinuses, including the septum (Fig. 10).

The **external auditory canal** - usually matrix filled - can readily be recognized. Its bony part is mainly cylindrical with a 7 to 10 mm diameter in modern man, while, in Homo erectus (Pithecanthropus IV) (Wind 1984), in Australopithecus robustus (TM 1517), and in A. africanus (STS 5, MLD 37/38) it

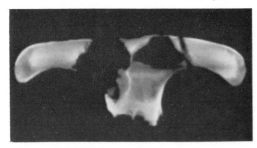

Fig. 10.
Coronal scan of OH9.
Note the difference
in density between
the bone and the
matrix, demonstrating
the outline of the
frontal sinuses and
the septum in between.

appears to have a funnel shape with its most medial diameter
being 7 to 9 mm and its most lateral diameter 12 to 15 mm (Fig.
11) - unlike Rak and Clarke (1979) who found this shape only in
A. robustus (SK 46, SK 848, SK 52, SKW 18) but not in A.
africanus (STS 25, STS 67).

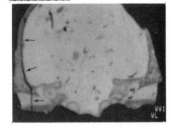

Fig. 11.
Coronal scan of MLD 37/38 demonstrating the matrix-filled endocranium
and funnel-shaped external auditory
canals. Note the separation line
between MLD 37 and 38 (arrows).

CT imaging allows **exact measurements** making it an appropriate tool for morphometry. In sagittal scans, for example,
we compared the diameter of the descending part of the facial
nerve canal in modern man, Broken Hill 1, and OH9, and found
values of 1.4, 1.7, and 2.7 mm respectively (Fig. 12). (Of
course, for an exact assessment cross-sectional diameters are
also required). Enamel thickness in A. robustus molars was
found to a maximum of 2.6 mm in the M3 of SK 48, and 3.3 mm in
the M2 of TM 1517 (Fig. 13).

The **angle** between the plane of the lateral semicircular
canal and other planes in the skull, e.g. the posterior surface
of the clivus, may be of importance in assessing function, as a
result of the relationship between the position of the head and
the canal with respect to the direction of gravitational force
(Girard 1923; Delattre 1951; Fenart, Empereur-Buisson 1970).
It has also been suggested that the head has a sort of gyro-
scopic axis (Beauvrieux 1934) which appears to be parallel to
the posterior surface of the clivus (Caix, Outrequin 1976) and

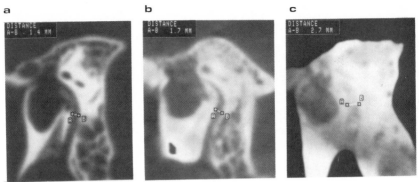

Fig. 12. Measurement of the antero posterior diameter of the facial canal in modern man (left), Broken Hill (centre), and OH9 (right).

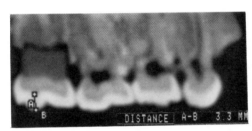

Fig. 13. Sagittal scan of P3, P4, M1 and M2 in TM 1517. Note the clear difference between enamel and dentine. A maximum enamel thickness of 3.3 mm was found in M2.

hence to the axis of the cervical spine. CT scanning can assist in determining the direction of the planes, the axis, and the Frankfurt Horizontal. The plane of the canal can be deduced from a sagittal scan by connecting the canal's anterior and posterior limbs (Fig. 14a and c). The nasion-biauricular plane is parallel to the plane of the canal (Fusté 1956). The plane of the posterior surface of the clivus can be assessed from a mid-sagittal scan (Fig. 14b and d). The Frankfurt Horizontal can be determined from a lateral scanogram (Fig. 3) or, from sagittal scans, by registration of the coordinates of the inferior orbital margin and the superior margin of the external porus. Table 1 summarizes the angles that we measured.

Density measurements of the fossils' tooth enamel demonstrated that its range is the same as that of modern man.

a   b   c   d

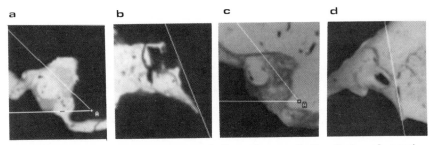

Fig. 14. Angle measurements of the plane of the lateral semi-circular canal and the posterior surface of the clivus. **a** and **b** show STS 5, **c** and **d** MLD 37/38.

Specimen	clivus/Frankfurt H.	clivus/lateral s.c. canal
STS 5	52°	25°
MLD 37/38	--	30°
Modern man	78°	53°

Table 1. Angles between the posterior surface of the clivus and, the Frankfurt Horizontal and the plane of the lateral semicircular canal respectively.

CONCLUSION

We have shown that, with a few adaptations, modern CT scanners allow assessment of the intracranial morphology of fossil hominid skulls. As even the small intratemporal structures, such as those of the labyrinth, can be distinguished and as quite exact measurements are possible, we feel that this CT scanning technique opens new perspectives for exact morphometry and densitometry in paleoanthropology.

ACKNOWLEDGEMENT

The authors are grateful to Dr. J. de Vos, Dr. C.B. Stringer, Prof. W. Maier, Prof. P.V. Tobias, Dr. C.K. Brain, Dr. R.J. Clarke and Dr. E.S. Vrba for allowing access to and transportation of the fossils, Dr. D.E. Ricklan, Dr. A. Turner, Mr. P.C. Panagos, and Mr. T. Gill for assistance, and Dr. R. Hoare, Prof. P.F.G.M. van Waes and Prof. A.T. Scher for providing examination time on the CT scanners.

REFERENCES

Beauvrieux J (1934). Recherches anatomiques sur les canaux semi-circulaires des vertébrés. Thesis. Bordeaux.

Caix M, Outrequin G (1976). "Les canaux semi-circulaires osseux". Monographie du College medical français des professeurs d'anatomie. Saint-Yrieix: Fabrègue.

Damsma H, de Groot JAM, Zonneveld FW, van Waes PFGM, Huizing EH (1984). CT of cochlear otosclerosis (otospongiosis). Radiol Clin North Am 22(1):37.

Dart RA (1962). The Makapansgat pink breccia australopithecine skull. Am J Phys Anthropol 20:119.

Delattre A (1951). "Du crâne animal au crâne humain." Paris: Masson.

Fenart R, Empereur-Buisson R (1970). Application de la methode "vestibulaire" d'orientation, au crâne de l'enfant du Pech de l'Azé, et comparaison avec d'autres crânes néandertalièns. Mém Arch Inst Paleont Hum. 33/3:89.

Fusté M (1956). Estudio comparativo sobre la rotacion de la región occipital en cráneos de Neandertal y Sapiens. Trabajos del Instituto Bernardino de Sahágún de Antrop. y Etnol., C.S.I.C., Bd XV, no. 1, Barcelona.

Girard L (1923). L'Attitude normale de la tête determinée par le labyrinthe de l'oreille. Soc Anthr Paris:79.

Heron IC (1923). Measurements and observations upon the human auditory ossicles. Am J Phys Anthropol 6:11.

Hounsfield GN (1973). Computerized transverse axial scanning (tomography) I. Description of system. Br J Radiol 46:1016.

Rak Y, Clarke RJ (1979). Aspects of the middle and internal ear of early South African hominids. Am J Phys Anthropol 51:471.

Wind J (1984). Computerized X-ray tomography of fossil hominid skulls. Am J Phys Anthropol 63:265.

Wind J, Zonneveld FW (1985). Radiology of fossil hominid skulls. In Tobias PV (ed.) "The past, present and future of hominid evolutionary studies". New York: Alan R. Liss.

Zonneveld FW (1980a). Computed tomography. Philips publication 422 984 53441.

Zonneveld FW (1980b). The scanogram: technique and applications. Medicamundi 25:25.

Zonneveld FW, van Waes PFGM, Damsma H, Rabischong P, Vignaud J (1983) Direct multiplanar computed tomography of the petrous bone. RadioGraphics 3:400.

Hominid Evolution: Past, Present and Future, pages 437–442
© *1985 Alan R. Liss, Inc.*

RADIOLOGY OF FOSSIL HOMINID SKULLS

Jan Wind, M.D.
Institute of Human Genetics, Free University,
P.O. Box 7161, Amsterdam, The Netherlands.

Frans Zonneveld, M.Sc.
Radiology Dpt., Utrecht University Hospital,
Catharijnesingel 101, Utrecht, The Netherlands

Until now, the internal morphology of fossil hominid skulls has remained largely unknown, because their proprietors do not generally appreciate the physical sectioning of the specimens. And even some of the usual external features may remain unknown after inspection, viz when they are covered by matrix; this applies equally to their spatial interrelationships because these may be only partly simultaneously visible, e.g. in one plane. Because these problems largely coincide with those faced by clinicians, and humankind so far has been ready to spend more resources on medicine than on paleoanthropology, clinical methods have been devised to meet these difficulties. The most widely used method is, of course, radiological examination. However, in the case of fossil hominid skulls this method has never, for reasons to be explained, become very popular.

Among the most intriguing - and at the same time least explored - parts of fossil hominid skulls are the temporal bones. Their morphology, for example, may be relevant for finding the origin of one of the most diagnostic properties of man: speech. While we lack, of course, recordings of extinct hominids' voices, we do find parts of the microphones of recorders that were, in fact, operating during the hominids' lifetime and that were adapted to their vocalizations (Wind 1976). These microphones, naturally, are the ears (the sound spectrum of which in most species coincides with the spectrum of their vocalizations - the latter being most intensively used for intraspecific communication). This means that if we could reconstruct the whole of the middle and inner ear of

fossil hominids, we could reconstruct several physical qualities of their hearing. Furthermore, the plane of the lateral
semicircular canal - having in most vertebrates the same position relative to the direction of the forces of gravity - can
be used as an indicator for the usual position of our ancestors' head in space (Fenart, Empereur-Buisson 1970). Third, the
size of the facial canal may reflect the size of the facial
nerve and, hence, the refinement of facial mobility. Some
other paleoanthropologically interesting features of internal
cranial morphology include (i) cranial base inclination which
assists in assessing airway anatomy (Laitman et al. 1979) and
taxonomy (Kimbel et al. 1984); (ii) paranasal sinus size which
may be a climatic adaptation; (iii) cranial capacity which can
be computed on the basis of cross-sectional areas; (iv) osseous and dental pathology.

What problems face one in implementing the imaging of
the internal morphology of fossil hominid temporal bones? In
addition to the usual difficulties (only few specimens are
available most of which show postmortem damage and are geographically widely distributed), distinguishing parts of
intratemporal structures is hampered by their complicated
pattern, their close proximity, their superimposition in
traditional radiographs, and their small size. The clinical
solution consists in using X-ray tomography in which slices
of the body part to be examined are imaged. This technique,
however, suffers from the contrast degradation by the superimposition of neighbouring structures, and is focused only in
the plane of interest. Therefore it is applicable only to
objects containing large contrasts, such as exist between
bone and soft tissue or air.

Tomography of fossil hominid skulls is often rather dis-
appointing, viz on account of the skulls' high degree of
mineralization and the filling of their cavities with matrix.
This turns the specimens radiographically into something like
one piece of stone with little contrast (Fig.1) - unless the
specimen has been acid-treated (Fig.2). Computed Tomography
(CT) (or CT scanning) appeared to be able to partly overcome
the difficulties (Wind 1980), because it provides a truly two-
dimensional representation of a cross-sectional plane without
the disturbing influence of superimposition. While the first
types of CT scanners provided a rather crude resolution (Fig.
3), the latest scanners provide an improved resolution (Zonneveld, Wind 1985)permitting the recognition of ossicles and
other small structures; a first report is given in Wind (1984).

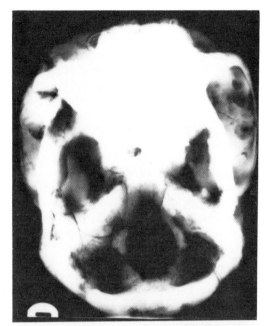

Fig.1.
A regular X-ray of the
Sts 5 skull (or "Mrs.
Ples"). The view is
from above. Mineral-
ization and superimpo-
sition hamper analyz-
ing intratemporal and
other structures.

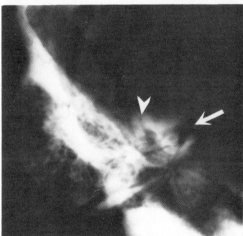

Fig.2.
X-ray (Stenvers inciden-
ce) of the left temporal
of SK 46. The specimen
has been acid-treated by
which minerals are washed
out and bony structures
are better visualized.
Arrow indicates internal
auditory meatus, arrow-
head superior semi-
circular canal.

Fig.3.
A coronal section of the Modjokerto infant
cranium (filled with matrix) made with a
first-generation CT scanner and showing the
crude picture elements not allowing the dis-
crimination of intratemporal structures.

We have recently introduced a new technique, by adapting the scanner, enabling the visualization of heavily mineralized skulls (Zonneveld, Wind 1985).

Having thus mainly resolved the imgaging difficulties, we are left only with the 'Mohammed-and-Mountain' problem: it is almost as difficult to move the skull as the CT-scanner (a machine which is quite heavy, can hardly be moved, and is only present in few, very well-equipped X-ray departments). Hence, implementing CT-scanning of fossil hominid skulls is largely a matter of logistics.   So far only a limited number of skulls has been investigated in this way.

Specimens that were examined with normal X-raying include: Taung 1, Sts 5 (or "Mrs. Ples"), SK 46, La Quina 5, and Modjokerto. Specimens that were examined with traditional tomography include Sts 5, SK 46, Pithecanthropus IV, and La Quina 5 (Figs. 4 and 5). Specimens that were CT-scanned with a non-

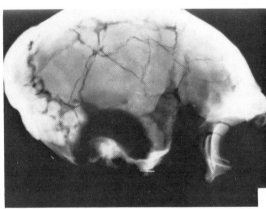

Fig.4.
A lateral normal X-ray of the neandertal La Quina 5. Insert represents a normal tomogram of the right temporal bone showing that in this, little mineralized, specimen labyrinthine details can be recognized.

Fig.5.
Coronal, normal, tomogram of the right petrous bone of La Quina 5. Arrow indicates the internal auditory meatus, arrowhead the superior semicircular canal.

adapted CT scanner include Pithecanthropus IV (Fig. 6), Modjokerto (Fig. 3), and La Quina 5. Skulls that were CT scanned
with the newest technique include Taung 1, Sts 5, SK 48,
Gibraltar 1 (Fig. 7), Broken Hill 1 (Fig. 8), Wadjak 1, Olduvai
Hominid 9, STW 53, MLD 37/38, and TM 1517. In addition, CT-
scans were made of chimpanzee, gorilla, and orang utan skulls.
For a detailed description of the CT-scanning of Pithecanthropus IV, see Wind (1984).

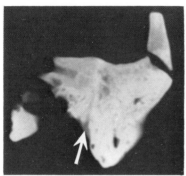

Fig. 6.
Coronal section of Rt temporal of Pithecanthropus IV.
It shows the vertical part
of the facial nerve canal.

Fig. 7.
Coronal section of the right
temporal of Gibraltar 1. It
shows the cochlear spiral in
a heavily mineralized area
(white); black spot indicates
CT number overflow (see
Zonneveld & Wind 1985).

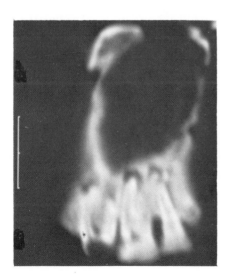

Fig. 8.
A sagittal CT scan of the
left maxilla of Broken
Hill 1. It shows the max-
illary sinus and teeth.
Around the latter's apices
bone has been resorbed
testifying to a periapical
infection, probably as a
result of the root canals
having been exposed due to
dental wear.

The results so far indicate that the temporal bones, including the labyrinth, have largely been similar to those in modern man. This indication is supported by comparative anatomy and physiology indicating that both the organ of balance and the inner ear - in their function as well as in their morphology - are largely similar in humans and apes. In conclusion, CT-scanning is able to provide more satisfactory imaging of the internal structures of fossil hominid skulls than methods used so far; and it may well become a standard craniometric tool in paleoanthropology.

ACKNOWLEDGEMENTS

We thank C.K. Brain, R.J. Clarke, E. Claus, J. Cook, E. Desmedt, Elscint Ltd., R. Fenart, P. Gerlings, H. Hacker, B.A. den Herder, R. Hoare, F. Hotton, A. Hughes, H.R. von Ilberg, H. Jackson, the late G. von Koenigswald, F. Kroon, W. Maier, R.D. Martin, G.R. Nep, E. Pantak, Philips U.K., Philips S.A., M. Plhak, M. Sakka, C. Stringer, P.V. Tobias, J.C. de Valois, J. Vignaud, J. de Vos, E.S. Vrba, G. Wheeler, and B.A. Wood.

REFERENCES

Fenart R, Empereur-Buisson R (1970). Application de la méthode "vestibulaire" d'orientation, au crâne de l'enfant du Pech de l'Azé, et comparaison avec d'autres crânes néandertaliens. Mém Arch Inst Paléont Hum. 33/3:89.

Kimbel WH, White TD, Johanson DC (1984). Cranial morphology of Australopithecus afarensis: a comparative study based on a composite reconstruction of the adult skull. Am J Phys Anthropol 64:337.

Laitman, JT, Heimbuch RC, Crelin, ES (1979). The basicranium of fossil hominids as an indicator of their upper respiratory systems. Am J Phys Anthropol 51:15.

Wind J (1976). Phylogeny of the human vocal tract. Ann N Y Acad Sci 280:612.

Wind J (1980). X-ray analysis of fossil hominid temporal bones. Antropol Contemp 3:299.

Wind J (1984). Computerized X-ray tomography of fossil hominid skulls. Am J Phys Anthropol 63:265.

Zonneveld F, Wind J (1985). High-resolution computed tomography of fossil hominid skulls: a new method and some results. In Tobias PV (ed.) "The past, present and future of hominid evolutionary studies". New York: Alan R. Liss.

Hominid Evolution: Past, Present and Future, pages 443–454
© *1985 Alan R. Liss, Inc.*

COMPARATIVE PRIMATE DENTAL ENAMEL THICKNESS: A
RADIODONTOLOGICAL STUDY

Geoffrey H. Sperber

Department of Oral Biology
University of Alberta
Edmonton, Alberta, T6G 2N8, CANADA

"The teeth more than any other single feature removes
Australopithecus from apes and assembles him with men. As
in the jaws of primitive men we find here the dentition of
a man in the jaws of an ape".   R.A. Dart (1934)

     Radiographic examination of mineralized tissues -
bones and teeth - provides a non-invasive technique of
analysis of the internal morphology of these organs that
can be as revealing as their external structure. This
form of investigation is particularly applicable to
irreplaceable fossil material that provides much of the
evolutionary record, and is most appropriate for dental
examination of both extant and fossilized forms. For it
is a peculiarity of teeth that their highly mineralized
nature - so susceptible to decay in modern man - is
pristinely retained in fossilized specimens as the only
original remnants of long-decayed bodies. In reality,
enamel is fossilized during life by becoming so heavily
mineralized during its ontogeny that it is the only part
of the body not replaced by minerals during fossilization.
Palaeoradiodontology, the radiographing of fossil teeth,
thus provides a means of direct comparison of living and
deceased dentitions.

     Much attention has been paid in the palaeo-
anthropological literature to variations of teeth as
whole organs (Ashton and Zuckerman 1950, 1951, 1952 a, b;
Bennejeant 1936; Broom 1929, 1939; Dart 1934; Robinson 1956;
Senyürek 1941; Sperber 1975; Weidenreich 1937;
Zuckerman 1951 a, b). Only recently have the individual

tissues constituting the teeth received consideration by
palaeo-anthropologists (Gantt 1977, 1979, 1983; Grine
1981; Molnar, Gantt 1977). In particular, the enamel,
being the most exposed of the dental tissues and the most
susceptible to dietary variation, has received detailed
attention as part of the recent surge of interest in
primate and early hominid diets (Gordon KD 1984; Gordon KR
1984; Jolly 1970; Puech 1984; Puech et al 1983; Rensberger
1973; Teaford, Walker 1984; Walker 1981; Walker et al
1978; Wallace 1973, 1975). Dental enamel, being the
product of a rigidly genetically controlled set of cells -
the ameloblasts - can be a very specific interpreter of
the genotype of a species and of the selective pressures
to which they are subjected. Apart from enamel histology -
which constitutes a study in itself - enamel thickness,
both in its pristine form and in its later worn condition,
can be very revealing of the habits and habitat of various
exemplars of hominid evolution (Hand 1968; Huszar 1971).

METHODS

In this study the author measured dental enamel
thickness on radiographs of samples of fossil hominids,
extant apes and modern man. The limitations of accuracy
of this indirect form of measurement of a minute
dimension are well recognized - from variations arising
from distortion by magnification to different X-ray
angulations casting variable "thicknesses" of enamel
shadows; but since a standardized procedure of radiography was practised wherever possible, comparisons of
similarly derived shadows were felt to have some validity.
The radiographic technique consisted in applying
periapical dental radiographic plates as closely to the
teeth as possible and directing X-rays at right angles to
the imaginary plane bisecting the vertical axis of the
tooth and the plane of the radiographic plate, a standard
dental radiographic procedure. The amount of radiation
exposure was consistent with obtaining the sharpest
photographic image possible, and varied with the radio-
density of the material. Heavily fossilized specimens
required the greatest exposure, while modern material
required the least.

Maximum and minimum thicknesses of enamel on the
occlusal surfaces only were measured at least 3 times,
with the mean of each of these dimensions reported as
representative.

MATERIAL

The enamel of postcanine teeth was studied in three primate groups: fossil hominids, extant apes and modern man.

The extensive collection of radiographs taken by Sperber (Skinner, Sperber 1982) of early South African hominids, specifically the australopithecines, was used. Because it was necessary to measure unworn enamel, the heavy attrition of most australopithecine dentitions precluded measurement of the enamel of the majority of these hominids, accounting for the small numbers cited. Teeth showing any evidence of occlusal wear were excluded, thus greatly reducing the available number of teeth and confining enamel measurement, in many instances, to unerupted teeth.

The small numbers of extant apes included were chosen for their unworn dentitions, greatly reducing the sample size. The extraordinary thinness of gorilla enamel revealed by this study diminished the accuracy of minimum thickness measurements, accounting for their deletion from the results.

The samples of modern human dentitions included in this study were drawn from radiographs of patients attending the University of Alberta Dental School Clinic. Measurements were confined largely to unerupted teeth.

RESULTS

Data for the three primate groups are revealed in Tables 1 and 2. The maximum and minimum measurements cited represent the means of these two dimensions for each of the samples.

The enamel of A. robustus was generally thicker than that of A. africanus; both outranked modern man's enamel thickness, which in turn exceeded that of the extant apes.

DISCUSSION

Degeneration of ameloblasts that deposit enamel prior to dental eruption determines the end of enamel deposition, and, moreover, precludes any natural repair of this

TABLE 1
<u>Greatest & Least Occlusal Surface Enamel Dimensions (in mm)</u>

	n	MAX.	± SD	MIN.	± SD	MIDRANGE

Tooth P$_3$

A. africanus Sterk.	1	1.50	----	0.60	----	1.05
A. robustus Swtkrns	3	3.00	.721	1.83	.586	2.42
H. sapiens	12	1.33	.249	0.78	.295	1.05
Pan troglodytes	2	0.85	.071	0.65	.071	0.75
Gorilla gorilla	1	0.40	----	----	----	----

Tooth P$_4$

A. africanus Taung	1	2.80	----	1.40	----	2.10
A. africanus Sterk.	1	2.30	----	1.00	----	1.65
A. robustus Swtkrns	8	2.81	.617	1.73	.682	2.27
H. sapiens	12	1.46	.168	0.88	.234	1.17
Pongo pygmaeus	1	1.70	----	0.80	----	1.25
Pan troglodytes	3	0.70	.100	0.63	.058	0.67
Gorilla gorilla	3	0.90	.100	----	----	----

Tooth M$_1$

A. africanus Taung	1	2.60	----	1.10	----	1.85
A. africanus Sterk.	1	3.00	----	1.40	----	2.20
A. robustus Swtkrns	5	3.54	.391	1.82	.691	2.68
H. sapiens	12	1.88	.312	0.96	.375	1.42
Pongo pygmaeus	4	1.75	.129	0.80	.082	1.28
Pan troglodytes	6	1.20	.395	0.55	.055	0.88
Gorilla gorilla	1	1.00	----	----	----	----

Tooth M$_2$

A. africanus Sterk.	3	2.50	.265	1.17	.153	1.83
A. robustus Swtkrns	6	3.08	.508	1.57	.266	2.32
H. sapiens	12	2.02	.313	1.18	.347	1.60
Pongo pygmaeus	2	1.65	.071	0.70	----	1.18
Pan troglodytes	6	0.85	.152	0.53	.121	0.69
Gorilla gorilla	4	1.43	.435	----	----	----

Tooth M$_3$

A. africanus Sterk.	4	2.58	.287	1.60	.216	2.09
A. robustus Swtkrns	3	3.14	.643	1.20	.265	2.17
H. sapiens	12	2.32	.260	1.40	.274	1.86
Pongo pygmaeus	3	1.50	.100	0.73	.058	1.12
Pan troglodytes	5	0.96	.207	0.50	0.71	0.73
Gorilla gorilla	3	1.33	.252	----	----	----

TABLE 2
Greatest & Least Occlusal Surface Enamel Dimensions (in mm)

	n	MAX.	+ SD	MIN.	+ SD	MIDRANGE
Tooth P^3						
A. africanus Sterk.	4	2.03	.330	1.20	.739	1.61
A. robustus Swtkrns	2	2.95	.778	1.75	1.20	2.35
H. sapiens	12	1.74	.168	1.03	.347	1.39
Pan troglodytes	1	0.60	----	0.40	----	0.50
Gorilla gorilla	1	0.60	----	----	----	----
Tooth P^4						
A. africanus Sterk.	1	2.90	----	2.10	----	2.50
A. robustus Swtkrns	2	3.00	.566	1.60	.849	2.30
H. sapiens	12	1.75	.268	1.16	.183	1.45
Pan troglodytes	3	0.86	.153	0.53	.115	0.70
Gorilla gorilla	3	0.97	.153	----	----	----
Tooth M^1						
A. africanus Taung	1	2.40	----	1.40	----	1.90
A. africanus Sterk.	2	2.55	.495	1.15	----	1.85
A. robustus Swtkrns	2	2.70	.707	1.30	.990	2.00
H. sapiens	12	1.93	.299	1.16	.340	1.55
Pongo pygmaeus	3	1.70	.200	0.77	.115	1.23
Pan troglodytes	4	0.85	.238	0.55	.058	0.70
Gorilla gorilla	3	1.13	.321	----	----	----
Tooth M^2						
A. africanus Sterk.	2	2.80	.283	1.05	.495	1.93
A. robustus Swtkrns	1	3.10	----	1.60	----	2.35
H. sapiens	12	2.28	.234	1.23	.177	1.75
Pongo pygmaeus	2	1.30	----	0.75	.212	1.03
Pan troglodytes	8	0.96	.192	0.58	.149	0.77
Gorilla gorilla	4	1.30	.294	----	----	----
Tooth M^3						
A. africanus Sterk.	1	3.10	----	1.90	----	2.50
A. robustus Swtkrns	4	3.83	.544	2.13	.386	2.98
H. sapiens	12	2.18	.289	1.08	.349	1.63
Pongo pygmaeus	3	1.77	.208	1.13	.058	1.45
Pan troglodytes	4	1.05	.300	0.58	.050	0.81
Gorilla gorilla	4	1.18	.386	----	----	----

Figure 1:   Taung (<u>A. africanus</u>)

Skiagram of right maxillary first molar (centre) and second deciduous molar (right and partially obscured). The hypotaurodont pulp chamber, the thick dentinal wall and thick enamel layer are evident.

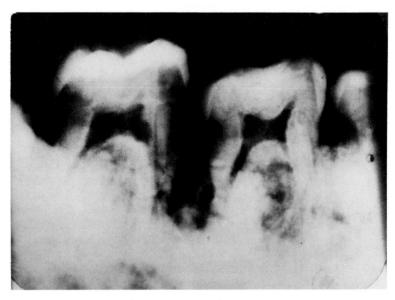

Figure 2:   Sts 52b (<u>A. africanus</u>)

Skiagram of mandibular right first and second molars revealing the worn enamel on the first molar (right) and the thick unworn enamel on the second molar (left). The hypotaurodont pulp chambers are conspicuously clear.

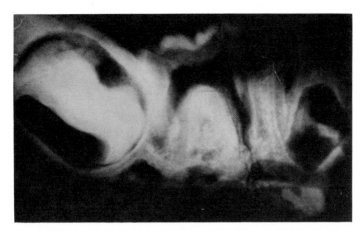

Figure 3:   SK 64 (<u>A. robustus</u>)

Skiagram of child's right mandible revealing the deciduous molars (to the right) and the unerupted first molar (to the left). The moderately thick enamel on the second deciduous molar and the very thick enamel on the forming crown of the first molar are evident. The large hypotaurodont pulp chamber of the second deciduous molar is noteworthy.

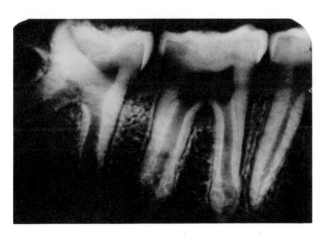

Figure 4:   Za 93 <u>Pongo</u> <u>pygmaeus</u> (orang-utan)

Skiagram of right mandibular molars revealing slightly worn enamel on the first and second molars and the moderate thickness of the unworn enamel on the third molar. The hypotaurodont pulp chambers resemble those of the australopithecines.

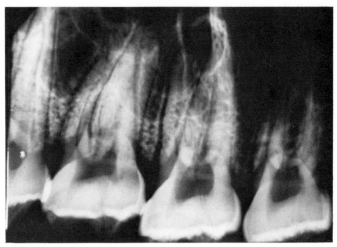

Figure 5: Za 1041 <u>Pan troglodytes</u> (chimpanzee)

Skiagram of left maxillary molar teeth revealing the thin enamel layer covering the thick dentine, surrounding moderately large hypotaurodont pulp chambers.

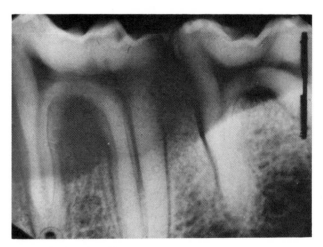

Figure 6: Za 1290 <u>Gorilla gorilla</u>

Skiagram of the enormous mandibular second and third molars, revealing the extraordinarily thin enamel layer. The attenuated cynodont pulp chambers and the extraordinarily long roots relative to the constricted crowns are noteworthy.

specialized, highly mineralized tissue (Ten Cate 1985).
Determination of the thickness of enamel is the
consequence of genetic programming of the life and death
of ameloblasts, that ultimately is responsive to the
evolutionary pressures based upon the utility of enamel.

The enamel thicknesses in Tables 1 and 2 indicate a
general gradation from thickest to thinnest in the
following order: A. robustus, A. africanus, H. sapiens,
Pongo pygmaeus (orang-utan), Pan troglodytes (chimpanzee)
and G. gorilla. Apart from Pan troglodytes, the order of
increasing enamel thickness is inversely related to the
general body size of these primates. This curious
negative allometric relationship of enamel thickness and
body mass invites conjecture on its significance. The
duration of odontogenesis, the age of full dental eruption,
the longevity of the species and the diet are all factors
of significance in determining dental enamel thickness
(Gantt 1977).

While the quality of enamel is undeniably a factor
in determining the rate of attritional wear of this
tissue, the abrasivity of the diet and the longevity of
the species are the final determinants of enamel persistence on the occlusal surfaces (Walker et al 1978). Once
the enamel has been worn away, dental degeneration ensues
fairly rapidly due to the softer nature of underlying
dentine and the possible onset of dental caries.

The thick australopithecine enamel may be correlated
with their presumed increase in longevity over extant apes.
The extremely worn teeth found even in adolescent
australopithecine specimens (by virtue of unerupted third
molars) suggest a highly abrasive diet, the nature of
which is the subject of much study (Wallace 1973, 1975).
The thin enamel of the chimpanzee and gorilla has been
correlated with their highly folivorous diet, while the
thicker enamel of the orang-utan has been correlated with
its frugivorous diet (Kay 1977). The diet of Australopithecus remains speculative, but microwear studies
suggest that it is neither exclusively graminivorous nor
carnivorous (Grine 1981).

Understanding the feeding habits of the australopithecines would allow profound insights into their body size,
locomotor behaviour, ecological niches, breeding

strategies and social milieu. One of the rarest of the
body's tissues, dental enamel is capable of reflecting all
these extraordinarily diverse aspects of primate life and
evolution.

ACKNOWLEDGEMENTS

Appreciation is expressed to Virginia Putnam and
Christopher Barratt of the University of Wales Dental
School for their contributions to this study and to
Jennifer Leckelt for her typing of the manuscript.

REFERENCES

Ashton EH, Zuckerman S (1950). Some quantitative dental
    characters of fossil anthropoids. Philos Trans R Soc
    Lond 234B:485.
Ashton EH, Zuckerman S (1951). Some dimensions of the
    milk teeth of man and the living great apes. Man 51:23.
Ashton EH, Zuckerman S (1952a). Overall dental dimensions
    of hominoids. Nature 169:571.
Ashton EH, Zuckerman S (1952b). The measurement of
    anthropoid teeth, with particular reference to the
    deciduous dentition. Man 52:64.
Bennejeant Ch (1936). La dentition de l'Australopithecus
    africanus (Dart). Mammalia 1:8.
Broom R (1929). Note on the milk dentition of Australo-
    pithecus. Proc Zool Soc Lond 1929:85.
Broom R (1939). The dentition of the Transvaal Pleisto-
    cene anthropoids, Plesianthropus and Paranthropus.
    Annals Transv Mus 19:303.
Dart RA (1934). The dentition of Australopithecus
    africanus. Folia Anat Japon 12:207.
Gantt DG (1977). Enamel of primate teeth: its thickness
    and structure with reference to functional and phyletic
    applications. Ph.D. dissertation, Washington Univ,
    St. Louis.
Gantt DG (1979). Taxonomic implications of primate dental
    tissues. J Biol Buccale 7:149.
Gantt DG (1983). The enamel of Neogene hominoids:
    structural and phyletic applications. In Ciochon RL,
    Corruccini RS (eds): "New Interpretations of Ape and
    Human Ancestry", New York: Plenum Press.
Gordon KD (1984). Hominoid dental microwear: complica-
    tions in the use of microwear analysis to detect diet.
    J Dent Res 63:1043.

Gordon KR (1984). Microfracture patterns of abrasive wear striations on teeth indicate directionality. Am J Phys Anthropol 63:315.

Grine FE (1981). Trophic differences between "gracile" and "robust" australopithecines: a scanning electron microscope analysis of occlusal events. S Afr J Sci 77:203.

Hand WL (1968). Enamel dentin thickness of young maxillary permanent incisors. J N Carolina Dent Soc 51:30.

Huszar G (1971). Observations of the thickness of the enamel. Bull Group Int Rech Sc Stomat 14:155.

Jolly CJ (1970). The seed eaters: a new model of hominid differentiation based on a baboon analogy. Man 5:5.

Kay R (1977). Diets of early Miocene African hominoids. Nature 268:628.

Molnar S, Gantt DG (1977). Functional implications of primate enamel thickness. Am J Phys Anthropol 46:447.

Puech PF (1984). Acidic food choice in Homo habilis at Olduvai. Curr Anthropol 25:349.

Puech PF, Albertini H, Serratrice C (1983). Tooth microwear and dietary patterns in early hominids from Laetoli, Hadar and Olduvai. J Hum Evol 12:721.

Rensberger JM (1973). An occlusion model for mastication and dental wear in herbivorous animals. J Paleont 47:515.

Robinson JT (1956). The dentition of the Australopithecinae. Transvaal Mus Mem 9, Pretoria:Transvaal Mus.

Senyürek MS (1941). The dentition of Plesianthropus and Paranthropus. Annals Transv Mus 20:293.

Skinner MF, Sperber GH (1982). "Atlas of Radiographs of Early Man". New York: Alan R Liss.

Sperber GH (1975). "Morphology of the Cheek Teeth of Early South African Hominids". Ann Arbor: Xerox University Microfilms.

Teaford MF, Walker A (1984). Quantitative differences in dental microwear between primate species with different diets and a comment on the presumed diet of Sivapithecus. Am J Phys Anthropol 64:191.

Ten Cate AR (1985). "Oral Histology", 2nd ed. St. Louis: Mosby.

Walker A (1981). Diet and teeth. Philos Trans R Soc Lond B 292:57.

Walker A, Hoeck HN, Perez L (1978). Microwear of mammalian teeth as an indicator of diet. Science 201:908.

Wallace JA (1973). Tooth chipping in the australopithecines. Nature 244:117.

Wallace JA (1975). Dietary adaptations of Australopithecus
and early Homo. In Tuttle HR (ed): "Paleoanthropology,
Morphology and Paleoecology". The Hague: Mouton,
pp 203-233.

Zuckerman S (1951a). Comments on the dentition of the
fossil australopithecines. Man 51:20.

Zuckerman S (1951b). The dentition of the Australopithecinae. Man 51:32.

Sizing Up Brains and Human Nature

Hominid Evolution: Past, Present and Future, pages 457–464

QUANTITATIVE MODELS ON PHYLOGENETIC GROWTH OF THE HOMINID BRAIN

O.-J. Grüsser and L.-R. Weiss

Department of Physiology, Freie Universität, Arnimallee 22, 1000 Berlin 33, Berlin (West), Germany

INTRODUCTION

The human brain can be considered as the result of a 'biohistorical' process starting about 4 to 5 million years ago, when hominids ancestral to <u>Homo sapiens</u> deviated from the developmental branch of other higher primates. Despite the uncertainties which definitely exist about dating and exact values of cranial capacity CC and body height H or body weight W, the data available from palaeoanthropological research can be treated as randomly found samples representing the phylogenetic growth of the human brain. Several quantitative models describing the phylogenetic growth function of the hominid brain have been discussed. Lestrel and Read (1973) and Lestrel (1975), using estimates of CC and the age of fossil hominids, plotted the data on double logarithmic diagrams and found a non-linear relationship between CC and the geological age (time past T_p). From the strong reduction in the slope in their diagram they concluded that about 300,000 years ago a 'stasis' in further brain growth began. Considering the log-log transformation of these data, this statement is, of course, misleading. When an anthropologist living half a million years from now would plot the data in the manner Lestrel did for his time T_p, he would be led to a totally different conclusion.

Cronin et al. (1981) used a simple linear regression function to describe the relationship between CC and T_p:

$$CC = 1376.6 - 352\,T_p \quad cm^3 \qquad (1)$$

whereby T_p is expressed in units of millions of years. One of the present authors (Grüsser 1982) used a hyperbolic function:

$$CC = (1.658 \times 10^9)/(T_p + 1.128 \times 10^6) \quad cm^3 \qquad (2)$$

whereby T_p is expressed in units of years.

Hofman (1983) applied an exponential function

$$CC = 1446\,e^{0.41\,T_p} \quad cm^3 \qquad (3)$$

to describe a similar data base as that used in the present report. T_p is in units of million of years.

Eqs. (1) to (3) assume a continuous temporal growth, i.e. functions which are not particularly satisfactory, since unlimited phylogenetic growth of an organism or an organ goes against biological evidence. In the following we have therefore included a formalism which automatically limits phyloge-netic brain growth. The present paper is the extension of a talk given by one of the authors at the Nice International Congress of Palaeoanthropo-logy (Grüsser 1982) and a paper presented at the Xanthi Palaeoanthropo-logical Symposium (Grüsser and Weiss 1983).

MATERIAL AND METHODS

From the relevant literature, estimates on dating, average brain weight or endocranial volumes and the body height of different hominids and pre-sent-day man were collected (Tab. 1, Holloway 1968, 1970, 1974, 1982; Tobias 1971, 1979; Cronin et al. 1981; Hofman 1983; Pakkenberg and Voigt 1964). Relative brain size or relative cranial capacity RC was com-puted by dividing the absolute values by body height H as stated in Szil-vassy and Kentner (1978), since the following allometric function was found between brain weight CC and body height from the data of adult Danes published by Pakkenberg and Voigt (1964):

$$\log CC = .711 \log H + 1.569 \qquad (4)$$

while no systematic allometric function exists for the same CC-data cor-related with body weight.

For computing the functions described in RESULTS a digital computer (HP 1000) was applied. More recently Dr. K. Jacobs from the Dept. of Anthropology, University of Texas at Austin, provided us with a list of individual data on CC and T_p from 92 specimens from <u>Australopithe-cus afarensis</u> to <u>Homo sapiens</u>. Eq. (5) was tested with this data base and some of the results are included in the present report (for details see Grüsser, Jacobs and Weiss 1985).

RESULTS

Models of phylogenetic hominid brain growth

Our new phylogenetic brain-growth function contains limiting factors. We have combined two exponential growth functions, one representing all phylogenetic factors leading to an increase in brain size, the other lumping together all factors limiting the phylogenetic growth of the hominid brain. When these exponential growth functions were coupled in a multiplicative manner, a very good description of the data base was obtained:

$$CC = a + c \exp\left[\alpha(4.5 - T_p)\right] / (1 + d \exp\left[\beta(4.5 - T_p)\right]) \; cm^3 \quad (5)$$

a, c, d, α and β are constants.

Fig. 1 shows results with the data base of Tab. 1. Since the data on the average cranial capacity of modern man seemed to be more reliable than

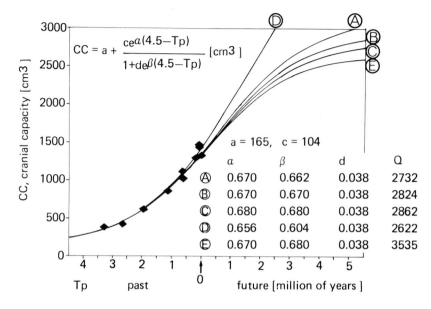

Fig. 1 : Brain growth function (CC, eq. 5). Curve A fits data best (Tab. 1).
Curve D without weight on data from the present.

Tab. 1 Data (averages) used for model computation of phylogenetic growth of hominid brain

| | CC, cranial capacity |cm3| | | | Tp, time past, | H, estimated average | RC, relative cranial |
	mean	estimated	95 percent limits	millions of years	body height |cm|	capacity |cm2|; CC/H
Australopithecus afarensis	401			3.3	140	2.864
A. africanus	450	385	525	2.7	145	3.103
Homo habilis	646	420	875	1.95	155	4.168
H. erectus er.	883	640	1130	1.1	162	5.451
H. erectus pekin.	1043	730	1355	0.6	162	6.440
H. soloensis	1135	1055	1283	0.63	162	7.006
H. praeneand. Eur.	1314	1170	1460	0.2	165	7.964
H. sap. neanderth.	1487	1300	1641	0.055	165	9.012
H. sap. europ. Wu.	1460	1293	1748	0.04	170	8.588
H. sapiens sap.	1345	955	1850	0	172	7.820

Tab.2. Increase in brain weight (CC/1.08) and number of cortical nerve
cells per generation (25 years) according to the different brain growth
functions. 45 000 nerve cells per mg neocortex;weighting 0.6.

equation	1	2	3	5 curve A	
weight	8.1	30.2	13.7	10.8	mg/generation
nerve cells	218	815	370	292	$\times 10^3$ nerve cells/generation

those available on our ancestral candidates, in computing the curves of Fig. 1 the average cranial capacity of recent man (1345 cm^3, Voigt and Pakkenberg 1964) was given the weight 7, while all other data of Tab. 1 were applied with the weight 1. Curve A corresponds to the optimal fit, but curves B and C located below the optimally fitting curve are also realistic descriptions of the experimental data. Thus from the data representing the last 3 million years of hominid brain development a maximum growth to less than 3,000 cm^3 on the average would be predicted for the next 4 million years of hominid brain development. When the constant α in eq. (5) is larger than the constant β, brain growth is not limited (curve D in fig. 1). This is, however, an unreasonable biological assumption. When the constant β is larger than the constant α, the brain growth function goes through a maximum. This also does not seem very probable in the light of other phylogenetic evidence (curve E).

Changes in relative brain size

Part of the phylogenetic increase in CC is caused by an increase in body height H, but the relative brain size also increased during the last 3 million years. For computation of relative cranial capacity RC we used an equation for RC corresponding to eq. (5)

$$RC = a' + c' \exp\left[\alpha'(4.5 - T_p)\right] / (1 + d' \exp\left[\beta'(4.5 - T_p)\right]) \quad cm^3 \quad (6)$$

Fig. 2 shows the function best fitting the data base of Tab. 1. It is interesting that in computing eq. (6) optimum values were obtained when $\alpha' = \beta'$.

Computation with individual data

The data base provided by Dr. Jacobs on individual specimens (listed in Grüsser, Jacobs, Weiss 1985) was used in a further step of computation. Hereby uncertainties increase even more, of course, than when average values are used and the individual weight given to the specimens representing different phylogenetic periods depends on the number of specimens found. Therefore Homo sapiens CC-values younger than 150,000 years had a greater impact on the computation results than early hominids. Since Homo sapiens neanderthalensis had higher CC-values than modern man, eq. (5) led to a much stronger accelerating function than for the average data, even when the weight of the CC-values for modern man was attributed a factor 30 (fig. 3 A). This was also the case when all Neandertal data were deleted (fig. 3, curve B). Nevertheless, the function suggested by the average data is still a good fit (fig. 3, curve C).

Predictions on human brain growth in the near future

Assuming that no dramatic changes occur in the average human living conditions on earth in the near future - an assumption which is perhaps unrealistic considering the danger of world-wide nuclear war - we computed the predictions for human brain growth in the near future from the phylogenetic growth functions (eqs. 1, 2, 3, 5) by linear interpolation between computed values of time past T_p = 50,000 years and future time T_f = 50,000 years. In addition, the data of average brain growth per generation were computed, whereby an average generation change every 25 years was arbitrarily chosen. Assuming that further phylogenetic growth in the human brain would be caused predominantly by an enlargement in cortical grey matter, an estimated 60 % of the growth would be due to an increase in the number of cortical nerve and glial cells and the remaining 40 % to an increase in axon space, subcortical glial cell space and extracellular fluid space. Then one can predict how many cortical nerve cells each generation would gain on the average in the near future (tab. 2).

Continuous or discontinuous brain growth functions?

The idea of 'stasis' and punctuated rapid increase in brain growth (Gould and Eldridge 1977) can be tested on the basis of individual data available. If this idea was correct, the differences in CC-data from historical neighbours in the samples should lead to clustered peaks when plotted as a function of T_p. This was not the case. Thus the conjecture of a sequence of phylogenetic periods of stasis and a punctuational graduated increase in brain growth can be refuted or at least cannot be deduced from the present data base.

DISCUSSION

Overall brain weight as well as relative brain weights are, of course, rather dull standards of measurements when one wishes to understand the phylogenetic growth of human brain. Therefore the present paper cannot provide more than a simple approach to a quantitative description of an existing data base. This base is - compared to other data available on human brain functions - rather fuzzy and includes errors of at least 20 percent. Certainly much more important than the overall increase in brain size is the phylogenetic appearance of new cytoarchitectonic neocortical areas and the corresponding transformation of corpus callosum connections and thalamic relay nuclei. Especially important for the difference in the brain of man as compared to other living primates are the development and the enlargement of isocortical integration areas in prefrontal-orbital, parietal and temporal cortical regions. In addition, neuropsychological findings obtained from brain lesional patients (cf. Grüsser, Grüsser-Cornehls, Kiefer and Kirchhoff 1985) indicate that the man-specific functions are characterized by a very strong lateralisation to one of the two hemispheres. This is true not only for the isocortical integration areas used by modern man for language (in particular area 6, 40, 41, 42, 44), but also for constructive praxia, tool-making, musical rhythms, melodies and harmony, reading, writing, graphical design, painting, chess playing, etc.

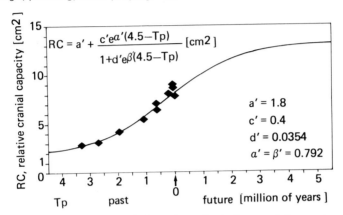

Fig. 2: Phylogenetic growth function of relative cranial capacity RC.

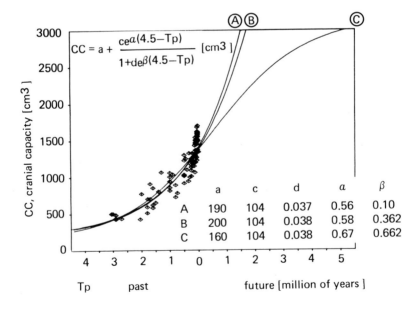

Fig. 3: Phylogenetic growth function of cranial capacity CC, data from individual specimens.

SUMMARY

Despite the fact that palaeoanthropological data on brain and body size of hominids are rather provisional, enough material is available to apply quantitative models of phylogenetic brain growth. Eqs. (5) and (6) were proposed for describing cranial capacity CC or relative cranial capacity RC as a function of time past T_p. In both equations facilitating and inhibiting phylogenetic factors were combined, leading to the prediction of a further increase in CC and RC until a plateau or maximum is reached in the far-distant future. It should be emphasized, however, that CC and RC represent only a rather crude aspect in the complex phylogenetic evolution of the human brain, which deviates from the brain of the other primates in the development of new isocortical cytoarchitectonic areas and in a strong functional left-right asymmetry for man-specific higher brain functions.

REFERENCES

Cronin JE, Boaz NT, Stringer CB, Rak Y (1981). Tempo and mode in hominid evolution. Nature 292:113.

Gould SI, Eldridge N (1977). Punctuated equilibria: tempo and mode of evolution reconsidered. Paleobiol 3:115.

Grüsser O-J (1982). The phylogenetic development of the human brain. Congrès International de Paléontologie Humaine Resumées des Communications. p.216.

Grüsser L-J, Weiss L-R (1983). A quantitative description of the phylogenetic growth of the hominid brain. Xanthi Symposium of Palaeoanthropology (in press).

Grüsser O-J, Jacobs KH, Weiss L-R (1985). The phylogenetic growth of the hominid brain: quantitative models. Brain Behav Evol (submitted)

Grüsser O-J, Grüsser-Cornehls U, Kirchhoff N, Kiefer R (1985). Averbal social communication of hominids is phylogenetically older than verbal communication. Some evidence from studies in brain lesioned patients. Abstract Taung Diamond Jubilee International Symposium, Johannesburg.

Hofman MA (1983). Encephalization in hominids: evidence for the model of punctuationalism. Brain Behav Evol 22:102.

Holloway RL Jr (1968). The evolution of the primate brain: some aspects of quantitative relations. Brain Res. 7:121.

Holloway RL Jr (1970). New endocranial values for the australopithecines. Nature 227:199.

Holloway RL (1974). The casts of fossil hominid brains. Scient Am 231:106.

Holloway RL (1982). Homo erectus brain endocasts: volumetric and morphological observations with some comments on cerebral asymmetries. Congrès Internationale de Paléontologie

Humaine, Nice, Vol. 1:355.

Lestrel PE (1975). Hominid brain size versus time: revised regression estimates. J Hum Evol 5:207.

Lestrel PE, Read DW (1973). Hominid cranial capacity versus time: a regression approach. J Hum Evol 2:405.

Pakkenberg G, Voigt J (1964). Brain weight of the Danes. Forensic material. Acta Anat 56:297.

Szilvassy J, Kentner D (1978). "Anthropologie. Entwicklung des Menschen, Rassen des Menschen". Naturhistorisches Museum Wien 150p

Tobias PV (1971). "The Brain in Hominid Evolution". New York: Columbia University Press.

Tobias PV (1979). "Evolution of human brain, intellect and spirit". Adelaide, South Australia: University of Adelaide. 70p.

Supported in part by a grant of the Deutsche Forschungsgemeinschaft (Gr. 161)

Hominid Evolution: Past, Present and Future, pages 465–473
© *1985 Alan R. Liss, Inc.*

THE STUDY OF HUMAN EVOLUTION AND THE DESCRIPTION OF HUMAN NATURE

Leonard O. Greenfield, Ph.D.

Department of Anthropology
Temple University
Philadelphia, PA 19122

Traditionally, students of human evolution have tried to discover the evolutionary events in our lineage since its divergence from the lineages leading to Gorilla and Pan. Their intention has been to find evidence to test the Darwinian hypothesis (1859, 1871) about our species' origin, to reconstruct the paleobiology of extinct hominoids, and to supplement the data base of the social sciences in their attempt to describe human nature.

On this last intention, Washburn (1983) wrote

I do not claim that evolutionary studies will give final answers to the question of human nature, only that such investigations will alert us to new problems and help us to see unsuspected facts and biases.

If this is an accurate assessment, and I believe it is, one could conclude from this quotation that currently the study of human evolution plays a limited role in the construction of a description of human nature.

Perhaps many reasons could be put forth to argue why the study of human evolution has had and can have only a limited impact on a description of human nature. For example, a contributing factor might be that relatively few people are doing this kind of work.

However, the major reason for the limited impact evolutionary studies have had and can have on a description of human nature is that those who try to make statements about human nature concentrate on describing the novel and unique evolutionary changes in our lineage over the past few million years (for example, the acquisition of language, tool use, sharing and cooperation). I believe that this is an inadequate approach to the phenomena about which they wish to make statements.

Perhaps most workers would agree that the paleontological record can provide only a very incomplete picture of our lineage since its divergence from the lineages leading to Pan and Gorilla. Many systems, and their evolution over this span of time, cannot be interpreted from the fossil record. However, this is not the sense in which I use the term inadequate. By inadequate I mean that, even if we were to know everything about the evolutionary events in our lineage since divergence, this would still be an inadequate source for making statements about human nature.

The purpose of this paper is threefold. First, to analyze critically two commonly held assumptions about the origin of human nature upon which this traditional approach is based. Second, to provide more appropriate assumptions about the origin of human nature. And third, based on the more appropriate assumptions, to suggest a theoretical framework (which goes beyond the scope of traditional human evolutionary studies) necessary for the construction of a description of human nature.

TWO COMMON ASSUMPTIONS ABOUT THE ORIGIN OF HUMAN NATURE

In a single quotation Washburn (1983) made two assumptions about the origin of human nature which are commonly found in evolutionary studies that attempt to contribute to the description of human nature. He said "our nature is the summary of the evolutionary events of the past few m.y." The two assumptions about the origin of human nature are 1) that human nature is a consequence of evolutionary events and 2) that specifically those evolutionary events of the past few million years are the ones which are necessary and sufficient to know about in order to construct a description of human nature. But, is human nature the summary of the past few million years of evolutionary events? The following critical evaluation of both assumptions shows that they help to define the boundaries of a data base which, under no circumstances, can be an adequate source for statements about human nature.

The second assumption can be more easily dealt with and it is discussed first.

Evolutionary Events of the Past Few Million Years

Over the past few m.y. the human lineage has, without doubt, been modified by the process of natural selection (as well as by mutation, random genetic drift and gene flow). Since our divergence from Pan and Gorilla there have been many novel changes in the form and behavior of individuals and societies that have contributed to our uniqueness as a species. But would a "list" of these changes contain all of the evolutionary events that have contributed to human nature? For example, would this "list" include the attainment of a homeothermic system or a system of parenting based on lactogenic capacities? While these systems undoubtedly have changed in humans to some extent over the past few m.y., there are also invariant (or only slightly changed) organization relations within them which were realized long before a few m.y. BP and perhaps as long ago as the therapsid radiation (Dobzhansky et al. 1977). The invariant organizational relations of these systems are still in us today. These persistent patterns of organization we generally refer to as homologies, and most importantly they would not appear on our "list" of evolutionary events of the past few m.y. except perhaps to note that they were unchanged, or nearly so, and under stabilizing selection during that time.

Surely, if we are to operationally describe living people, we would want to include the fact that they, like other mammals, attempt to maintain an internal temperature favorable to the maintenance of cellular activity. The organizational relations of the homeothermy system, established prior to our divergence from the African apes and which remained invariant, or which changed only slightly, since our divergence from the extant African apes, are those which, by means of negative feedback organization, contribute to the perception and correction of disturbances in body temperature. Precisely how this is done has changed somewhat since our lineage's divergence from the lineages leading to Pan and Gorilla. For example, there has been, among others, the novel addition of clothing, fire and shelters to our behavioral repertoire of responses. But, these changes have occurred without disturbing, and in fact they were subordinated to, the invariant organizational relations of the homeothermy system. In sum, this example shows that humans are the product of more than a few m.y.

of evolutionary events.

    This line of inquiry can be expanded to a more general question. If it is insufficient to base a description of human nature on the past few m.y. of evolutionary events, then what are the necessary evolutionary events to study in order to make statements about human nature? I argue that the products of <u>all</u> events (evolutionary and otherwise – see below) which have contributed to the present form and behavior of humans are necessary to study.

    For example, wouldn't it be necessary to describe people as living entities? What formal relations constitute this emergent property we call "aliveness" and when did these relations evolve? Students of human evolution cannot seek an answer to this profound question in the last few m.y. of evolutionary events. Rather, they must look at biogenesis- a series of events that occurred more than three billion years ago (Dickerson 1978). Man's aliveness is a consequence of this series of events and a subsequent maintenance of that organization we call living till the present time. Biogenesis, like the origin of the homeothermic system in our therapsid ancester, is not just an obscure event of the distant past. In humans, as in all entities we describe as living, it is the series of events which produced our present "aliveness" (Maturana, Varela 1980; Varela 1979). Thus, a knowledge of this <u>product</u> of biogenesis is therefore relevant to the process of constructing a description of human nature.

    Thus the first assumption, that human nature is the result of the evolutionary events of the past few m.y., is clearly false. While it is necessary to know about these events they are not a sufficient source of information for making statements about human nature. Thus, even if we were able to find out <u>everything</u> about the products of the evolutionary events which occurred in our lineage over the past few m.y., this knowledge would still be insufficient. This is so because living men also retain systems that evolved prior to a few m.y.BP and as long ago as biogenesis and upon which the products of the last few m.y. of evolutionary events were built. Thus, whatever systems evolved and were subsequently retained, in whatever form, from the time of biogenesis to the present must be included in a description of human nature. As Bateson (1979) suggests "there is something relevant in every step of phylogeny and between the steps."

The next question to ponder is whether one can construct a description of human nature on the basis of a knowledge of the products of evolutionary events. I conclude below that it is not possible. There is another class of events which contribute to human nature and which therefore must also be included in a description.

Evolutionary Events

By evolutionary events I presume Washburn was referring to those changes in our form and behavior that have occurred as a consequence of the operation of natural selection, mutation, random genetic drift, and gene flow. However, all of these phenomena occur (and are defined) as a consequence of many interactions between many organisms within complex environments from one generation to the next. Given that these evolutionary processes can be used to explain the presence, persistence, and disappearance of systems on an intergenerational time scale, one could ask, "What are the causes of changes in the form and behavior of an individual(s) or a population(s) within a generation and in particular on a moment by moment basis?" At present the best answer, which I will try to explain below, is that it is always, in all organisms, the operation of autopoietic negative feedback circuits in relation to environmental variables that "triggers" their operation (Powers 1973; Maturana and Varela 1980; Varela 1979).

What are autopoietic negative feedback circuits and how is their operation "triggered" by environmental variables? For the moment I shall concentrate on negative feedback circuits. According to Powers (1973), a feedback circuit is "a closed unidirectional causal chain (loop) relating a system and its environment." A negative feedback circuit "is a feedback situation in which a disturbance acting on any variable in the feedback loop gives rise to an effect at the point of disturbance which opposes the effect of the disturbance."

I illustrate these definitions with a well known example of a negative feedback circuit in a bacterium. Many properties of this circuit can be found in negative feedback circuits in all organisms, including man. Empirical observations supporting this contention are to be found in Guyton (1976) who notes that all physical systems in organisms are negative feedback circuits.

The empirical example of a negative feedback circuit referred to above is the lac operon of the bacterium, E. coli which was first analyzed by Jacob and Monod (1961) (the description here was taken, nearly verbatim, from Elseth and Baumgardner, 1984). The lac operon (fig. 1) possesses three structural genes (Z, Y and A) that code for enzymes that aid in the metabolism of lactose. Lactose is a source of energy as well as a source of carbon for E. coli.

As the diagram shows, in the absence of lactose, the I gene is transcribed and the repressor protein is produced from translation of MRNA for the I gene. This repressor protein sits on the operator and blocks the transcription of

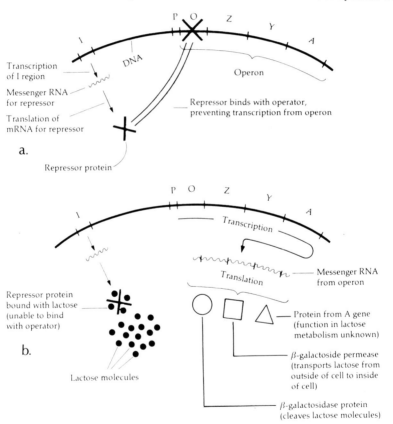

Fig. 1. The lac operon (modified from Hartl, Our Uncertain Heritage (1977)).

the Z, Y and A genes. In the presence of lactose the repressor protein's ability to bind with the operator and block the transcription of the Z, Y and A genes is greatly diminished. This occurs because the repressor protein has an affinity for lactose. This bonding between the repressor protein and lactose changes the conformation of the repressor protein so that it is much less able to sit on the operator. When this happens, transcription of the Z, Y and A genes gradually occurs more freely. The Y gene product brings more lactose into the cell, while the Z gene product metabolizes the incoming lactose, thus reducing this sugar's concentration in the cell. As the external supply of lactose disappears, so does the amount of lactose inside the cell, as the intracellular supply is degraded into galactose and glucose by the Z gene product. As the intracellular concentration of lactose decreases, the concentration of unbound repressor protein molecule, which is continuously manufactured, increases. The unbound repressor protein gradually regains its ability to sit on the operator, thus again preventing the transcription of the Z, Y and A genes. The moment by moment operation of the lac operon is caused by its interaction with an environmental variable. Whether or not the reproduction of an individual with this system is enhanced, relative to the reproduction of other individuals with different forms of the homologous system, is a consequence or effect (measurable in the next generation) of the loop's operation, not the cause of it.

Finally, let me consider the relationship between this negative feedback loop and E. coli as a whole. It has been noted that the lac operon provides a source of energy and carbon for E. coli. Some of this energy and carbon will be used in the very same loop that secured their supply. This recursive property is the basis of living organization not only in E. coli but also in all organisms.

This is the emergent property that arose during biogenesis and has been maintained until the present time. Thus, all systems in humans as in E. coli and all other organisms have functional significance only in relation to the maintenance of their own organization. This emergent property of living entities has been called autopoiesis. Autopoietic entities are homeostatic systems that "have their own organization as the variable they maintain constant"(Maturana and Varela 1980).

A GENERAL FRAMEWORK

I have attempted to outline this second set of causes
for changes in the form and behavior of individuals and
populations on an intragenerational time scale. The table
below is a partial comparison of this set of causes with
the evolutionary causes for changes in form and behavior. I
suggest (as did Bateson, 1979) that the combination of both
sets forms a necessary framework (or paradigm) for the understanding of human nature. That this new framework is <u>suffi-
cient</u> for deriving a description of human nature is not
claimed, i.e., I do not know if there are any other sets of
causes for changes in human form and behavior.

One could conclude from these arguments that traditional
human evolutionary studies provide necessary but not sufficient data to contribute to a description of human nature.
There are at least two sets of causes for human form and
behavior, evolutionary and autopoietic, and both must be part
of any framework which seeks to explain the nature of living
humans.

I see two major choices of study for students of human
evolution based on one's declared objective. If one aims to
document the events of human phylogeny and the paleobiology
of the hominoids, this constitutes one choice which could
perhaps be referred to as human paleontology. However, if

Natural Selection and Adaptation vs. Autopoiesis	
intergenerational time	intragenerational time
changes in form and behavior are the effect of competitive process-- selective component is the differ- ential fertility and mortality of genotypes	changes in form and behavior due to oper- ation of systems within individual(s) and population(s) in relation to environmen- tal variables which trigger operation-- selective component is changes in systems contribute to the maintenance of autopoiesis.
direction of change --adaptation	direction of change--maintenance of autopoietic organization
random component--mutation	random component--occurrence of environ- mental variables--independent of organ- ism's circuits.
explains--the presence, persistence and disappearance, of systems across generational time	explains--operation of systems within generational time
intergenerational changes in cognitive domains=evolution	intragenerational changes in cognitive domains while maintaining autopoiesis= ontogeny

one's objective is to describe human nature and living people and societies, relevant phenomena to understand are 1) evolutionary events in the human lineages since biogenesis, and 2) the operational rules for, and the actual operation of, negative feedback systems in living humans - i.e., how autopoiesis is maintained in humans. This second choice of study, given the declared objective, should be referred to as anthropology.

ACKNOWLEDGMENTS

I gratefully acknowledge M. Kirkpatrick and L. Ecker for technical assistance and Temple University and the Taung Symposium organizers for financial assistance.

Bateson G (1979). "Mind and Nature," New York: Bantam.
Darwin C (1859). "The Origin of Species," New York: Mentor, reprinted 1958.
Darwin C (1871), "The Descent of Man," New York: Hurst and Co., reprinted 1971.
Dickerson RE (1978). Chemical evolution and the origin of life. Sci Am 239:70.
Dobzhansky T, Ayala FJ, Stebbins GL, Valentine JW (1977). "Evolution," San Francisco: Freeman.
Elseth GD, Baumgardner KD (1984). "Genetics," Reading, MA: Addison Wesley.
Guyton A (1976). "Textbook of Medical Physiology," 5th ed, Phila: Saunders.
Jacob F, Monod J (1961). Genetic regulatory mechanisms and the synthesis of proteins. J.Molecular Bio 3:318.
Maturana HR, Varela FJ (1980). Autopoiesis and cognition. Boston Studies in the Philosophy of Science 42:1.
Powers W (1973). "Behavior: The Control of Perception," Chicago: Aldine.
Varela FJ (1979). "Principles of Biological Autonomy, New York: Elsevier-North Holland.
Washburn SL (1983). Human nature. Anthroquest (LSB Leakey Foundation News) 26:9.

*Present and Future Prospects for
Dating the Past*

Hominid Evolution: Past, Present and Future, pages 477–484
© *1985 Alan R. Liss, Inc.*

PROSPECTS FOR DATING OF FOSSIL HOMINID MATERIALS

JPF Sellschop, A Zucchiatti* and HJ Annegarn

Wits-CSIR Schonland Research Centre for Nuclear Sciences, University of the Witwatersrand, Johannesburg 2000, Republic of South Africa.

INTRODUCTION

The need for direct dating of actual hominid materials has become increasingly necessary as advances in other fields of study relative to these materials have revealed anomalies and discrepancies in interpretation. No truly appropriate method has been available, in the sense that questions relating to distortion of data through losses of volatiles have been obvious. Even if current methods are above question in so far as methodology is concerned, there has still been an understandable reluctance to apply any of them to actual hominid materials. The reason for this is not hard to find: such fossil materials are extremely precious, so that on the one hand samples made available for such analyses tend to be very small, and on the other hand there is a legitimate demand that any method used should be truly non-destructive.

There is a pressing need for methods of dating that are ultra-sensitive, non-destructive, capable of calibration, matrix independent, and covering the range 0.5 to 3.5 million years before present. Recent developments in ultra-sensitive isotope analysis have made it appropriate to re-examine radioactive isotope dating; for isotopes with suitable half-lives it is possible to speculate on mechanisms of source production, transfer and incorporation into the material of interest.

* On leave from INFN, Genova, Italy.

MECHANISM FOR COSMIC RAY ISOTOPE PRODUCTION AND LOCALIZATION

The composition of the primary cosmic radiation, i.e. the specific isotopes, their intensity and energy spectra, is the obvious starting point. These characteristics change quite dramatically as the atmosphere is first of all reached, and then penetrated. As the consequential cosmic rays penetrate the earth, the composition is again starkly changed. The primary and secondary cosmic radiation may be influenced by various perturbations, such as solar flares. For this varying incident flux, five target areas can be identified (Lal, Peters 1967):

(i) upper 70% of atmosphere - stratosphere, with residence times for cosmic ray produced nuclides of 7 to 8 years; (ii) lower 30% - troposphere, with residence times of 30 to 90 days; (iii) near surface region with residence times of 4 to 14 days; (iv) surface region of the earth; (v) sub-surface region of the earth.

The primary cosmic radiation consists of charged particles ($\sim$90% protons, $\sim$10% alphas and 1% heavy particles) and uncharged neutrinos. From measurements of the ^{10}Be to ^{9}Be ratio, the flux of these primaries can be concluded to have been constant to within ±6% back to not less than 7.9 m.y. ago, when it appears that the primary flux increased (Raisbeck, Yiou 1984). This primary radiation striking the atmosphere acts as a high energy nuclear laboratory within which occurs cascade development of myriad subnuclear 'elementary' particles (secondary cosmic radiation), production of a range of isotopes and attenuation of both primary and secondary radiation. Initially as the atmosphere is penetrated the total flux of particles increases. With decay of very short-lived species and absorption, the flux reaching the earth's surface is reduced 1000-fold. The composition of the radiation has changed also, so that what strikes the lithosphere comprises charged particles (muons up to TeV energies, and electrons at high altitudes only) and uncharged particles (neutrons up to MeV energies and neutrinos of both electron and muon type over a wide energy spectrum).

The atmosphere as a target (% composition indicated) produces the following isotopes, ignoring those that are stable or have half-lives less than 10 years:

from N (80%) and O (19.8%) - ^{10}Be, ^{14}C, ^{3}H (tritium); from Ar(.2%) and Kr(.001%) - ^{26}Al, ^{36}Cl, ^{81}Kr, ^{32}Si, ^{39}Ar.

The earth as a target is quite different in composition from the atmosphere; major target nuclides of relevance now include Fe, Si and Ca. While such electrons as survive to the earth's surface are instantly damped and neutrinos are essentially unaffected in their penetration of the earth, the other two components, muons and neutrons, are most significant in production of radioactive isotopes in the first few meters of the crust. Ignoring again nuclides with half-lives less than 10 years and those that are stable, we find no fewer than 11 are produced (Table 1).

Table 1. Isotopes ($T_{\frac{1}{2}} > 10$ years $\neq \infty$) produced by cosmic radiation in the atmosphere and the lithosphere

Isotope	Half-life (years)	Atmospheric production: target nuclide	Lithospheric production: target nuclide
^{129}I	1.57×10^7	-	Te, Ba, La, Ce
^{53}Mn	3.70×10^6	-	Fe
^{10}Be	1.60×10^6	N,O	Be, O, Mg, Si
^{26}Al	7.16×10^5	Ar	Al, Si
^{36}Cl	3.00×10^5	Ar	Ca, Fe
^{81}Kr	2.10×10^5	Kr	Sr, Zr
^{41}Ca	1.30×10^5	-	Ca
^{59}Ni	7.50×10^4	-	Fe, Ni
^{14}C	5.73×10^3	N,O	O, Mg, Si
^{32}Si	6.50×10^2	Ar	-
^{39}Ar	2.70×10^2	Ar	K, Ca, Fe
^{3}H	1.23×10^1	N,O	O, Mg, Si

We note that 7 of the 8 atmospherically produced isotopes are produced also in the lithosphere, the sole exception being ^{32}Si. This presents us with a problem: if any of these isotopes are to be used as chronometers the dual source nature must be incorporated in the model. The simple source/sink model for atmospherically produced isotopes is not applicable to lithospherically produced isotopes - the modelling will have to address continuous production, mechanisms for removal from production sites and incorporation into hominid skeletal materials. The group of lithospherically produced isotopes will have to be considered in terms of the characteristics of each isotope.

CANDIDATES AS CHRONOLOGICAL TOOLS

There are potentially five: ^{53}Mn, ^{10}Be, ^{26}Al, ^{36}Cl and ^{41}Ca.

(i) ^{53}Mn; production purely terrestrial; half-life 3.7 m.y.; most probable production reactions ^{56}Fe(92%)($\mu^{\pm}$,^{3}He)^{53}Mn, ^{57}Fe(2%)(μ^{+},^{4}He)^{53}Mn, ^{54}Fe(6%)(n,d)^{53}Mn; production rate at 3840 m altitude estimated as 0.1 atoms/min/kg for granites or basalts (Yokoyama et al. 1977); concentration of ^{53}Mn thus low, but detectable; eligible chronometer.

(ii) ^{10}Be; dual source production; half-life 1.60 m.y.; atmospheric production -- spallation product with residence time of a few years, attaches to aerosols and rains out; production rate $\sim$4.5x10^{-2} atoms/cm^{2}/sec (cf. ^{14}C - 2.5 atoms/cm^{2}/sec); terrestrial production -- most probable reactions ^{9}Be(100%)(n,γ)^{10}Be and ^{14}N(99%)(μ^{-},^{4}He)^{10}Be; production rate very small, $\sim$10^{4}x less than atmospheric; conclude that atmospheric source only is effective. Many examples of ^{10}Be detection exist, thus an eligible chronometer for fossils.

(iii) ^{26}Al; dual source production; half-life 0.716 m.y.; atmospheric production -- spallation product; production rate $\sim$1.4x10^{-4}atoms/cm^{2}/sec; terrestrial production; most probable reaction ^{28}Si(92%)(n,t) ^{26}Al; production rate very small. Conclude that atmospheric contribution $\sim$10^{2}x greater than lithospheric, thus it remains an eligible chronometer for hominid materials. Could be especially attractive if one measures the ratio ^{26}Al/^{10}Be because of the similar origins and chemistry of these two isotopes.

(iv) ^{36}Cl; dual source production; half-life 0.31 m.y.; atmospheric production -- spallation product from Ar; production rate $\sim$1.1x10^{-3} atoms/cm^{2}/sec; terrestrial production -- most probable reactions ^{35}Cl(76%)(n,γ)^{36}Cl, ^{40}Ca(97%)(μ^{-},^{4}He)^{36}Cl; estimate of terrestrial production rate shows it is significant compared with atmospheric. Conclude there is interplay of two production systems and that experiments will be required to establish if the isotope is sufficiently unique for fossil studies.

(v) ^{41}Ca; produced terrestrially only; halflife 0.130 m.y.; most probable reaction ^{40}Ca(97%)(n,γ) ^{41}Ca, strongly exothermic, Q = +8.363 MeV, cross section for thermal neutrons 0.44 barns. Assuming a time greater than the half-life and a thermal neutron flux of 3x10^{-3}n/cm^{2}/sec, saturation concentration estimated as ^{41}Ca/^{40}Ca = 8x10^{-15} (Raisbeck, Yiou 1979). This ratio implies a decay rate of 1x10^{-3} dpm/gm of ^{41}Ca i.e. $\sim$10^{4}x smaller than for contemporary ^{14}C. Besides having a low activity, the

decay by electron capture produces the ^{41}K 3.3 keV x-ray
line, which is very difficult to detect. This is an
interesting prospect - the bones of the living hominid
will have been irradiated by cosmic ray neutrons, but we
may accept that a saturated ^{41}Ca component, ingested from
food and water, will have dominated. After death, the
isolation of the corpse from the neutron flux is
important; if coverage was several meters deep then the
flux would have been greatly attenuated and the
circumstances would satisfy chronometer requirements.
With regard to the ultimate limits, the uranium and
thorium content of the coverage materials would also have
to be considered. ^{41}Ca is therefore a challenging case,
which in most instances may be eligible as a chronometer.

We conclude that of the five isotopes, not one should
yet be excluded. There are questions that remain open as
noted above, and others of exchange and erosion, but if
measurements of sufficient accuracy and precision can be
made, there are prospects for sensible modelling and dating.

TWO RELEVANT EXPERIMENTS

Measurement of a ^{36}Cl Profile through Limestone

The mechanisms which may participate include: direct
spallation, mainly on Ca; $^{40}Ca(\mu^-,^4He)^{36}Cl$; $^{35}Cl(n,\gamma)^{36}Cl$;
$(\mu^-,n) \rightarrow {}^{35}Cl(n,\gamma)^{36}Cl$; U, Th fission neutrons $\rightarrow {}^{35}Cl(n,\gamma)^{36}Cl$.
These have all been modelled and calculated to give the
ratio $^{36}Cl/^{40}Ca$ for comparison with experiment (Kubik et al.
1984). Below a few meters depth, the shape of the curve
agrees with a model based on $^{40}Ca(\mu^-,^4He)^{36}Cl$. The situation
is thus relatively unperturbed by mixing. Near the surface
the situation is more complex and may include mixing/erosion
effects. The ^{36}Cl case considered here presents a more
complicated situation (as anticipated earlier in the
arguments for each potential isotope). For interpretation of
such data detailed background traverses and measurement of
other isotopes will be necessary.

Non-destructive, Small Spot/Volume Analyses of Fossil Bones

In the chronology endeavours proposed, complex
situations must be expected and single isolated measurements
will be unreliable or meaningless. Dual isotope approaches

will be desirable or even essential. It will be appropriate
to complement dating-orientated analyses with broad spectrum
elemental analyses of the micro-volumes being dated, to
establish whether there is any aspect which is unusual, or
which might bear on questions such as exchange.

A non-destructive method which meets these needs is
proton induced X-ray emission (PIXE), with few-MeV protons
being used to irradiate the sample spot. We have carried out
such measurements on a vertical traverse through the
Sterkfontein hominid site, investigating fossil and matrix
materials. The following elements are found in most
materials: P, K, Ca, Ti, Cr, Mn, Fe, Ni, Cu, Zn and Sr. The
inter-element ratios vary greatly and serve to characterise
fossil as distinct from matrix materials. A key comparison
which we have made, for example, is of a finger bone and the
cave mud inside (Fig. 1). We will attempt to establish
inter-element correlation coefficients for the full traverse
through all sites investigated. Such data will be valuable
for the elucidation of perturbative effects in chronology
measurements, and for the establishment of changes in the
matrix relevant to possible exchange corrections.

PRAGMATIC CONSIDERATIONS WITH REGARD TO PROPOSED MEASUREMENTS

To measure any of the five isotopes proposed,
considering the concentration estimates, implies the most
demanding analytical techniques. There has been however an
important development during the past decade of a new form
of ultra-sensitive mass spectrometry using particle
accelerators, generally of the Tandem DC configuration
(Purser et al. 1977). This approach has attractive features
of unique Z and A specificity, is minimally destructive in
that a small volume only of the sample material is sputtered
to produce a beam, and has the additional claim that all
five isotopes of interest to the hominid problem have been
determined in various laboratories. Without the recent
developments of accelerator mass spectrometry it is
difficult to see how any of the five isotopes could be
exploited for hominid chronology purposes.

CONCLUSIONS

The need for more and reliable information on the range
of ages of materials from hominid sites is becoming

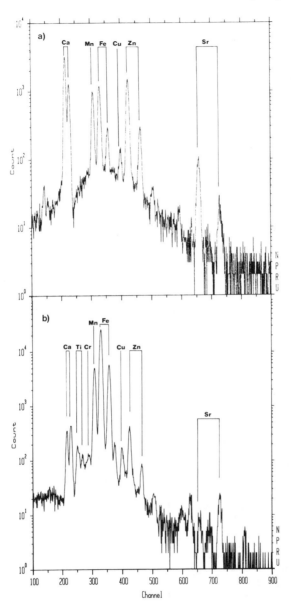

Fig 1. Proton Induced X-ray Emission (PIXE) spectra of:
a) outer surface of a finger bone from Sterkfontein; b) cave
mud within the cavity of the same bone. Ca dominates the
bone spectrum. Mn in the bone, relative to Fe, is enhanced
over the crustal ratio of Mn/Fe as observed in the cave mud.

increasingly essential. Five isotopes, produced in either the atmosphere or the lithosphere or both, have appropriate radioactive half-lives, but occur at extremely low concentrations. Because of the complexity of the modes of formation, and of removal from the cosmic ray production sources, it will be necessary at each site to establish that there is a meaningful chronology. We are attempting this in collaboration with colleagues from the ETH, Zürich (^{10}Be and ^{36}Cl) and from the TU, München (^{36}Cl and ^{41}Ca), by using precisely located materials in vertical traverses, starting with the numerous non-hominid fossils (e.g. antelope, baboon). At all three centres the techniques of Tandem van de Graaff accelerator mass spectrometry are being employed. With these techniques, there are good prospects for at least narrowing the window of ages with which the paleontologist is currently confronted.

The three physics groups involved in exploring the prospects of hominid dating express their appreciation to the Paleo-anthropological Research Group at the University of the Witwatersrand for the provision of materials. The authors thank the CSIR's Foundation for Research Development for continuing support.

Kubik PW, Korschinek G, Nolte E, Ratzinger U, Ernst H, Teichman S, Morinaga H, Wild E, Hille P (1984). Accelerator mass spectrometry of ^{36}Cl in limestone and some paleontological samples using completely stripped ions.   Nucl Instr and Meth in Phys Res B5:326.
Lal D, Peters B (1967).   Cosmic ray produced radioactivity on the earth.   Handbuch Phys 46(2):551.
Purser KH, Liebert RB, Litherland AE, Beukens RP, Gove HE, Bennett CL, Clover MR, Sondheim WE (1977).   An attempt to detect stable N$^-$ ions from a sputter ion source and some implications of the results for the design of Tandems for ultra-sensitive carbon analysis.   Rev Phys Appl 12:1487.
Raisbeck GM, Yiou F (1979).   The possible use of ^{41}Ca for radioactive dating.   Nature 277:42.
Raisbeck GM, Yiou F (1984).   Production of long-lived cosmogenic nuclei and their applications.   Nucl Instr and Meth in Phys Res B5:91.
Yokoyama Y, Reyss J-L, Guichard F (1977).   Production of radionuclides by cosmic rays at mountain altitudes. Earth Planet Sci Lett 36:44.

Index